GASEOUS MATTER

GASEOUS MATTER

Editor
Braxton Henryson

scitus
academics

Gaseous Matter

Edited by **Braxton Henryson**

This Book is Printed on Acid-Free Paper. ♾

ISBN: 9781681178547
Library of Congress Control Number: 2017948881

SCITUS Academics LLC,
616, Corporate Way, Suite 2, 4766,
Valley Cottage, NY 10989

www.scitusacademics.com

Notice

Printed in United States of America.

Preface

Gas is a state of matter that has no fixed shape and no fixed volume. It was the early-seventeenth-century Flemish chemist-physician Jan Baptista van Helmont who coined the word "gas" in order to convey the idea that a gas had an indefinite shape and size. The terms gas and vapor are both used to describe the gaseous state of a substance. However, gas is primarily for a pure substance or mixture that exists in gaseous state under normal conditions. Vapor is used to describe a substance that is in gaseous state, which exists in liquid or solid state under normal conditions. A gas can be compressed above the atmospheric pressure and even nontoxic gases can be lethal when their concentrations are high enough to displace too much oxygen in the air. While too much vapor will result in the phase change from gas to liquid by condensation. Gases have lower density than other states of matter, such as solids and liquids. There is a great deal of empty space between particles, which have a lot of kinetic energy. The particles move very fast and collide into one another, causing them to diffuse, or spread out, until they are evenly distributed throughout the volume of the container.

This book aims to describe some basic properties of gases, which are applicable to not only air but also gaseous air emissions. Kinetic theory of gases attempts to explain macroscopic properties of gases, such as pressure, temperature, or volume, by considering their microscopic compositions and motion. It includes a critical analysis of existing methods as well as hands-on practical recommendations. This book will be ideal for graduate students and researchers at the cutting edge of investigations into the fundamental nature of gaseous matter; this text presents the basic methods used in the treatment of gaseous flows.

Table of Contents

CHAPTER 1

Gas Hydrate—Properties, Formation and Benefits

Azeez G. Aregbe

Institute of Oil and Gas Engineering, University of Lagos, Lagos, Nigeria

ABSTRACT

There are numerous gas hydrate reserves all over the world, especially in permafrost regions and ocean environments. The abundance of gas hydrate reserves is estimated to be more than twice of the combined carbon of coal, conventional gas and petroleum reserves. These hydrate deposits hold significant amount of energy which can make hydrate a sustainable energy resource. The comprehensive research on the properties and formation of methane hydrates is paramount to ensure efficient and effective exploration and development of hydrate reserves. Natural gas is mostly distributed for different purposes through pipelines or pressure vessels such as dry gas, compressed gas or liquefied gas, which means transporting natural gas creates serious safety concerns because methane is highly flammable and almost impossible to detect any leak without using odorant. Alternatively, natural gas can be stored and transported as gas hydrate turns solid or slurry. Gas hydrate can be stored at equilibrium conditions with either saturation temperature or pressure. The equilibrium conditions are influenced by the cost and weight of storage vessel. Hydrate can be transported either as slurry or solid depending on the location of target or destination. The slurry form is usually a better option for distance of approximately 2500 mile or less while the solid form can be used for distances of roughly 3500 miles or more. The paper examines the properties, formation and benefits of gas hydrate. The suitability of gas hydrate as a sustainable energy resource and the possibility of using gas hydrate for the transportation and storage of natural gas (methane) are also stated. Natural gas transportation and storage as gas hydrate will create effectively and efficiently alternative bulk gas transportation and storage for future use of the gas.

Keywords: Natural Gas, Hydrate, Storage, Energy, Transportation

1. INTRODUCTION

The term "gas hydrates" refers to crystalline compounds that are composed of water and any of the following light molecules: methane, ethane, propane, iso-butane, normal butane, nitrogen, carbon dioxide, and hydrogen sulfide. It is known that some polar components between the sizes of argon (0.35 nm) and ethyl cyclohexane (0.9 - 1) can also form hydrates. Hydrate formation usually occurs when water molecule exists in the vicinity of these molecules at temperatures above or below the ice point and relatively high pressure. The water molecules enclose these host molecules and form cage-like structures which are stable at these conditions.

Natural gas hydrates commonly called gas hydrates the crystalline compounds formed when water molecule and methane gas exist together at the right temperature and pressure. Methane hydrate is stable at a temperature slightly above or below 0°C with high pressure. The pressure and temperature of most ocean environments offer appropriate conditions for methane hydrate stability but considerable amounts of hydrate are formed at the continental shelf due to the effect of geothermal gradient [1] . Geothermal gradient makes it possible for the temperature at some depths to be more than the equilibrium temperature at the in situ pressure, which creates stability for methane hydrate. The cage-like structure formed is stabilized by the interactions that exist between the water molecules and the methane gas. These interactions create a relatively strong and stable lattice. Gas hydrates have high capacity to store methane gas—1 cubic metre of a typical hydrate contains approximately 180 standard cubic metres of methane gas at standard temperature and pressure [1] .

Natural gas which is mainly methane is used as fuel because the technologies for handling it directly. As fuel for heating and indirectly through energy conversion, technologies are readily available and they also cost money effective. In addition to that, methane gas contains more hydrogen atoms for each carbon atom than the other hydrocarbon fuels. This implies less production of carbon dioxide during combustion of methane. Methane gas is also pure and easy to purify, which makes its use as fuel more environmentally friendly than other hydrocarbon fuels such as oil and coal. Methane gas is also surprisingly be found to produce much less carbon dioxide than alcohols, and even much lesser than in liquid petroleum and oil based fuels.

Natural gas (Methane) produced from conventional oil and gas reserves is enormous, which is easily distributed to the consumers through pipeline, and also has abundant heat content. The exploration and development of gas hydrate reserves, with a prospect of almost indefinite supply of methane will definitely influence the growth of the gas-based energy economy to overcome the oil-based energy economy. Figure 1 shows the comparison between hydrate deposits and other sources of fuel, using a unit of 10^{15} g of carbon.

Natural gas is mostly distributed for different purposes through pipeline system or pressure vessels either as dry gas, compressed gas or liquefied gas. These means of transporting of natural gas poses serious safety concerns due to the fact that methane is highly flammable and almost impossible to detect any leak without using odorant.

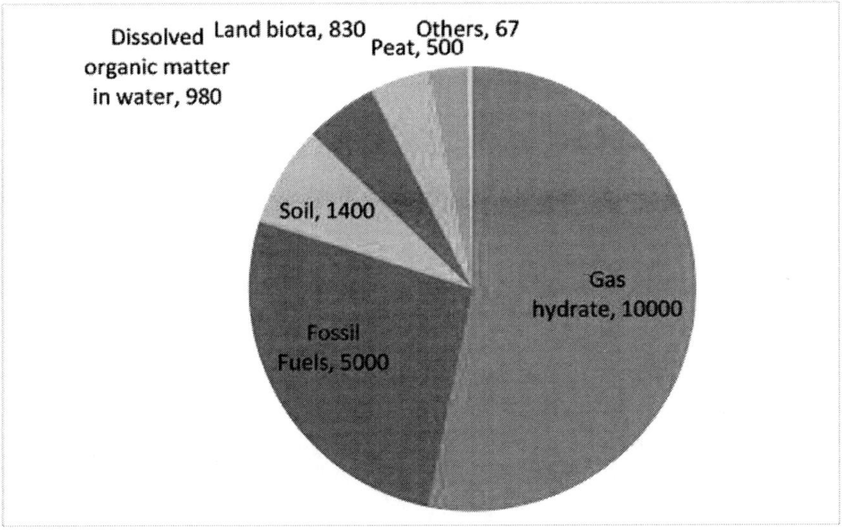

Figure 1. Gas hydrates deposits compared with other fuel resources, units = 10^{15} g of carbon, [2] .

Hydrates have been one of the flow assurance problems in gas production and transportation. Different models and approaches have been adopted to solve the gas hydrate problems. Hydrates block the conduit of oil and gas pipelines and transportation systems with significant economic impacts. In recent years, hydrate production has been studied and investigated as potential for sustainable energy resource. The abundance of gas hydrate is estimated to be more than twice the combined carbon of coal, conventional gas and petroleum reserves [3] . Gas hydrate reserves contain enormous source of energy in form of pure methane gas which can serve as sustainable energy resource.

Nature of Gas Hydrate Deposits

The hydrates of CH_4 and C_3H_8 gases occur naturally all over the world. The formation of Arctic hydrates is usually under the permafrost layer. Oceanic hydrates are also discovered to be deposited at the continental shelves. In natural settings, such as the ocean bottom, where organic matter are buried and decomposed to methane, the methane formed is dissolved in water to form methane hydrates at temperatures around 277 K. Biogenic methane usually dissolves slowly in water to form hydrate because of mass transfer limitations.

Over geologic time, the estimated methane hydrate in the ocean is approximately 2.1×10^{16}standard cubic metres—twice the total energy of all other fossil fuels on earth. The amount of methane hydrate in the northern latitude permafrost is approximately 7.4×10^{14} standard cubic metres. Hydrates deposits are usually determined by indirect methods such as the seismic reflection method known as bottom simulating reflectors (BSRs). This method uses seismic signals which are caused by velocity inversion because the presence of gas below some high-velocity barriers such as a hydrate deposit.

Significant hydrate sediments in the ocean beds can jeopardize the foundation of subsea platforms, manifolds or even pipelines. Water molecules form frameworks of large cavities as a result of the hydrogen bond within the water molecule which enclose the guest gas molecule. The gas hydrate is thermodynamicaly stable because of the interaction between the water and the gas molecules [4] . Gases hydrates samples have been discovered from many places—19 or more areas worldwide and are believed to be deposited at roughly 77 areas including the Antarctica and Siberia. Generally, they are discovered in continental shelf sediments and onshore polar areas beneath the permafrost due to the fact that the pressure-temperature conditions in these regions are within the hydrate stability conditions [5] . An artificially produced sample of methane hydrate is shown in Figure 2.

Hydrate deposits are generally found in waters deeper than 300 m—their zones of existence is from seafloor to a depth of few hundred metres and this is also a function of the local thermal gradient. Based on research and investigation, huge amount of methane gas is believed to be trapped in the hydrate deposits as well as the sediments beneath these deposits. Exploration and development of gas hydrate reserves is the key solution to unlimited supply of energy. This source of energy is pure and environmentally friendly. The investigation of the properties and how these hydrate deposits are formed is one of the objectives of this paper. The study of the formation of hydrate is also crucial for the transportation and storage of natural gas as hydrates which is the focus of this paper. A better understanding of the hydrate formation and dissociation mechanism is vital for proper design and optimization of separation processes involving hydrate equilibrium and the understanding of natural gas hydrate formation, accumulation and destabilization in nature.

2. LITERATURES/RECENT WORKS ON GAS HYDRATES

There has been several research works on gas hydrate to mitigate its formation and also to explore gas hydrate reserves. Although a significant amount of research on gas hydrates has been carried out in both industrial and academic laboratories, several study areas such as the formation mechanism and the hydrate growth are still poorly understood, especially for complex systems. However, a better understanding of the hydrate formation and dissociation mechanism is required for both the design and optimization of separation processes involving hydrate equilibrium and the understanding of natural gas hydrate formation, accumulation and destabilization in nature.

Gas Hydrate was discovered by Sir Humphrey Day in 1810. The study and research on hydrate became of interest to the oil and gas industry in 1934 when the first pipeline blockage was observed by Hamrnerschmidt. This was due to the crystalline, non-flowing nature of hydrates. Different experiments have been performed to determine the structures of gas hydrates:

Jin et al. [7] used micro focus X-ray computed tomography (CT) to study the structure of natural gas sediments with and without gas hydrates. A newly developed high pressure vessel was used for the analysis to observe the sediments at temperatures above 273 K and relatively high pressures. They

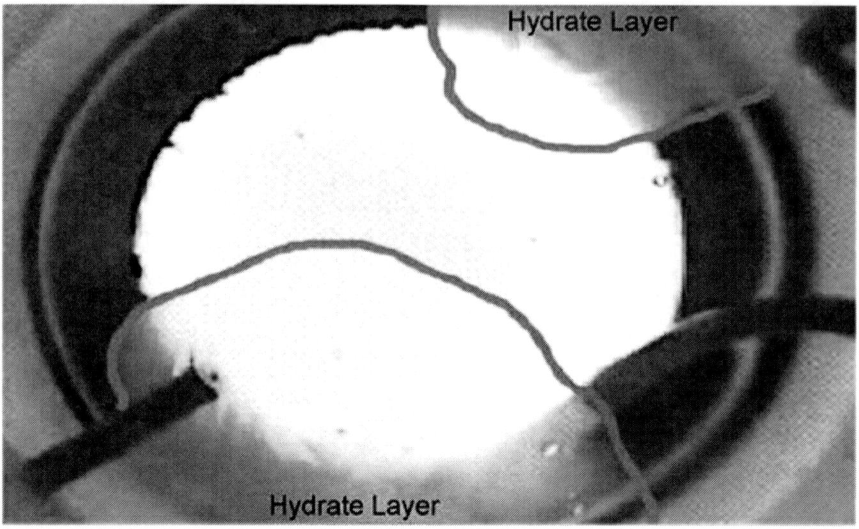

Figure 2. Sample of methane hydrate [6] .

measured the spatial distribution of the free gas, sand particles, liquid water and solid hydrate phases. They also assessed the permeability of the sediments through the correlation between the absolute permeability and the pore networks of the sediment. The results showed that the proportion of horizontal continuous pore channel is a major determinant of the absolute permeability.

Minagawa et al. [8] used proton nuclear magnetic resonance measurements combined with a permeability measurement system to characterize methane hydrate sediments based on their permeability and pore size distribution. They compared the effective permeability of the sediments and different effective porosities, which had been measured by using water flow based on Darcy's law, with the permeability calculated by NMR spectra based on SDR (Schlumberger-Doll Research) model. The results of the permeability from both methods were similar with a minimal difference of less than factor of 2. The results were used to describe the relationship among pore size distribution, porosity and effective permeability.

Santamarina et al. [9] conducted series of experiments to determine the mechanical, thermal, electrical and electromagnetic properties of hydrate-bearing soils using standardized geotechnical devices and test protocols. They experimented with a range of grain sizes subjected to an effective stress of up to 2 MPa and with well-controlled saturations of synthetic hydrates.

Stoll and Bryan [10] conducted an experimental work on the thermal conductivity and acoustic wave velocity of hydrates as well as sediments containing hydrates. The most significant part of their results was that the formation of hydrates decreases with the thermal conductivity of the sediments. This behavior was found to be different from what should be expected of the behavior of frozen sediments. Their research work was also able to confirm that both pure water and water-bearing sediments are converted to a stiff elastic mass by formation of a sufficiently high amount of hydrate. This led to their conclusion that as a basis for using sharp acoustic impedance contrast at the boundary of sediment containing hydrates to locate hydrate deposits.

Pearson et al. [11] were able to predict the physical properties of sediments containing hydrates in order to include their effects on production models and develop geophysical exploration as well as reservoir characterization techniques. They incorporated empirical relationships between the composition of ice and seismic velocity, electrical resistivity, density and heat capacity for frozen soils to estimate the physical properties of the deposits. They stipulated that the resistivity of laboratory permafrost samples followed a variation of Archie's equation. They measured the parameters and also calculated these parameters for variety of lithological types. They also concluded that the compressional wave velocities of the partially frozen sediments are related to the velocities of the matrix, liquid and solid phases present in the pores of the sediments.

Winters et al. [12] performed experimental to measure a wide range of acoustic-wave velocities in coarse grain sediments for different pore space occupants. The measured values from the experiments ranged from less than 1000 m/s for gas-charged sediments to 1770 - 1940 m/s for water saturated sediments, 2910 - 4000 m/s for other sediments with varying degree of hydrate saturations and 3880 - 4330 m/s for frozen sediments. The sediments shear strength was increased as a result of the presence of gas hydrate, ice and other solid pore-filling substances. The magnitude of the shear strength increment was found to be a function of the amount of hydrate in the pore spaces and cementation characteristics between the hydrate and grains of the sediments. They also concluded that the presence of free gas in pore spaces influenced pore pressure response and during shearing and also the strengthening effect of gas hydrate in sands.

There has been a significant improvement in characterizing the relationship between hydrates and sediments containing hydrates. Tinivella et al. [13] conducted a research work to quantify the concentrations of gas hydrate in pore spaces of the sediments by travel-time inversion modeling of the acoustic properties of these sediments. Such analysis has created rooms for the identification of free gas distribution in pore spaces, likely patterns of fluid migration and the physical properties of these sediments.

Tohidi et al. [14] performed series of visualization experiments using two-dimensional transparent glass micro-models and stipulated that hydrates can be formed form either free gas or dissolved gas in the system. They concluded that hydrates usually form at the center of pore spaces with a thin film of water covering the grains instead of nucleating around the grains.

Kingston et al. [15] conducted an experimental work by using a specially manufactured laboratory porous medium, gas hydrate resonant column (GHRC) to explore different methods hydrate synthesis and also measure the properties of the sediments formed. The experiments were performed with different water saturation conditions of the porous medium. In low water saturated condition, where there is gas saturation; the growth of the hydrate was on the water location. Thus, the water saturation was considered the limiting factor. In partially saturated sands, water was collected at grain contacts and coated individual sand grains. As the content of hydrate increased, the filling of the pore spaces began but the increased stiffness was produced by the increased quantity of cement at grain contacts.

Moridis and Kowalsky [16] researched on Gas Production from Unconfined Class 2 Hydrate Accumulations in the Oceanic Subsurface. Hydrate accumulations in oceanic sediments are characterized by mobile saline water zones encaging the hydrate deposits and by the absence of impermeable layers to vertical flow. They evaluated the gas production potential of such hydrate deposits using both single-well and five-spot well configurations. Single-well production is based on depressurization-induced dissociation of the hydrates, whereas the five-spot configuration involves both depressurization at the production wells and thermal stimulation at the injection wells. The study showed that unconfined Class 2 hydrate accumulations are among the most challenging targets for gas production because of the absence of confining boundaries, water production with the gas, and thermal stimulation to enhance production requires substantial energy inputs.

Different international R&D programs have been conducted on gas hydrates all over the world. The review on International Gas Hydrate Research conducted by USGS—science for a changing world in 2014 and the reports of the review showed that different countries all over the world conducted R&D programs on gas hydrate exploration and production. Some of the R&D programs are [17] :

I. Japan

- 2012/13: R&D on both Arctic and Marine projects.
- 2013: A week deepwater production test.

- 2014/15: "Extended" deepwater production test.
- New Japan Sea project.

II. Korea

- 2007 & 2010: UBGH-1 & UBGH-2 expeditions.
- 2015: Marine production test.

III. China

- 2007 & 2013: GMGS-1 & GMGS-2 expeditions.
- 2007 through 2011: Onshore "tests".

IV. Norway (Statoil)

- Onshore long-duration production test.
- Gas hydrates global screening.

V. Canada

- Onshore Mallik Project 1998, 2002, 2007-2008.
- Beaufort Shelf hazard and climate research.
- Pacific and Atlantic marine gas hydrate studies.

VI. New Zealand

- Gas hydrates on the Hikurangi Margin, GNS and University of Auckland.
- Energy focus, marine surveys and drilling.

VII. Germany

- SUGAR Energy Assessment Project, BGR plus others.
- GEOMAR marine gas hydrates research, marine surveys.
- MARUM MeBo (sea floor drilling rig) drilling research.

3. FORMATION OF GAS HYDRATE

When water molecules come in contact with gas molecules at low temperature and high pressure, different geometric structures contrary to that of a hexagonal ice are formed. The water molecules serve as host molecules and create cage lattices that can hold gas molecules as guest molecules. These cage-like crystalline structures are less dense than crystalline water structure because of the presence of the gas molecules. The gas hydrate formed is held together by the hydrogen bonds of the water molecules and also stabilized by Vander Waals forces holding the gas and water molecules together. The Vander Waals force is responsible for the stable nature of the gas hydrate and even makes the hydrate more stable than normal ice formed by water. There are different structures of gas hydrate and they are characterized by the shape of their cages. Natural gas composed mainly of methane gas and the complete combustion of methane gas gives water, carbon dioxide, and energy, as shown in the Equation (1.0).

$$CH_4 + O_2 \Rightarrow CO_2 + H_2O + \text{Energy} \uparrow \tag{1.0}$$

The energy liberated from this process can serve different purposes. This makes natural gas more environmentally friendly than other fossil fuels because more energy is liberated and less CO_2produced. The pictorial view of the lattice structure of gas hydrate is shown in Figure 3. Methane gas is the guest in the middle (Green) while water molecule is the host (Pink).

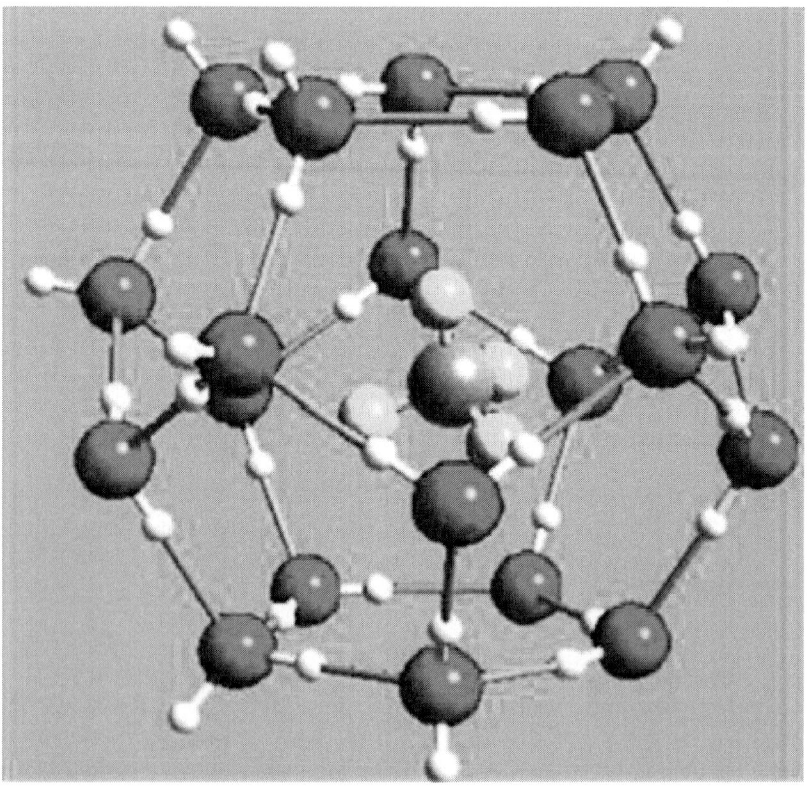

Figure 3. Cage-like structure of gas hydrate [18] .

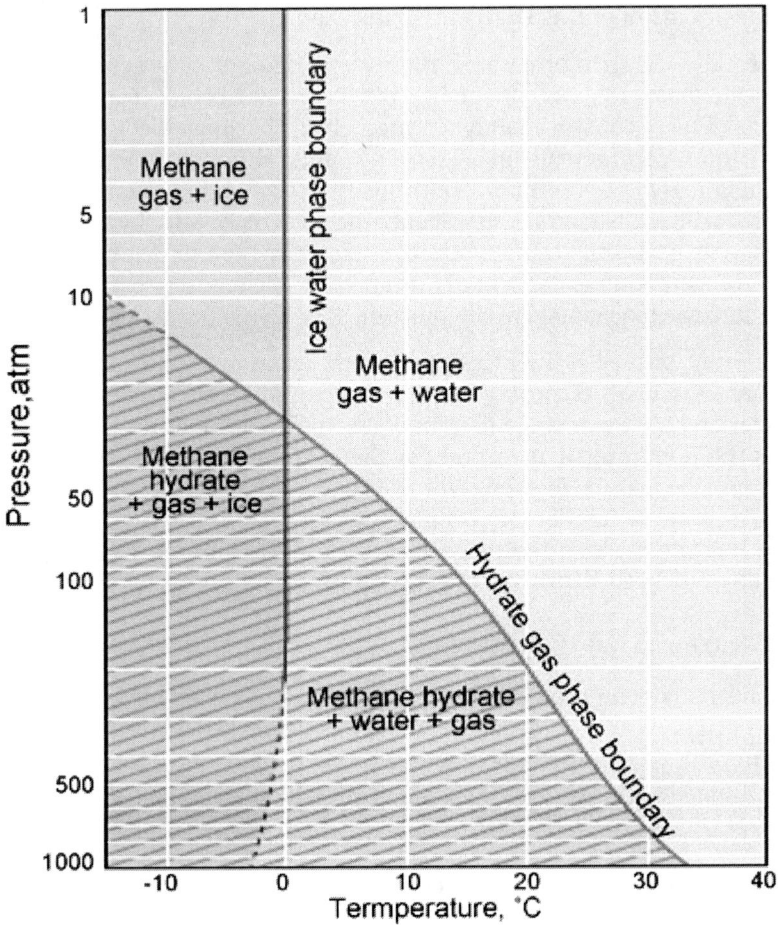

Figure 4. Phase diagram of gas hydrate stability [18] .

Gas hydrate can be stored or transported at equilibrium conditions with either its saturation temperature or pressure as shown in Figure 4. At the saturation temperature and pressure, hydrates are usually stable. Some factors affect the saturation pressure and temperature of the hydrate.

Factors such as cost and weight of material for hydrate storage vessel as well as the environment of the sediments containing the hydrate deposits. Hydrates are usually stable at moderate temperatures and pressures when compared to the conditions required for LNG and CNG.

3.1. Classification of Gas Hydrate Structures

For naturally-occurring hydrates, the gases originate either from biogenic or thermogenic sources. Biogenic gases are microbial transformation products which consist mainly of methane. Thermogenic gases usually come from deep petroleum reservoirs and although the methane is overwhelmingly represented in those gases, the fraction of C_2 - C_5 hydrocarbons is very significant. In the case of hydrates from thermogenic gases, the change in the vapor phase composition during that process of formation is particularly interesting because it provides valuable insight into the mechanism of the reaction.

The different classes of hydrates are formed when methane gas is trapped. They are classified according to the structures of the cages which are characterized by the arrangement of water molecules in the crystals. When the gas (CH_4) is trapped, it rearranges the host molecules (H_2O) and a cage that traps the gas is formed. Not every water molecule attracts a gas molecule; some remain empty even when a cage appears. But immediately the gas hydrate melts, methane gas is released. The properties of the cages formed by the three structures of gas hydrate crystals are shown in Table 1.

3.1.1. Structure I Gas Hydrates (sI)

These are body-centered cubic structures formed from small gas molecules and are usually found in deep ocean environments. They are formed from two different sizes and shapes of cage; small cages and large cages. The compositions of the two cages are:

I. Small Cages

They are composed of 9 dodecahedrons formed from 12 pentagons (5^{12}). They are not in contact with one another but are bonded to the larger cages.

II. Large Cages

They are made of 12 tetradecahedrons formed from 12 pentagons and 2 hexagons ($5^{12}6^{2}$). The small cages are attached to the large cages to form cubic structures.

3.1.2. Structure II Gas Hydrates (sII)

These hydrates have diamond lattices within cubic frameworks and are formed when natural gas or oil containing molecules larger than ethane but smaller than pentane. The structure II hydrates are usually found in oil and gas production and processing systems. They are also formed from two different shapes and sizes of cages. These are:

Table 1. Characteristics of the structures I, II, and H gas hydrates.

S/N	Properties of the Cages	Structure I		Structure II		Structure H		
1	Cavity	Small	Large	Small	Large	Small	Medium	Large
2	Description	5^{12}	$5^{12}6^{2}$	5^{12}	$5^{12}6^{4}$	$5^{12}6^{4}$	$4^{3}5^{6}6^{3}$	$5^{12}6^{8}$
3	Cavities/unit Cell	2	6	16	8	3	2	1
4	Density, kg/m^3	912		940		1952		
5	Unit cell parameters, nm	a = 1.2		a = 1.7		a = 1.21, c = 1.01		
6	Average Cavity Radius, Å	3.95	4.33	3.91	4.73	3.91	4.06	5.71
7	Variation in radius, %	3.4	14.4	5.5	1.73	-	-	-
8	Lattice type	cubic		Face-centered cubic		Hexagonal		
9	Coordination number,	20	24	20	28	20	20	36
10	Water molecules/unit Cell	46	-	136	-	34	-	-

I. Small Cages

The small cages comprise 16 dodecahedrons which are formed from 12 pentagons (5^{12}).

II. Large Cages

The large cages consist of 8 hexadecahedrons which are formed from 12 pentagons and 4 hexagons ($5^{12}6^{4}$).

3.1.3. Structure H Gas Hydrates (sH)

This is the newest structure of gas hydrates discovered and was also found to occur in the Gulf of Mexico. The structure H hydrates have hexagonal framework and also possess cavities large enough to hold molecules as big as naphtha and gasoline.

Table 1 shows the properties of the different structures of gas hydrates. The variation in distance of oxygen atoms from center of cage is the cavity radius and the number of oxygen molecules at the periphery of each cavity signifies the variation in radius. The estimation of structure H cavities was based on geometric models. The three different crystal structures of gas hydrates are cubic structure type I (sI), cubic structure type II (sII) and hexagonal structure (sH) as shown in Figure 5. The unit cell of the cubic structure type I requires 46 water molecules as host to form small and large cages. There are two small cages and six large cages in the unit cell.

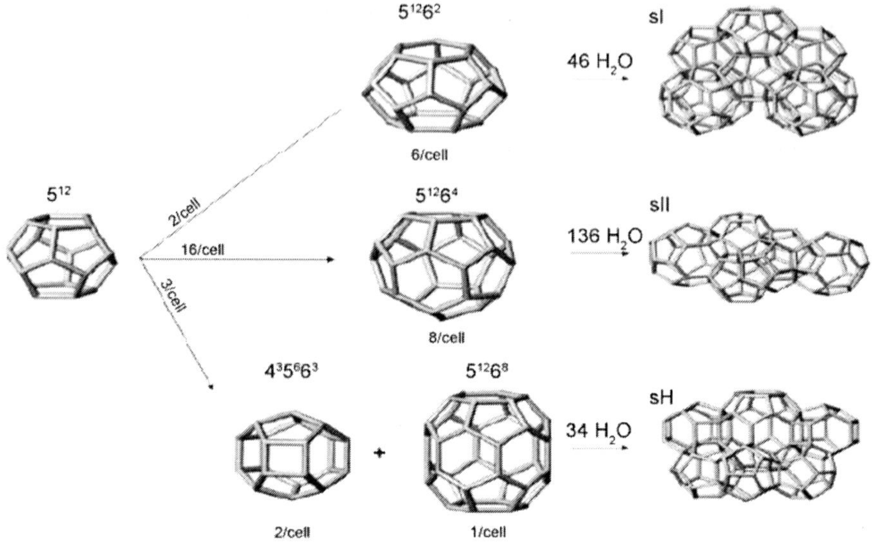

Figure 5. The structures of the classes of gas hydrate.

The small cage has the shape of a pentagonal dodecahedron and the large cage has hexagonal truncated trapezohedron shape. Typical guests molecules that form the type I hydrates are carbon dioxide and methane gas. The cubic structure type II requires 136 molecules of water to form small and large cages. There are sixteen small cages and eight large cages in a unit cell. The small cage has the shape of pentagonal dodecahedron while the large one has hexadecahedron shape. The structure type II hydrates are formed by oxygen and nitrogen gases. The hexagonal structure (sH) requires 34 molecules of water to form two small cages of different types and a huge cage. The formation of structure type H hydrate requires the guest molecules to be stable. The size of the large cavity allows large molecules (i.e. butane) to fit into the structure type H hydrates in the presence of other small gases to support the remaining cavities.

3.2. Properties of Gas Hydrates

Gas hydrates structures are in three different crystal structures: cubic structure type I (sI), cubic structure type II (sII), and hexagonal structure (sH) as shown in Figure 5. The type I (sI) gas hydrate is formed with those gas molecules having diameters between 4.2 and 6 Å, i.e. methane, ethane, carbon dioxide, and hydrogen sulfide. Other small molecules like nitrogen ($d < 4.2$ Å) form the type II (sII) gas hydrates. Molecules with diameters: 6 Å $< d < 7$ Å, i.e. propane and iso-butane also form sII gas hydrates. Larger molecules with diameters: 7 Å < d < 9 Å, i.e. iso-pentane and 2, 2-dimethylbutane form sH gas hydrates in the presence of smaller molecules like methane or nitrogen [4] as shown in Table 2.

Flow assurance problems caused by gas hydrates occur as a result of slow cooling of oil and gas in pipeline or rapid cooling due to depressurizing across the valves installed with the pipeline or distribution systems. Recent studies on gas hydrate showed that there are three primary conditions that influence hydrate formation in oil and gas pipelines and in petrochemical processes and they are:

1) The presence of water and gas components,

2) Low temperatures and,

3) High pressures.

Table 2. Properties of Ice, structures I and II gas hydrates.

S/N	Properties (Unit Cell)	Ice	Structure I (sI)	Structure II (sII)
1	Water molecules number	4	46	136
2	Lattice parameters at 273 K, nm	a = 0.452, c = 0.736	1.20	1.73
3	Dielectric constant at 273 K	94	~58	58
4	Water diffusion correlation time (*μsec*)	220	240	25
5	Water diffusion activation energy (kJ/m)	58.1	50	50
6	Shear Velocity (Vs), m/s	1949	1963.6	2001.1
7	Compressional Velocity (Vp), m/s	3870.1	3778.0	3821.8
8	Refractive index, 638 nm, −3°C	1.3082	1.3460	1.350
9	Density, kg/m³	916	912	940
10	Poisson's Ratio	0.33	~0.33	~0.33
11	Bulk Modulus (272 K)	8.8	5.6	-
12	Shear Modulus (272 K)	3.9	2.4	-
13	Velocity Ratio (comp./shear)	1.99	1.92	1.91
14	Linear thermal expn., K^{-1} (200 K)	56×10^{-6}	77×10^{-6}	52×10^{-6}
15	Heat capacity, J/kg-K	3800	3300	3600
16	Thermal conductivity, W/m-K (263 K)	2.23	0.49 ± 0.02	0.51 ± 0.02

There are also secondary influencing factors which favor hydrate formation such as high fluid velocities, agitation, pressure, pulsations or any source of fluid turbulence, the presence of CO_2 and H_2S [4] . Understanding the properties of gas hydrates is crucial to the exploration and production of these enormous deposits of methane gas. The presence of hydrate in the marine sediments significantly influences some of the properties of the sediments. This can be capitalized on and detected by field measurements and down hole logs.

3.3. Rate of Hydrate Formation

The formation of hydrate has two major steps-nucleation and growth and is a time-dependent process. The knowledge of the dynamics of hydrate formation and accumulation of hydrate crystals is important in determining the parameters for mass production of gas hydrates, and in understanding plug conditions in the gas pipeline and other equipment [17] . The former has an advantage for natural gas storage as hydrate and the later has an advantage for preventing of hydrate blockage in offshore processing, production and transportation of oil and gas. The rate of hydrate formation can be calculated using the Equation (2.0) [17] :

$$R = 4\pi K \mu_2 \left(f - f_{eq}\right) \tag{2.0}$$

where R is the rate of hydrate formation, K is the empirical kinetic parameter obtained by Englezos 1987, μ_2 is the second moment of the particle size distribution for the hydrate crystals, f is the fugacity of the gas at the given conditions, and f_{eq} is the fugacity of the gas at equilibrium conditions. The fugacities required can be calculated using the CNGA equation of state for natural gas mixtures where the compressibility factor (Z) is:

$$Z = \frac{1}{1 + \frac{344400 \times P \times 10^{1.785SG}}{T^{3.825}}} \tag{3.0}$$

P is the pressure (psig), SG the specific gravity of the gas relative to air, and T the temperature (°R). The second moment of the particle size distribution (μ_2) can be calculated by the following equations (Englezos et al. 1987):

$$\mu_2 = 4r^2 \mu_o \tag{4.0}$$

$$\mu_o = \frac{3M\left(N - N_{eq}\right)}{4\pi V \rho r^3} \tag{5.0}$$

M is the molecular mass of the hydrate, N is the number of moles of gas in solution at the given conditions, N_{eq} is the number of moles of gas in solution at the equilibrium conditions, V is the volume of water for the above number of moles of gas, ρ is the density of the hydrate, and r is the mean particle radius for the hydrate crystals.

4. BENEFITS OF GAS HYDRATE

Methane hydrate, commonly known as gas hydrate holds enormous amount of energy which cancan be used for different purposes. Natural gas contains mainly methane gas which is colorless, odorless and combusts completely to generate carbon dioxide, water and significant amount of energy. There are many gas hydrate deposits in permafrost and deepwater marines. Gas hydrate deposits can be explored to produce methane gas which can serve as a sustainable energy resource. Gas hydrate can also serves as a means of storing and transporting natural gas. This will serve as alternative solution to gas flaring of useful natural resource.

4.1. Gas Hydrate for Natural Gas Storage

Natural gas can be effectively and efficiently stored as gas hydrate. This can be achieved through proper study and understanding of the processes involved to convert the gas to hydrate and the processes required to prevent the hydrate from dissociating. The storage of natural gas as hydrates will require the synthesis of the hydrate and its regasification. This process is beneficial because the density of natural gas hydrates reduces the space requirements for the storage of natural gas. The stored gas can be used in the future and for peak-shaving applications to obtain a higher price for the natural gas as well as to ensure adequate natural gas supplies during periods of peak usage. Peak-shaving application is storing natural gas when natural gas demand is low, then selling the natural gas during periods of high demand.

Gas hydrate can be stored at equilibrium conditions with either its saturation temperature or pressure. The main factors that determine the optimum/limiting pressure and temperature are cost and weight of material, i.e. steel needed for hydrate storage vessel. There are basically three approaches to this operation to make it an economical process:

1) The first is the initial formation of large quantity of gas hydrates to avoid high pressure recompression on recycle,

2) In addition, reproducible, rapid conversion of the gas and water to hydrate to reduce the amount of water transported, and

3) Lastly, the transportation of gas hydrates to locations with small amount of refrigeration and dissociation units needed for the whole operation.

4.2. Sustainable Energy Resource

The estimate of the gas hydrate reserves in the world vary widely and a range between 10^{15} - $10^{18}m^3$ has been widely reported [19] . These estimates are based on research investigation and data obtained by different researchers on conventional hydrocarbon resources. The total quantity of gas in hydrate deposits exceeds, by a factor of two, the total energy content of fuel fossil recoverable reserves by conventional methods [19] . The magnitude of this resource can make gas hydrate reserves a sustainable energy resource. The potential benefit of gas hydrates is also supported by the environmentally friendly nature of natural gas compared to other fossil fuels [14] . Gas can be produced from hydrate deposits through dissociation of hydrate. The three methods of dissociation of gas hydrate are:

1) Depressurization: the pressure of the system containing the hydrate is reduced to a value lower than the hydration pressure at the corresponding temperature,

2) Thermal stimulation: the temperature of the system is allowed to increase above the hydration temperature at the corresponding pressure, and

3) The use of hydrate inhibitors: inhibitors such as salts and alcohols can be used to shift the equilibrium temperature and pressure of hydrate to the region of instability of the hydrate for effective dissociation of the hydrate to its gas and water components.

The most producible hydrate reserves are those directly adjacent gas reserves, such that gas production causes hydrate dissociation by reducing the reservoir pressure below hydrate stability pressure. A similar technique was used in a Siberian permafrost reservoir (Messoyakha) to produce for almost a decade in the 1970s. Gas production from hydrate deposits in the permafrost and ocean sediments containing hydrates is costly but technically feasible. In 2002, Mackenzie delta of Canada used depressurization and thermal stimulation to produce gas from hydrate reserves. The objective of the well was extended to other findings to produce gas from leaner hydrate reserves.

4.3. Gas Hydrate as a Transportation Means

Natural gas can be shipped or transported to different locations where there is complex terrain for pipeline and the quantity of the gas is insufficient to justify liquefaction. It may be economical and viable to transport by converting the gas to hydrate. Hydrate concentrates the gas by a factor that ranges from 150 - 180 and without the cost of liquefaction/compression or transporting the gas at high pressure. In hydrate transportation or shipment, as shown in Figure 6, preservation of the hydrate is vital for preventing losses due to dissociation of the hydrate [17] . Recent research works on hydrates have shown that only a minute amount of refrigeration, i.e. 20 °F is required to prevent dissociation of hydrate with rates that are several magnitudes less than those stipulated for lower or higher temperatures. The reason for the anomalous behavior of the hydrate is due to the presence of outer ice barriers that prevent inner-particle dissociation. The ice protective shell is as a result of re-freezing of water from the melted hydrate surface caused by endothermic process.

Research experiments and calculations on gas hydrate have shown that hydrate-gas transportation is economical for distances greater than 400 m. British Gas Ltd., an energy company in the United Kingdom has a pilot facility that converts natural gas to hydrate. The Japanese National Marine Research Institute also started a similar project to transport natural gas as hydrate.

Hydrate can be transported either as slurry or solid blocks depending on the location of target or destination [17] . Hydrate transportation in slurry form has a faster loading, unloading and also ease of handling. This form of transporting hydrate loses storage capacity to anti-freezing agents and extra equipment required to remove anti-freezing agents. The solid form of transporting hydrate uses simplified regasification facility and offers slightly more capacity per storage than that of the slurry form. In this form of transporting hydrate, there is slower loading, unloading and additional solid handling equipment is needed. The slurry form of transporting hydrates is usually a better option for distances of approximately 2500 miles or less while the solid form is suitable for distances of roughly 3500 miles or more [17] . The economics of the two forms of hydrate transportation is shown in Table 3.

The transportation of hydrate at high pressure can be highly expensive because the storage vessel must have thicker walls increasing the cost of transportation and decreasing the amount of hydrate transported.

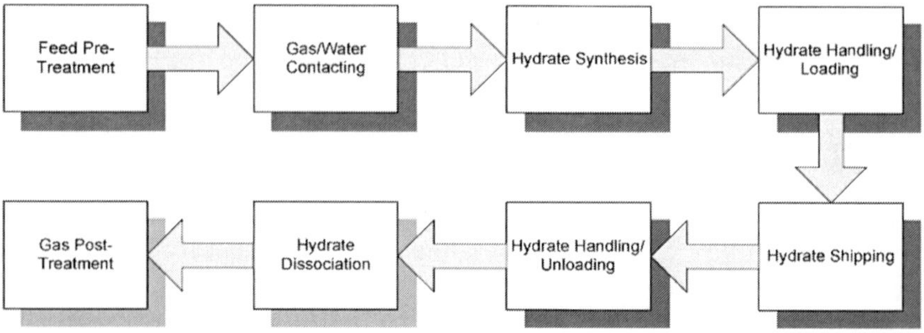

Figure 6. Schematic of a gas hydrate formation, transportation and dissociation system.

Table 3. Economics of hydrate transportation in slurry and solid forms.

S/N	Distance (miles)	Transit Time, days (18 Kts)	Total Trip Time, days		Percent time saved with slurry form, %
			Slurry	Solid	
1	1000	4.6	7.8	9.1	14.3
2	2000	9.3	12.5	13.8	9.4
3	3000	13.9	17.1	18.4	7.1
4	4000	18.5	21.7	23	5.7
5	5000	23.1	26.3	27.6	4.7

Unlike CNG, shipping gas hydrates at elevated pressures provides no increase in capacity due to the incompressibility of the gas hydrates. Also, polyurethane insulation is relatively more efficient at reducing hydrate losses due to dissociation than increasing the pressure rating of the vessel in order to transport hydrates at elevated temperature. However, at atmospheric pressure, the stable temperature for hydrates is below the freezing point of water which can make the hydrate to freeze any solid in the vessel. Therefore, to effectively handle the hydrate as fluid, ethanol can be added to depress the freezing point of the water. Alternatively, the hydrate slurry mixture can be frozen into manageable blocks and loaded onto the vessel for transportation. The ethanol-water-hydrate system will be more easily loaded and unloaded by the use of pumps, which will reduce the amount of time spent while loading and unloading. But, additional equipment will be required to separate the ethanol vapor from the natural gas. Also, the ethanol displaces natural gas hydrate, thereby reducing the effective capacity of the vessel.

Natural gas transportation as LNG is better economically because the shipping costs for gas hydrates are much higher than the shipping costs for LNG for high transportation distances. The costs for shipping gas hydrates are high because for every ton of natural gas shipped, 6.5 tons of water, which generates no revenue, is also shipped [17] . But for low transportation distances and low production capacities, the costs for shipping gas hydrates are lower than the costs for the entire LNG process. However, as transportation distance increases and capacity increases, LNG becomes a better economic option than gas hydrates in transporting natural gas.

Peak-shaving application is storing natural gas when natural gas demand is low, then selling the natural gas during periods of high demand. For peak-shaving application, shipping of the hydrate is not applicable because they are designed for the temporary storage of natural gas. Production and regasification of these hydrates is cheaper than that of LNG. The production of LNG requires very low temperatures with the use of expensive refrigeration systems while gas hydrates are produced at moderate temperatures with the use of less expensive refrigeration systems. This production condition of gas hydrates makes it economically more favorable than LNG for the temporary storage of natural gas.

5. CONCLUSIONS

The abundance of gas hydrate reserves can ultimately make gas hydrate to be a sustainable energy resource all over the world. These hydrate reserves hold significant amount of energy that is estimated to be more than twice the combined carbon of coal, conventional gas and petroleum reserves. The properties and formation of hydrates were studied and examined because they are paramount to efficient and effective exploration and development of hydrate reserves.

Gas hydrate can be stored at equilibrium conditions with either its saturation temperature or pressure. The factors that determine the optimum/limiting pressure and temperature are cost and weight of hydrate storage vessel. The stored gas can be used in the future and for peak-shaving applications to obtain a higher price for the natural gas as well as to ensure adequate natural gas supplies during periods of peak usage. Peak-shaving application is storing natural gas when natural gas demand is low, then selling the natural gas during periods of high demand. The possibility of storing and transporting natural gas as either gas hydrate solid or slurry was also considered. The advantages and disadvantages of solid and slurry forms of transporting gas hydrate were also examined. The slurry form of transport is usually a better option for distances of approximately 2500 miles or less while the solid form is suitable for distances of roughly 3500 miles or more.

Hydrate can serve as a sustainable energy resource and a means of storing and transporting natural gas from one end to the other. This will create an effective and efficient alternative for bulk gas transportation and storage for future use of the gas. Natural gas transportation as LNG is more economical because the shipping costs for gas hydrates are much higher than the shipping costs for LNG and for high transportation distances. The costs for shipping gas hydrates are high because for every ton of natural gas shipped, 6.5 tons of water, which generates no revenue, is also shipped. But for low transportation distances and low production capacities the costs for shipping gas hydrates are lower than the costs for the entire LNG process.

CITE THIS PAPER

Aregba, A.G. (2017) Gas Hydrate—Properties, Formation and Benefits. Open Journal of Yangtze Gas and Oil, 2, 27-44.

REFERENCES

1. Rempel, A.W. and Buffet, B.A. (1997) Formation and Accumulation of Gas Hydrate in Porous Media. Journal of Geophysical Research, 102, 151-164.
2. Tohidi, B. (2014) Advances in Avoiding Gas Hydrate Problems. Centre for Gas Hydrate Research & Hydrafact Ltd., Institute of Petroleum Engineering, Heriot-Watt University, 1-47.
3. Ruffine, L., Donvala, J.P., Charloua, J.L., Cremièrea, A. and Zehnderb, B.H. (2010) Experimental Study of Gas Hydrate Formation and Destabilisation Using a Novel High-Pressure Apparatus. Marine and Petroleum Geology, 27, 1157-1165.
4. Gabitto, J.F. and Tsouris, C. (2010) Physical Properties of Gas Hydrates: A Review. Journal of Thermodynamics, 2010, 1-12.
5. Atilhan, M., Aparicio, S., Benyahia, F. and Deniz, E. (2012) Natural Gas Hydrate. Advances in Natural Gas Technology, 10, 194-212.
6. Raz, S.B. (2012) A Study of Formation and Dissociation of Gas Hydrate. Master's Thesis, Office of Graduate Studies of Texas A&M University, College Station, 1-129.
7. Jin, S., Nagao, J. and Takeya, S. (2006) Structural Investigation of Methane Hydrates Sediments by Microfocus X-Ray Computed Tomography Technique under High Pressure Conditions. Japanese Journal of Applied Physics, 45, 714-716.
8. Minagawa, H., Nishikawa, Y. and Ikeda, I. (2008) Relation between Permeability and Pore-Size Distribution of Methane-Bearing Hydrates Sediments. Proceeding of the Offshore Technology Conference, Houston, 1-6.
9. Santamarina, J.C. and Ruppel, C. (2008) The Impacts of Hydrate Saturation on the Mechanical, Electrical and Thermal Properties of Hydrate-Bearing Sand, Silts and Clay. Proceedings of the 6th International Conference on Gas Hydrates, Vancouver, 1-12
10. Stoll, R.D. and Bryan, G.M. (1979) Physical Properties of Sediments Containing Gas Hydrate. Journal of Geophysical Research, 84, 1629-1634.

11. Pearson, C., Halleck, P.M., McGuire, P.L., Hermes, R. and Mathews, M. (1983) Natural Gas Hydrate Deposit: A Review of in Situ Properties. The Journal of Physical Chemistry, 87, 4180-4185.
12. Winters, W.J., Henry, P., Booth, J.S., Hovland, M. and Clennell, M.B. (1999) Formation of Natural Gas Hydrates in Marine Sediments. Conceptual Model of Gas Hydrate Growth conditioned by Host Sediments Properties, 104, 22985-23003.
13. Tinivella, U., Accaino, F., Guistiniani, M. and Loreto, M.F. (2009) Gas Hydrates and Mud Volcanoes Offshore Antarctic Peninsula: A Geophysical Study. Proceedings of the Goldschmidt Conference, Davos, Switzerland, 1-8.
14. Tohidi, B., Anderson, R., Clennell, M.B., Burgass, R.W. and Biderkab, A.B. (2001) Visualobservation of Gas Hydrate Formation and Dissociation in Synthetic Porous Media by Means of Glass Micro-Models. Journal of Geology, 29, 869-870.
15. Kingston, E., Clayton, C. and Priest, J. (2008) Gas Hydrate Growth Morphologies and Their Effects on the Stiffness and Damping of a Hydrate-Bearing Sand. Proceedings of the 6th International Conference on Gas Hydrates, Vancouver, Vancouver, Canada, 1-7.
16. Moridis, G.J. and Kowalsky, M. (2006) Gas Production from Unconfined Class 2 Hydrate Accumulations in the Oceanic Subsurface. Lawrence Berkeley National Laboratory, Earth Sciences Division, 1-8.
17. Mannel, D. and Puckett, D. (2008) Natural Gas Hydrate Transportation. University of Oklahoma, Norman, 1-54.[Citation Time(s):7]
18. USGS—Science for a Changing World (2014) Review on International Gas Hydrate R&D Research Programs. USGS-Science for a Changing World, 1-79.
19. Birchwood, R., Dai. J., Shelander, D., Boswell, R., Collett, T., Cook, A., Dallimore, S., Fujii, K., Imasato, Y., Fukuhara, M., Kusaka K., Murray, D. and Saeki, T. (2010) Developments in Gas Hydrates. Oilfield Review, 22, 18-33.

CHAPTER 2

Natural Gas Dehydration

Michal Netušil and Pavel Ditl

[1] Czech Technical University in Prague, Faculty of Mechanical Engineering, Department of Process Engineering, Prague, Czech

1. INTRODUCTION

The theme of natural gas (NG) dehydration is closely linked with storage of natural gas. There are two basic reasons why NG storage is important. Firstly, it can reduce dependency on NG supply. With this in mind, national strategic reserves are created. Secondly, NG storage enables the maximum capacity of distribution lines to be exploited. NG is stored in summer periods, when there is lower demand for it, and is withdrawn in the winter periods, when significant amounts of NG are used for heating. Reserves smooth seasonal peaks and also short-term peaks of NG consumption. Underground Gas Storages (UGS) are the most advantageous option for storing large volumes of gas. Nowadays there are approximately 135 UGSs inside the European Union. Their total maximum technical storage capacity is around 10^9 m_s^3. According to the latest update, over $0{,}7 \cdot 10^9$ m_s^3 of additional storage capacity will come on stream in Europe by 2020 [1]. There are three types of UGSs: (1) Aquifers, (2) Depleted oil/gas fields, and (3) Cavern reservoirs (salt or hard rock). Each of these types possesses distinct physical characteristics. The important parameters describing the appropriateness of UGS use are storage capacity, maximum injection/ withdrawal performance, and gas contamination during storage. Generally, the allowable pressure of stored gas inside a UGS is up to 20 MPa. The pressure inside increases as the gas is being injected, and decreases when gas is withdrawn. The output gas pressure depends on further distribution. Distribution sites from UGS normally begin at 7 MPa. The temperature of the gas usually ranges from 20 - 35°C. The exact temperature varies with the location of the UGS and with the time of year.

1.1. Water in The Gas

A disadvantage of UGSs is that during storage the gas become saturated by water vapors. In the case of depleted oil field UGSs, vapors of higher hydrocarbons also contaminate the stored gas. The directive for gas distribution sets the allowable concentration of water and concentration of higher hydrocarbons. In the US and Canada, the amount of allowable water in the gas is specified in units: pounds of water vapor per million cubic feet (lbs/MMcft). This amount should be lower than 7 lbs/MMcft [2], which is equivalent to 0,112 g_{H2O}/m_S^3. In Europe, the concentration of water and higher hydrocarbons is specified by their dew point temperature (T_{dew}). T_{dew} for water is -7°C for NG at 4 MPa, and T_{dew} for hydrocarbons is 0°C for NG at the operating pressures [3]. This value for water is equivalent to roughly 0,131 g_{H2O}/m_S^3 of NG at 4 MPa. As was stated above, the distribution specifications depend on the geographic region in which they are applied. For example, in Nigeria water T_{dew} should be below 4°C for NG at 4 MPa, which means that the NG can contain more than twice as much water vapors as in Europe.

The water content of NG at saturation is dependent on temperature and pressure. With increasing pressure of the gas the water content decreases, and with increasing temperature the water content in the gas increases. This is well presented in Figure No. 20, Chapter 20, in the GPSA Data Book, 12thEdition. The water content of the gas can be calculated using the following equation [4, 5]:

$$w_{water} = 593{,}335 \cdot exp(0.05486 \cdot t_G) \cdot P_G^{-0.81462} \quad (1)$$

Where w_{water} is in kilograms of water per 10^6 m_s^3 of NG, t_G is temperature of NG in °C, and P_G is pressure of NG in MPa.

The average value of water in NG withdrawn from UGS is 2 - 5 times higher than required. An NG dehydration step is therefore essential before further distribution.

1.2. Problems With Water In The Gas

If the temperature of pipeline walls or storage tanks decreases below the T_{dew} of the water vapors present in the gas, the water starts to condense on those cold surfaces, and the following problems can appear.

- NG in combination with liquid water can form methane hydrate. Methane hydrate is a solid in which a large amount of methane is trapped within the crystal structure of water, forming a solid similar to ice. The methane hydrate production from a unit amount of water is higher than the ice formation. The methane hydrates formed by cooling may plug the valves, the fittings or even pipelines.
- NG dissolved in condensed water is corrosive, especially when it contains CO_2 or H_2S.
- Condensed water in the pipeline causes slug flow and erosion.
- Water vapor increases the volume and decreases the heating value of the gas.
- NG with the presence of water vapor cannot be operated on cryogenic plants.

2. DEHYDRATION METHODS

2.1. Absorption

The most widely-used method for industrial dehydration of NG is absorption. Absorption is usually performed using triethyleneglycol sorbent (TEG). Absorption proceeds at low temperatures and the absorbed water is boiled out from TEG during regeneration in a reboiler at high temperatures. Some physical properties of pure TEG are given in the following text.

Viscosity data vs. temperature is shown in Table 1 and is shown in a graph in Figure 1 [6].

It follows from Figure 1 that the kinematic viscosity of TEG increases dramatically with low temperatures. The temperature of TEG during a process should never decrease below 10°C. The reason is to prevent pump damage or even clogging of the flow. For temperatures above 100°C, the viscosity changes just slightly and the average kinetic viscosity value $5 \cdot 10^{-6}$ m^2/s can be used. For a description of the dependency of viscosity on

TABLE 1. Kinematic viscosity of TEG according to temperature

[°C]	4	10	16	21	27	32	38	43	49	54	60	66
[m²/s 10⁻⁵]	9,53	7,094	5,367	4,124	3,214	2,539	2,032	1,646	1,348	1,116	0,934	0,788
[°C]	71	77	82	88	93	99	104	110	116	121	127	132
[m²/s 10⁻⁵]	0,672	0,577	0,5	0,436	0,384	0,34	0,303	0,272	0,245	0,222	0,203	0,186
[°C]	138	143	149	154	160	166	171	177	182	188	193	199
[m²/s 10⁻⁵]	0,171	0,159	0,147	0,138	0,129	0,121	0,115	0,109	0,103	0,099	0,095	0,091

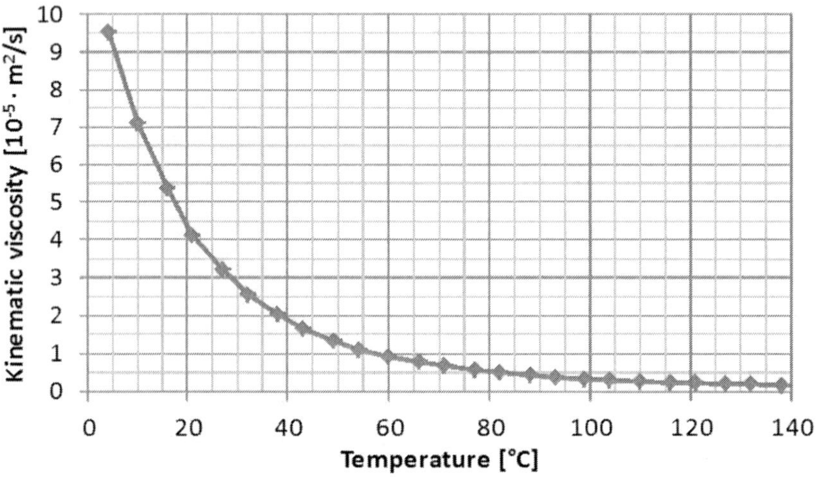

FIGURE 1. Kinematic viscosity of TEG as a function of temperature

temperature, the following polynomial interpolation coefficients have been calculated.

$$\nu = 1{,}159 \cdot 10^{-4} - 5{,}655 \cdot 10^{-6} t + 1{,}347 \cdot 10^{-7} t^2 + 1{,}872 \cdot 10^{-9} t^3 + 1{,}572 \cdot 10^{-11} t^4 - 7{,}820 \cdot 10^{-14} t^5 + 2{,}116 \cdot 10^{-16} t^6 - 2{,}393 \cdot 10^{-19} t^7 \quad (2)$$

The density of TEG at various temperatures is shown in the following table [7].

TABLE 2. Density of TEG according to temperature

[°C]	10	16	21	27	32	38	43	49	54	60	66	71	77
[kg/m³]	1132	1128	1124	1119	1114	1111	1106	1101	1098	1093	1089	1084	1080
[°C]	82	88	93	99	104	110	116	121	127	132	138	143	149
[kg/m³]	1076	1071	1066	1063	1058	1053	1050	1045	1041	1036	1032	1028	1023
[°C]	154	160	166	171	177	182	188	193	199	204	210	216	221
[kg/m³]	1019	1015	1010	1007	1002	997	993	989	984	980	975	972	967

Table 2 shows that in the range of working temperatures the density of TEG is a linear function of temperature. The following equation can be used for determining the density at a certain temperature.

$$\rho = -0{,}7831 \cdot t + 1140 \tag{3}$$

Finally, the thermal conductivity of TEG does not change in the range of working temperatures and has a value of 0,194 W/m²/°C.

For determining the physical properties of TEG solutions with water (concentrations c_{TEG} above 95 wt.%), the activity coefficient of water in TEG can be approximated by the following equation.

$$\gamma = -0{,}0585 \cdot c_{TEG} + 6{,}2443 \tag{4}$$

The industrial absorption dehydration process proceeds in a glycol contactor (a tray column or packet bed). In a contactor, a countercurrent flow of wet NG and TEG is arranged. During the contact, the TEG is enriched by water and flows out of the bottom part of the contactor. The enriched TEG then continues into the internal heat exchanger, which is incorporated at the top of the still column in the regeneration section of the absorption unit. It then flows into the flash drum, where the flash gases are released and separated from the stream. The TEG then runs to the cold side of the TEG/TEG heat exchanger. Just afterwards, the warmed TEG is filtered and then runs into the regeneration section, where is it sprayed in the still column. From there, the TEG runs into the reboiler. In the reboiler, water is boiled out of the TEG. The regeneration energy is around 282 kJ per liter of TEG. The temperature inside should not exceed 208°C, due to the decomposition temperature of TEG. Regenerated (lean) TEG is then pumped back through the hot side of the TEG/TEG and NG/TEG heat exchanger into the top of the contactor. The entire method is depicted in Figure 2 [8].

The circulation rate (l_{TEG}/kg_{H2O}) and the purity of the regenerated TEG are the main limiting factors determining the output T_{dew} of NG. The amount of circulating TEG is around 40 times the amount of water to be removed. The minimal TEG concentration should be above 95 wt.%, but the recommended value is higher. However, to obtain TEG concentration above 99 wt.% enhanced TEG regeneration has to be implemented.

The simplest regeneration enhancing method is gas stripping. Proprietary designs DRIZO®, licensed by Poser-NAT, and COLDFINGER®, licensed by Gas Conditioners International, have been patented as an alternative to traditional stripping gas units. The Drizo regeneration system utilizes a recoverable solvent as the stripping medium. The patent operates with iso-octant solvent, but the typical composition of the stripping medium is about 60 wt.% aromatic hydrocarbons, 30 wt.% naphthenes and 10 wt.% paraffins. The three-phase solvent water separator is crucial for this method. The Coldfinger regeneration system employs a cooling coil (the "coldfinger") in the vapor space of the surge tank. The cooling that takes place there causes condensation of a high amount of vapors. The condensate is a water-rich TEG mixture, which is led to a further separation process [9]. Figure 3depicts enhanced regeneration systems which replace the simple reboiler in the regeneration section shown in Figure 2.

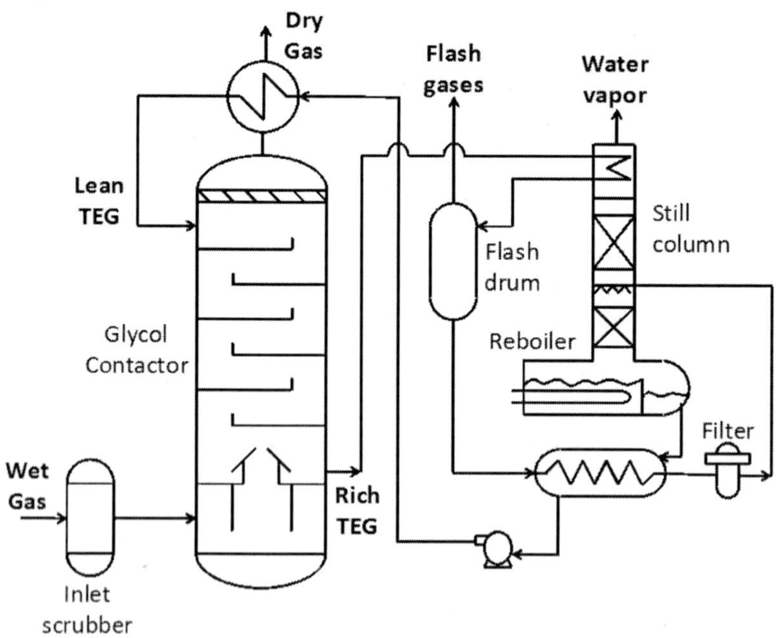

FIGURE 2. TEG absorption dehydration scheme

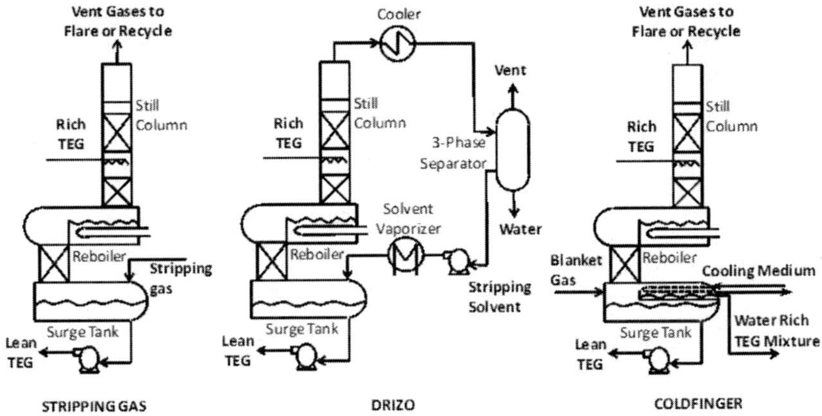

FIGURE 3. Enhanced TEG regeneration systems

2.2. ADSORPTION

The second dehydration method is adsorption of water by a solid desiccant. In this method, water is usually adsorbed on a mole sieve, on a silica gel or on alumina. A comparison of the physical properties of each desiccant is shown in Table 3 [4,10].

TABLE 3. Comparison of the physical properties of desiccants used for dehydration of NG

Properties	Silica gel	Alumina	Mol. sieves
Specific area [m²/g]	750 – 830	210	650 – 800
Pore volume [cm³/g]	0,4 – 0,45	0,21	0,27
Pore diameter [Å]	22	26	4-5
Design capacity [kg $_{H2O}$/100 kg $_{desiccant}$]	7-9	4-7	9-12
Density [kg/m³]	721	800 - 880	690 – 720
Heat capacity [J/kg/°C]	920	240	200
Regeneration temperature [°C]	230	240	290
Heat of desorption [J]	3256	4183	3718

The amount of adsorbed water molecules increases with the pressure of the gas and decreases with its temperature. These facts are taken into account when the process parameters are designed. Adsorption dehydration columns always work periodically. A minimum of two bed systems are used. Typically one bed dries the gas while the other is being regenerated.

Regeneration is performed by preheated gas, or by part of the dehydrated NG, as depicted in Figure 4.

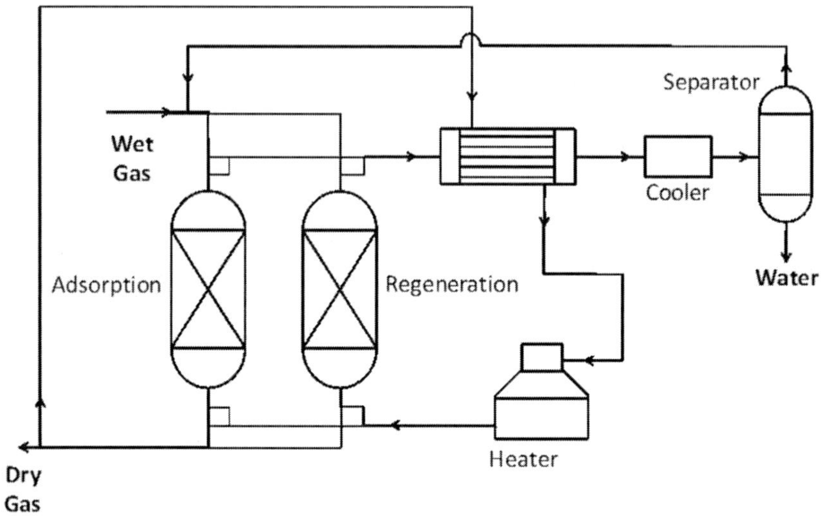

FIGURE 4. Scheme of the temperature swing adsorption dehydration process

This method is known as temperature swing adsorption (TSA). Regeneration can also be performed by change of pressure - pressure swing adsorption (PSA). However, PSA is not industrially applied for NG dehydration. Further details about PSA can be found in [11,12]. A combination of those two methods (PSA and TSA) seems to be a promising future option for adsorption dehydration of NG. This idea is still in the research process.

In classical applications, the TSA heater is realized as an ordinary burner or as a shell and tube heat exchanger warmed by steam or by hot oil. The regeneration gas warms in the heater and flows into the column. In the column passes through the adsorbent and the water desorbs into the regeneration gas. The water saturated regeneration gas then flows into the cooler. The cooler usually uses cold air to decrease the temperature of the regeneration gas. When the water saturated regeneration gas is cooled, partial condensation of the water occurs. The regeneration gas is led further into the separator, where the condensed water is removed.

A downstream flow of wet NG through the adsorption column is usually applied. In this way, floating and channeling of an adsorbent is avoided. Regeneration is performed by countercurrent flow in order to provide complete regeneration from the bottom of the column, where the last contact of the dried NG with the adsorbent proceeds. The typical temperature course for 12 h regeneration of molecular sieves is shown in Figure 5 [13].

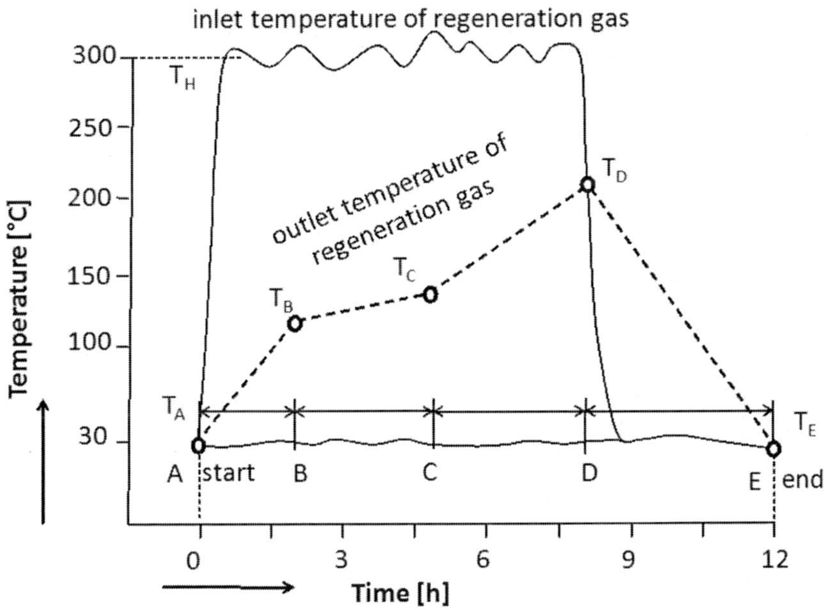

FIGURE 5. Typical temperature course for 12 h TSA regeneration of molecular sieves

The shape of the curve representing the course of the outlet regeneration gas temperature is typically composed of four regions. They are specified by time borders A, B, C and D with appropriate border temperatures T_A, T_B, T_C and T_D. Regeneration starts at point A. The inlet regeneration gas warms the column and the adsorbent. At a temperature around 120°C (T_B) the sorbed humidity starts to evaporate from the pores. The adsorbent continues warming more slowly, because a considerable part of the heat is consumed by water evaporation. From point C, it can be assumed that all water has been desorbed. The adsorbent is further heated to desorb C_{5+} and other contaminants. The regeneration is completed when the outlet temperature of the regeneration gas reaches 180 - 190°C (T_D). Finally, cooling proceeds from point D to point E. The temperature of the cooling gas should not decrease below 50°C, in order to prevent any water condensation from the cooling gas [13].

Part of the dehydrated NG is usually used as the regeneration gas. After regenerating the adsorbent the regeneration gas is cooled, and the water condensed from it is separated. After water separation, the regeneration gas is added back to the inlet stream or alternatively to the dehydrated stream.

The total energy used for regeneration is composed of heat to warm the load (30%), heat for desorption (50%) and heat going into the structure (20%). With proper internal insulation of the adsorption towers, the heat going to the structure can be minimized and around 20% of the invested energy can be saved.

So-called LBTSA (Layered Bed Temperature-Swing Adsorption) processes are an upgrade of the TSA method. Here, the adsorption column is composed of several layers of different adsorbents. Hence the properties of the separate adsorbents are combined in a single column. For example, in NG dehydration a combination of activated alumina with molecular sieve 4A is used. Alumina has better resistance to liquid water, so a thin layer is put in first place to contact the wet NG. This ordering supports the lifetime of the molecular sieve, which is placed below the alumina layer. The effect of adsorbent lifetime extension is shown for two cases in Figure 6. It can be seen that contact with liquid water dramatically decreases the lifetime of the molecular sieve [14].

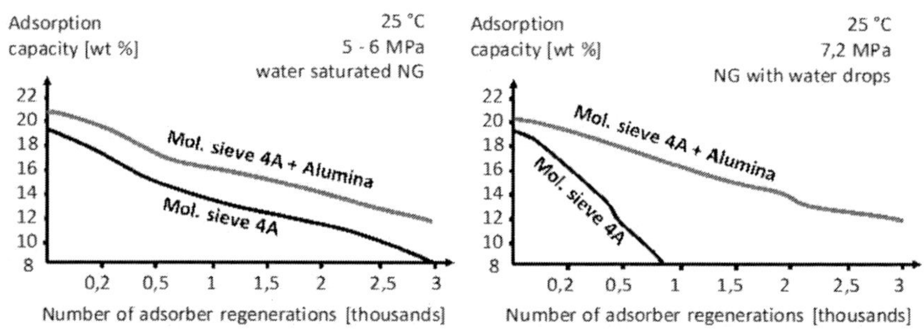

FIGURE 6. Effect of layered bed adsorption on the lifetime of the adsorbent

2.3. Condensation

The third conventional dehydration method employs gas cooling to turn water molecules into the liquid phase and then removes them from the stream. Natural gas liquids and condensed higher hydrocarbons can also be recovered from NG by cooling. The condensation method is therefore usually applied for simultaneous dehydration and recovery of natural gas liquids.

NG can be advantageously cooled using the Joule-Thompson effect (JT effect). The JT effect describes how the temperature of a gas changes with pressure adjustment. For NG, thanks to expansion, the average distance between its molecules increases, leading to an increase in their potential energy (Van der Waals forces). During expansion, there is no heat exchange with the environment, and no work creation. Therefore, due to the conservation law, the increase in potential energy leads to a decrease in kinetic energy and thus a temperature decrease of NG. However, there is another phenomenon connected with the cooling of wet NG. Attention should be paid to the formation of methane hydrate. Hydrates formed by cooling may plug the flow. This is usually prevented by injecting methanol or monoethylenglycol (MEG) hydrate inhibitors before each cooling. Figure 7 depicts an industrial application of dehydration method utilizing the JT effect and MEG hydrate inhibition.

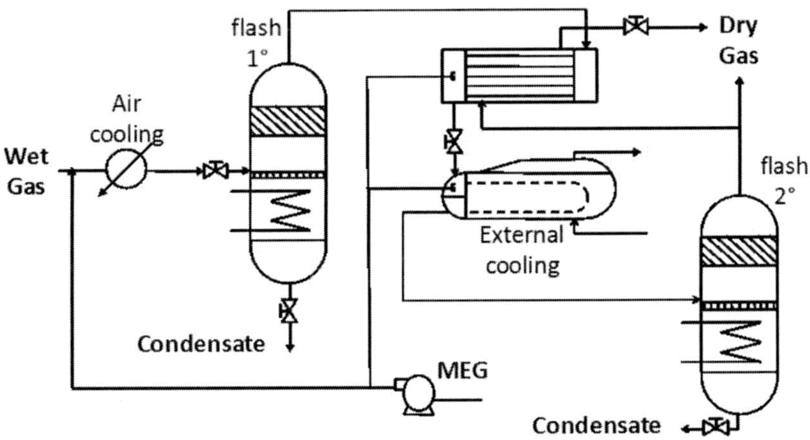

FIGURE 7. Dehydration method utilizing the JT effect and hydrate inhibition

The wet NG is throttled in two steps inside the flash tanks. The lower temperature (due to the JT effect) of the gas stream in the flash tanks leads to partial condensation of the water vapors. The droplets that are created are removed from the gas stream by a demister inside the flashes. In cases where cooling by the JT effect is insufficient (the usable pressure difference between the inlet and outlet of the gas is insufficient), the air pre-cooler and the external cooler are turned on. Since dehydration is normally applied to large volumes of NG, the external coolers need to have high performance, so this type of cooling is very energy expensive. For dehydration of low pressures NG the external coolers consumes up to 80% of total energy of dehydration unit. However, if the usable pressure difference is high, the JT effect inside the flashes is so strong that internal heating of the flashes is required to defreeze any methane hydrate or ice that may form. A condensation method is applied when suitable conditions for the JT effect are available.

2.4. Supersonic Separation

The principle of this method lies in the use of the Laval Nozzle, in which the potential energy (pressure and temperature) transforms into kinetic energy (velocity) of the gas. The velocity of the gas reaches supersonic values. Thanks to gas acceleration, sufficient temperature drops are obtained. T_{dew} of water vapor in NG is reached, and nucleation of the droplets proceeds. Figure 8 depicts the basic design of a supersonic nozzle [15].

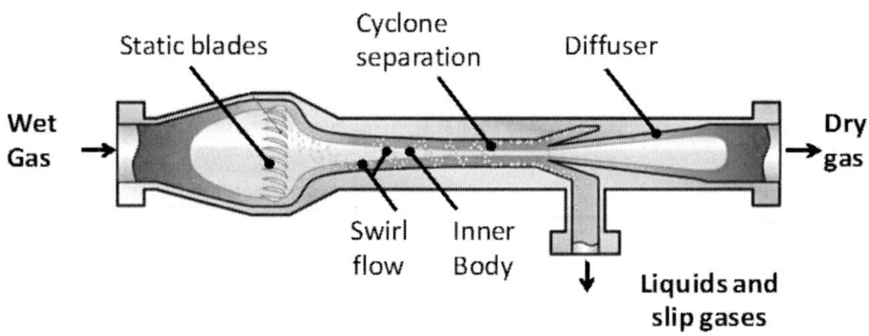

FIGURE 8. Design of a supersonic nozzle for NG dehydration

At the inlet to the nozzle there are static blades which induce a swirling flow of the gas. The water droplets that form are separated by the centrifugal force on the walls. The centrifugal force in the supersonic part of the nozzle can reach values up to 500 000 g [16]. The thin water film on the walls moves in the direction of flow into the separation channel. The separation channel leads into the heated degas separator. From here, the slip gas is returned back to the main stream and the water condensate is removed. After separation of the water it is important to recover the pressure of the gas from its kinetic energy. A shock wave is used to achieve this. Generally, shock waves form when the speed of a gas changes by more than the speed of sound. In supersonic nozzles, the shock wave is created by rapid enlargement of the nozzle diameter. This part of the nozzle is called the diffuser. Thanks to the diffuser 65 - 80% of the inlet pressure is recovered [17]. This section might also include another set of static devices to undo the swirling motion. The profile of pressure, temperature and velocity of a gas passing through the supersonic nozzle is depicted below.

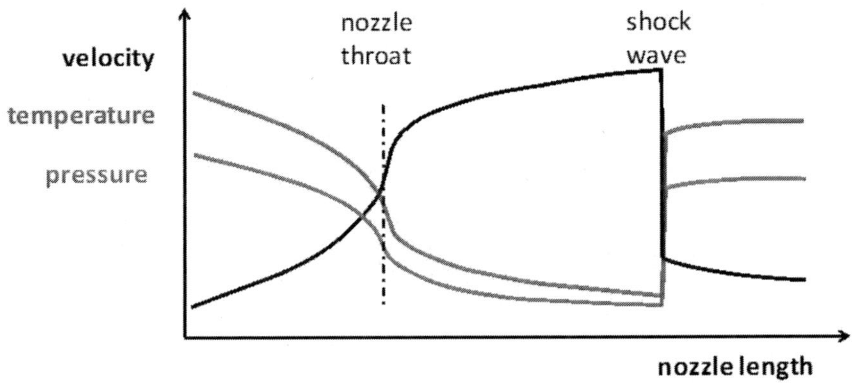

FIGURE 9. Profile of pressure, temperature and velocity of a gas passing through the supersonic nozzle

The scheme of a supersonic dehydration line working on the principle introduced here is depicted in Figure 10.

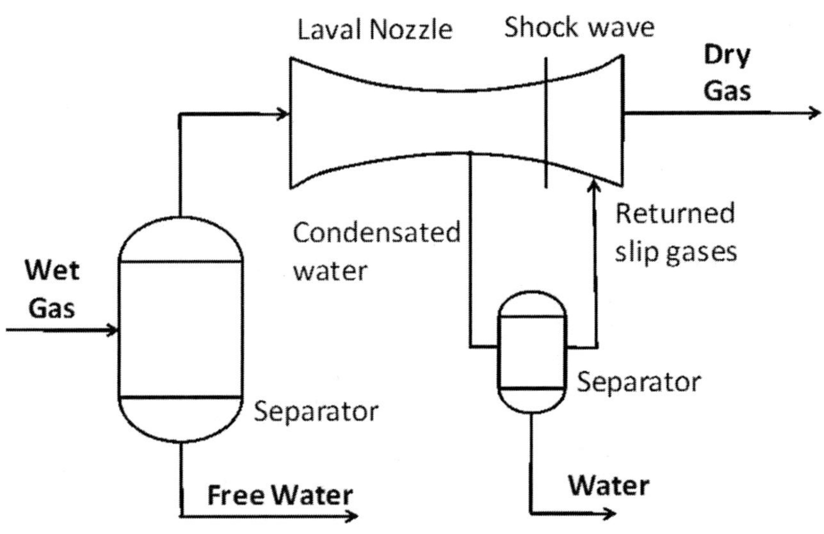

FIGURE 10. Scheme of a supersonic dehydration line

The gas residence time in the supersonic nozzle is below two milliseconds [18]. This time interval is too short for any methane hydrate formation, so no inhibitors are needed. To obtain supersonic velocity of the gas, the inlet diameter should be minimally $\sqrt{5}$ times higher than the nozzle throat. The geometry of the tapered section of the Laval nozzle is calculated by the following equations [16]

$$\frac{D-D_{cr}}{D_1-D_{cr}} = 1 - \frac{1}{X_m^2}\left(\frac{x}{L}\right)^3 \qquad \left(\frac{x}{L} \leq X_m\right) \tag{5}$$

$$\frac{D-D_{cr}}{D_1-D_{cr}} = \frac{1}{(1-X_m)^2}\left(1 - \frac{x}{L}\right)^3 \qquad \left(\frac{x}{L} > X_m\right) \tag{6}$$

Where D_1, D_{cr}, L, X_m stand for the inlet diameter, the throat diameter, the length of the tapered section, and the relative coordinate of tapered curve, respectively; x is the distance between an arbitrary cross section and the inlet, and D is the convergent diameter at an arbitrary cross section of x.

A model of a supersonic dehydration unit was analyzed with the use of numerical simulation tools, and the separation efficiency in respect to lost pressure was evaluated. The simulations were performed on water saturated NG at 30 MPa and 20°C. The results are presented in Table 4 [19].

TABLE 4. Supersonic water separation efficiency in respect to pressure lost in the nozzle

Pressure lost in nozzle (%)	17,3	20,0	27,6	49,0	51,5
Water separation efficiency (%)	40	50	90	94	96

The supersonic separation is a promising new technology. The main advantage of the method is the small size of the supersonic nozzle. For example, a nozzle 1,8 m in length placed in a housing 0,22 m in diameter was used for dehydrating 42 000 m_s^3 per hour of water-saturated NG at 25°C compressed to 10 MPa to output water $T_{dew} < -7$°C [20]. The corresponding absorption contactor would be 5 m in height and 1,4 m in diameter, and the corresponding adsorption line would be composed of two adsorbers 3 m in height and 1 m in diameter. A further advantage is the simplicity of the supersonic dehydration unit. The supersonic nozzle contains no moving parts and requires no maintenance. The operating costs are much lower than for other methods. The only energy-consuming devices are the pumps for removing the condensate and the heater for the degas separator. However, during supersonic dehydration a pressure loss occurs. However, if the same pressure loss were used for the JT effect, the temperature drop would be 1,5 – 2,5 times lower [21]. Supersonic separation enables simultaneous removal of water and higher hydrocarbons from the treated gas, and can be used as pretreatment method before NG liquefaction. This method could also be usable for other applications of gas separation.

The application of supersonic separation has some disadvantages. The most limiting condition of use is the need for stationary process parameters. Fluctuations in temperature, pressure or flow rate influence the separation efficiency. However, it is in many cases impossible to achieve constant process parameters. For example, this is the case when withdrawing NG from UGS. However, supersonic dehydration can be used even in this case. The problem is solved by arranging several nozzles into a battery configuration with a single common degas separator. The battery configuration enables an optimal number of nozzles to be switched on, depending on the inlet parameters of the gas. The scheme of a possible arrangement is depicted in Figure 11 below [22].

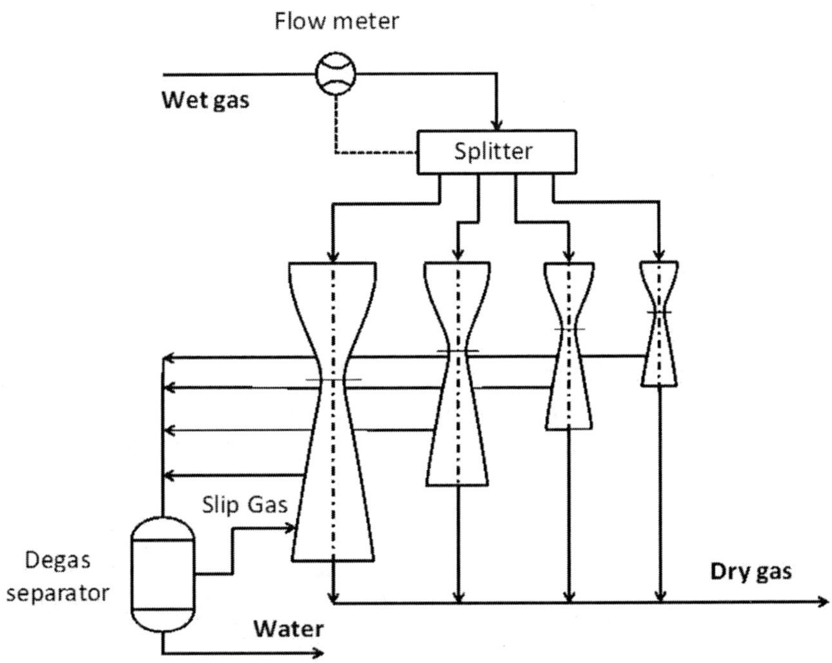

FIGURE 11. Arrangement of supersonic nozzles for unsteady inlet parameters of NG

Nozzle A is designed to process 80% of the nominal gas flow. Nozzles B, C, D are in the proportion 4:2:1 with respect to the processed gas flow. Nozzles B, C, D together can process 40% of the nominal gas flow. This arrangement therefore enables ± 20% deviation of the nominal flow to be covered. With appropriate switching of the nozzles, the maximal deviation between the real gas flow and the designed flow for the combination of nozzles is below 4%.

A further disadvantage of the supersonic dehydration is its novelty. The appropriate nozzle design is complicated, and "know how" is expensive. The geometry of the nozzle ranges in the order of micrometers. In addition, the construction material has to withstand abrasion and the impacts of a shock wave.

3. COMPARISON OF CONVENTIONAL DEHYDRATION METHODS

3.1. General Comparison

Each of the methods presented here has its advantages and disadvantages. Absorption by TEG is nowadays the most widely used method. Outlet T_{dew} around -10°C is usually reached and this water concentration is sufficient for pipeline distribution of NG. Indeed, with improved reboiler design (Vacuum Stripping, Drizo, Coldfinger), the outlet T_{dew} is even 2 - 3 times lower. However TEG has a problem with sulfur, and with gas contaminated with higher hydrocarbons. The TEG in the reboiler foams, and with time it degrades into a "black mud". BTEX emissions (the acronym stands for benzene, toluene, ethylbenzene and xylenes) in the flash gases and in the reboiler vent are a further disadvantage.

Adsorption dehydration can achieve very low outlet water concentration $T_{dew} < -50°C$, and contaminated gases are not a problem. Even corrosion of the equipment proceeds at a slower rate. However, adsorption requires high capital investment and has high space requirements. The adsorption process runs with at least two columns (some lines use three, four, or as many as six). Industrial experience indicates that the capital cost for an adsorption line is 2 - 3 times higher than when absorption is used [5]. In addition, the operating costs are higher for adsorption than for absorption.

Expansion dehydration is the most suitable method in cases where a high pressure difference is available between UGS and the distribution connection. However, the difference decreases during the withdrawal period and becomes insufficient, so that an external cooling cycle is needed.

A cycle for regenerating hydrate inhibitor from the condensate separated inside the flashes is also required.

The general overview of areas suitable for application of target dehydration method is depicted on the following Figure 12.

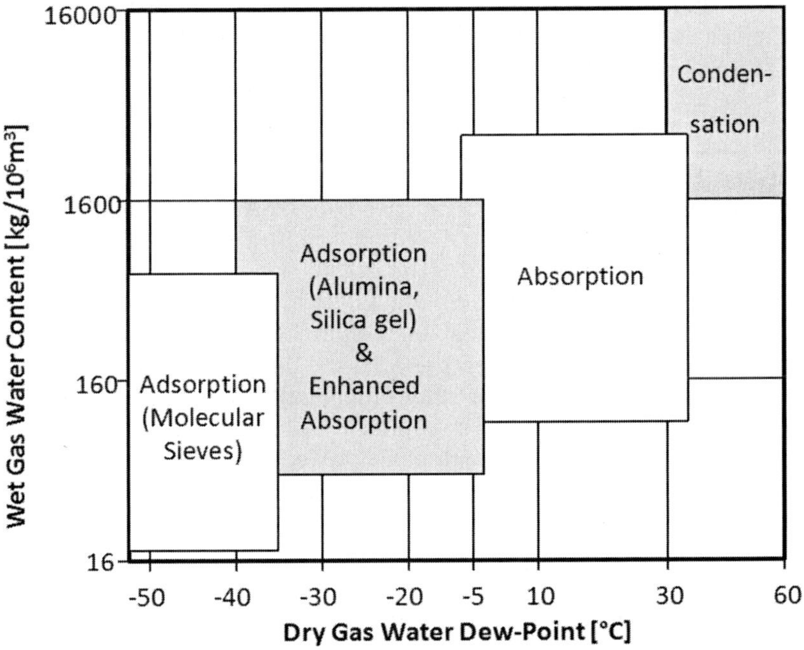

FIGURE 12. Overview of areas suitable for application of target dehydration method

3.2. Energy Comparison Based On Own Analyses And Available Data

The energy demand of the conventional methods presented here is compared on the basis of a model, where a volume of 10^5 m_s^3/h of NG from UGS is processed. The NG is water saturated at a temperature of 30°C. The pressure of the gas is varied from 7 to 20 MPa, but in the case of the condensation method the pressure range starts at 10 MPa. The required outlet water concentration of in NG is equivalent to dew point temperature -10°C at gas pressure 4 MPa [23].

The calculation of TEG absorption is based on GPSA (2004) [24]. The results are compared with the paper by Gandhidasan (2003) [5] and with industrial data provided by ATEKO a.s. The total energy demand is composed of heat for TEG regeneration in the reboiler, energy for the pumps, filtration and after-cooling the lean TEG before entering the contactor. Enhanced regeneration is not considered. The basic parameters for the calculation are: regeneration temperature 200°C, concentration of lean TEG 98,5 wt.%, and circulation ratio 35 l_{TEG}/kg_{H2O} [24].

For calculating adsorption dehydration, molecular sieve 5A is considered to be the most suitable adsorbent. The total energy demand is directly connected to the regeneration gas heater, and no other consumption is assumed. The calculations are again based on GPSA (2004) [24]. The results are compared with the paper by Gandhidasan (2001) [4] and also with the publication by Kumar (1987) [7]. The calculation procedure for GPSA and Gandhidasan arises from the summation of the particular heats, i.e. the heat for adsorbent warming, the heat for column warming (insulation of the adsorption towers is considered), and the heat for water desorption. Kumar's calculation procedure runs differently. The regeneration step is divided into four regions (as depicted in Figure 5). Afterwards, we determine what individual phenomena proceed in each region, what the border and average temperatures are, and how much energy is required to cover these phenomena. Finally, the demands for each region are added. The basic parameters for all procedures are: temperature of the regeneration gas 300°C, time of adsorption/regeneration 12 h, and two column designs.

The condensation method was calculated on the basis of industrial data provided by TEBODIN s.r.o. and supplementary calculations of the JT effect. The key parameter influencing energy demand is the pressure of NG. Because it is not feasible to apply this method for low pressures, and because the provided data starts at 10 MPa, the pressure range was adjusted. The total energy demand consists of the air pre-cooling unit, the external cooling, the pumps for MEG injection and condensate off take, the heat for MEG regeneration, and flash heating.

3.3. Results Of The Analyses

The results obtained for the TEG absorption method are the same for each of the calculation procedures, and good agreement with industrial data was also obtained. However, the calculation procedures for the adsorption method lead to different results. Hence the average energy demand value was taken as the reference. The maximum deviation from it is below 20% for all calculation procedures. The source of the deviation lies in the "loss factor and the non-steady state factor". In the case of the condensation method, the calculated values for the JT effect were in good agreement with the industrial data, but the amount of data was limited, resulting in limited representation of the condensation method.

The final energy consumption results for each dehydration method are summarized by the graph in Figure 13.

For low pressures (pressure of NG from UGS < 13 MPa), the condensation method is the most demanding. Its demand decreases linearly with pressure to a value of 145 kW for 13 MPa. At this point, the energy demand for the condensation method is roughly the same as for the adsorption method. When the NG pressure is further increased from 13 MPa to 16 MPa, the energy demand for the condensation method still decreases, but with a lowering tendency. For a high pressure of NG (> 16 MPa), the energy demand of the condensation method is at its lowest, and it remains nearly constant, with an average value around 36 kW.

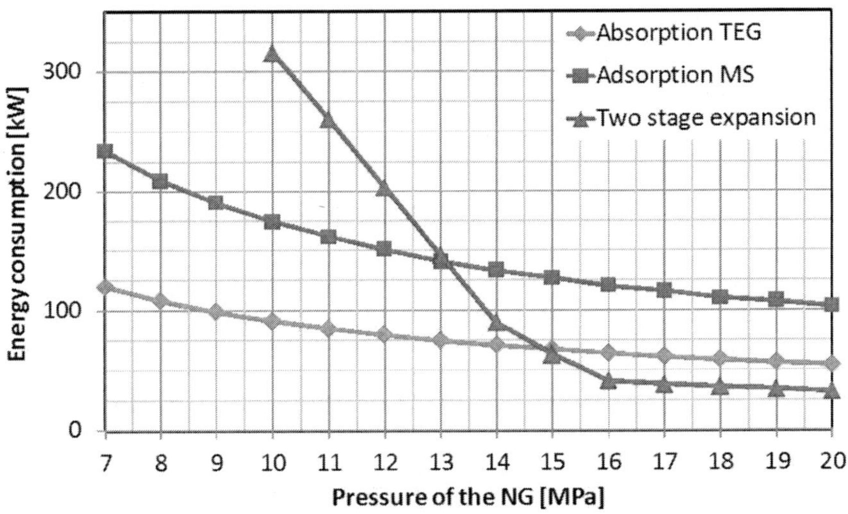

FIGURE 13. Energy consumption results for conventional dehydration methods

The course of the energy demand for the adsorption and absorption methods is quite similar: with increasing pressure of dehydrated NG the energy demand slowly decreases. The absorption method is less demanding on the whole pressure scale, and begins with consumption of 120 kW at 7 MPa. The adsorption method starts with 234 kW at 7 MPa, but the energy demand decreases slightly more as the pressure of NG in UGS rises. This leads to a gradual decrease in the difference between these methods, and the energy demand at the final pressure value of 20 MPa is equal to 54 kW for absorption and 103 kW for adsorption.

3.4. Conclusions Of The Analyses

By far the highest energy demand of the condensation method at low pressures of NG from UGS is due to the pressure being close to the distribution pressure, so that pressure cannot be used for the JT effect in flashes. Cooling is then compensated by the air pre-cooler and the external cooling device, which are unsuitable for large volumes of processed NG. However, as the pressure difference between UGS and the distribution site increases, the space for expansion rises and the JT effect proceeds with increasing impact. This is projected into a linear decrease in the energy demand of the air pre-cooler and the external cooling device. From the point where there is a pressure of NG > 14 MPa, flash heating is gradually turned on to prevent any freezing caused by the strong JT effect. The energy demand of flash heating is reflected in the total energy consumption. Finally, for pressures of NG > 16 MPa, total cooling and subsequent condensation is achieved by the JT effect. The total energy demand remains constant, and consists of flash heating and inhibitor injection and regeneration.

In case of the adsorption and absorption dehydration method, the similar falling course of the energy demand with increasing pressure of NG can be explained by the fact that with increasing pressure within a UDG the amount of water present in the NG decreases. The absorption method generally consumes less energy, because the regeneration of TEG is less demanding than adsorbent regeneration. The composition of the total energy demand of the adsorption method can be divided into three parts. The heat for water desorption is approximately 55%, for warming the adsorbent it is 31%, and for warming the column it is 14%. It also has to be assumed that just part of the heat in the regeneration gas transfers to the adsorbent, the column and heat loss leaves to the atmosphere, and the balance leaves with the hot gas.

In brief, in cases of high pressure the most appropriate dehydration method from the energy demand point of view is the stored NG condensation method. This holds for NG from UGS with pressure > 15 MPa and distribution pressure requirements 7 MPa. For lower pressures, the condensation method is used if the objective is to recover NGL and remove water simultaneously. However, this is usually not the case when storing NG in a UGS. In cases where insufficient pressure difference is available, the absorption method is therefore favored over the adsorption method in terms of energy demand. TEG absorption is almost twice less demanding. However, if a gas contaminated with sulfur or higher hydrocarbons is being processed, the TEG in the reboiler foams and degrades with time. This can occur when a depleted oil field is used as a UGS. Adsorption is preferred in cases where very low T_{dew} (water concentration lower than 1 ppm can be achieved) of NG is required, for example when NG is liquefied.

It is worth to note that the power comparison can be used as a measure of the technical excellence. From power data the specific energy consumption was calculated and its values indicate that the energy cost is much lower than the investment cost (depreciations). On the other hand the energy cost represents more than 60 % of the total operating cost.

4. CONCLUSIONS AND RECOMMENDATIONS

The chapter should help in the selection of a proper dehydration method and in calculating NG dehydration. The following methods are available as options: absorption, adsorption and condensation. Absorption is used in cases when emphasis is not placed on the water content of the output stream, and when low operating and capital investment are required. Adsorption is used in cases when bone dry NG is required. Low temperature separation employing the JT effect is used in cases where a sufficient pressure drop is available between the input and the output of the dehydration unit. Supersonic nozzles are a promising method that will in future displace these three conventional methods.

We have selected for citation here articles and procedures that we consider to provide reliable results.

ACKNOWLEDGEMENT

The authors are grateful for the financial support provided by Ministry of Industry and Trade of the Czech Republic (program TIP nr. FR-TI1/173).

REFERENCES

1. Gas infrastructure Europe2011Map Dataset in Excel-format Storage map. Available: http://www.gie.eu/maps_data/storage.html.Accessed 2011 Mar 8.
2. M. Foss, 2004Interstate Natural Gas Quality Specifications and Interchangeability. Center for Energy Economics.
3. NET4GAS2011Gas quality parameters. Available at: http://extranet.transgas.cz/caloricity_spec.aspx.Accessed 2011 Mar 8.

4. P. Gandhidasan, A. Al-Farayedhi, A. Al-Mubarak, 2001Dehydration of natural gas using solid desiccantsEnergy 26855868
5. P. Gandhidasan, 2003Parametric Analysis of Natural Gas Dehydration by Triethylene Glycol Solution.Energy Sources25189201
6. C. H. E. M. Group, Inc, 2012Triethylene Glycol- Liquid Density Data. Available at: http://www.chem-group.com/services/teg-density.tpl.Accesed 2012 Mar 6.
7. C. H. E. M. Group, Inc, 2012Triethylene Glycol- Kinematic Viscosity Data. Available at http://www.chem-group.com/services/teg-viscosity.tpl.Accesed 2012 Mar 6.
8. A. Bahadori, H. B. Vuthaluru, (2009) Simple methodology for sizing of absorbers for TEG gas dehydration systems. Energy 34 19101916
9. Hubbard R.A, Campbell J.M2000An appraisal of gas dehydration processesHydrocarbon Engineering57174
10. M. Tagliabue, D. Farrusseng, S. Valencia, S. Aguado, U. Ravon, C. Rizzo, 2009Natural gas treating by selective adsorption: Material science and chemical engineering interplayChemical Engineering Journal155553566
11. I. Roušar, P. Ditl, 1993Pressure swing adsorption: analytical solution for optimum purge Original Research, Chemical Engineering Science 48723734
12. I. Roušar, P. Ditl, M. Cekal, 1993Pressure swing adsorption- the optimization of multiple bed units, Precision Process Technology: perspectives for pollution prevention: 483492
13. S. Kumar, 1987Gas Production EngineeringHouston: Gulf Professional Publishing 239 p.
14. G. Jochem, 2002Axens Multibed Systems for the Dehydration of Natural Gas, PETEM.
15. P. Schinkelshoek, H. D. Epsom, 2008Supersonic gas conditioning- commercialization of twister technology. 87th Annual Convention. Grapevine, Texas.
16. C. Wen, X. Cao, J. Zhang, L. Wu, 2010Three-dimensional Numerical Simulation of the Supersonic Swirling SeparatorTwentieth International Offshore and Polar Engineering Conference. Beijing, China.
17. F. Okimoto, J. M. Brouwer, 2002Supersonic gas conditioningWorld Oil348991
18. Q. Hu, D. Jiang, J. Qiu, Z. , 2010Numerical study of the spontaneous nucleation of self-rotational moist gas in a converging-diverging nozzleInternational Journal of Computational Fluid Dynamics2936
19. A. Karimi, M. A. Abdi, 2006Selective dehydration of high-pressure natural gas using supersonic nozzlesChemical Engineering and Processing 48560568
20. Twister BV2012Twister supersonic separator- Experience. Available: http: //twisterbv.com/products-services/twister-supersonic-separator/experience/.Accesed 2012 Mar 7.
21. M. Betting, H. Epsom, 2007High velocities make a unique separator and dewpointer, World Oil, 197200
22. M. Netušil, P. Ditl, T. González, 2012Raw gas dehydration on supersonic swirling separator, 19th International Conference Process Engineering and Chemical Plant Design, Krakow.
23. M. Netusil, P. Ditl, 2011Comparison of three methods for natural gas dehydrationJournal of Natural Gas Chemistry20471476
24. GPSA2004Engineering Data Book. 12th ed. Tulsa: GPSA Press.

CHAPTER 3

Transport Properties of Superfluid Dipolar Fermi Gas at Low Temperatures

Maryam Khademi Dehkordi

Falavarjan Branch, Islamic Azad University, Isfahan, Iran

ABSTRACT

Transport coefficients of superfluid dipolar Fermi gas are calculated by using the Boltzmann equation approach. The interaction between Bogoliubov-Bogoliubov quasiparticles in the collision integral is considered in binary process. The shear viscosities $\eta_{xx\uparrow} = \eta_{yy\uparrow}, \eta_{xy\uparrow}$; $\eta_{xz\uparrow} \cong \eta_{yz\uparrow}$; and $\eta_{zz\uparrow}$ are proportional to T^{-6} ; T^{-8} ; and T^{-10} , respectively. Also, we have found the elements of the diffusive thermal conductivities $K_{xx\uparrow} = K_{yy\uparrow}$ with temperature dependency T^{-5} and $K_{zz\uparrow}$ which is proportional to T^{-7} and other components which are zero.

Keywords: Shear Viscosity, Diffusive Thermal Conductivity, Superfluid Dipolar Fermi Gas

1. INTRODUCTION

Fermi gases of dipolar particles present a different physical picture. Being magnetically polarized, these particles interact via anisotropic dipole-dipole forces. The potential dipole-dipole interaction is partially repulsive and partially attractive. These features of the dipole-dipole interaction have very important consequences not only for the scattering properties of the particles in ultra cold gas, but also for the stability of the system and for the variety of its properties. The recent success in observing quantum degeneracy in ultra cold atomic Fermi gas [1] -[4] stimulates a search for gaseous Fermi systems with an achievable temperature of super fluid phase transition, which is generally very low.

Fermionic superfluidity was obtained in 2003 and allowed for the exploration of BEC-BCS crossover physics, a theoretical scenario bridging the gap between the Bardeen-Cooper-Schrieffer mean-field theory describing the behavior of weakly attracting fermions (scattering length a small and negative), and the strongly attractive regime (a small and positive) where the system behaves as a Bose-Einstein condensate of tightly bound dimers. Remarkably, although the theoretical foundations of this crossover physics had been laid in the early 1980s by the pioneering works of Nozières, Schmitt-Rink and Leggett [5] [6] , its first experimental confirmation was only made possible by the possibility of tuning interactions in ultra-cold atom vapors using Fano-Feshbach resonances [7] -[8] .

Braby et al. [9] have computed the thermal conductivity and sound attenuation length of a dilute atomic Fermi gas in the framework of kinetic theory. Above the critical temperature for superfluidity, T_c, the quasiparticles are fermions, whereas below T_c, the dominant excitations are phonons. They have calculated the thermal conductivity in both cases. They have found that at unitarity the thermal conductivity K in the normal phase scales as $K\alpha T^{3/2}$.In the superfluid phase they have found $K\alpha T^2$.

The shear viscosity and the second viscosity of superfluid ^6Li have been formulated in the scope of Boltzmann equation approach [10] [11] . The value of the shear viscosity decreases as $(1-T/T_c)$ with temperature; and the second viscosity of superfluid ^6Li is temperature independent in the limit of low temperatures, and in the near transition temperature behaves as inverse of the gap temperature. Theoretical study show that a gas of spin-polarized atomic ^{6}Li becomes superfluid at densities and temperatures comparable with those at which the Bose-Einstein experiments are performed [12] .

The shear viscosity tensor of the A_1-phase of superfluid ^{3}He is calculated at low temperatures and melting pressure, by using the Boltzmann equation approach. They obtained that the elements of the shear viscosities η_{xy}, η_{xz} and η_{zz} are proportional to $(T/T_c)^{-2}$ [13] [14] . Shahzamanian [15] calculated the components of the diffusive thermal conductivity of superfluid ^{3}He-A using approximate collision integrals at low temperatures. The results are $K_{||} \propto T^{-4}$ and $K_{\perp} \propto T^{-2}$.

Transport coefficients of the superfluid p-wave Fermi gas with weak interaction were calculated using the Boltzmann equation approach and the quasiparticle relaxation rate [16] . They have found that the superfluid Fermi gases under their considerations are similar to $^{3}He-A_1$.

Diffusive thermal conductivity tensor of p-wave superfluid was calculated using the Boltzmann equation approach at low temperatures [17] . They have shown that their results are in good agreement with A_1-phases of superfluid ^{3}He at low temperatures [18] .

Here we present a calculation of the transport properties of a superfluid dipolar Fermi gas, based on the derivations of Abrikosov and Khalatnikov [19] (shear viscosity and diffusive thermal conductivity). We used a sufficiently high magnetic field so that all quasiparticles with spin up will go to superfluid state. Important collision processes are the binary scattering of quasiparticles, the decay processes in which one Bogoliubov quasi particle decays into three and coalescence processes in which three Bogoliubov quasi particles coalesce to produce one. We obtained the transition probabilities at low temperature. By using the Boltzmann equation, we calculated shear viscosity and diffusive thermal conductivity tensors.

2. TRANSITION PROBABILITIES

The interaction between quasiparticles in the superfluid dipolar Fermi gas was found by performing a Bogoliubov transformation on the normal state interaction:

$$H=\frac{1}{4}\sum_{1,2,3,4}\langle 3,4|T|1,2\rangle a_4^{\dagger}\, a_3\, a_1 a_2$$
$$=\frac{1}{4}\sum_{1,2,3,4}\langle 3,4|T|1,2\rangle\left(u_{4\uparrow\uparrow}\alpha_{4\uparrow}^{\dagger}-v_{4\uparrow\uparrow}^{*}\alpha_{-4\uparrow}\right)\left(u_{3\uparrow\uparrow}\alpha_{3\uparrow}-v_{3\uparrow\uparrow}^{*}\alpha_{-3\uparrow}\right)$$
$$\times\left(u_{1\uparrow\uparrow}\alpha_{1\uparrow}-v_{1\uparrow\uparrow}\alpha_{-1}^{\dagger}\right)\left(u_{2\uparrow\uparrow}\alpha_{2\uparrow}-v_{2\uparrow\uparrow}\alpha_{-2\uparrow}\right). \tag{1}$$

The Hamiltonian interaction contains the terms $\alpha_4^{\dagger}\,\alpha_3\,\alpha_{-2}\alpha_1$, $\alpha_4^{\dagger}\,\alpha_3\,\alpha_2\alpha_1$, $\alpha_4^{\dagger}\,\alpha_1\alpha_{-3}\alpha_2$, $\alpha_4^{\dagger}\,\alpha_3\,\alpha_{-2}\alpha_{-1}$ and $\alpha_{-4}\alpha_{-3}\alpha_2\alpha_1$. The last two processes are not allowed because in each process the total energy should be conserved.

At low temperatures Bogoliubov coefficients can be written as $u_p\cong 1$ and $v_p\cong 0$, because more particles are gathered in the nodes of gap, that gap in direction $\boldsymbol{p}$ on the Fermi surface is $\Delta_p=\Delta(T)\sin\theta_{p_i}$, where $\Delta(T)$ is the maximum gap and θ_p is the angle between the quasiparticle momentum and gap axis $\hat{l}$, which is supposed to be in the direction of z-axis, in this region of temperature $\sin\theta_{p_i}\cong 0(i=1,2,3,4)$ [20] . Using the bogoliubov transformation, transition probabilities in polarized Fermi gas are:

$$W_{22}(\uparrow\uparrow)=2\pi\left|\langle 4\uparrow 3\uparrow|H|1\uparrow 2\uparrow\rangle\right|^2=1/4\left|\left(V_{p_1-p_3}+V_{p_3-p_1}\right)-\left(V_{p_3-p_2}+V_{p_2-p_3}\right)\right|^2,$$
$$W_{31}(\uparrow\uparrow)=2\pi\left|\langle 4\uparrow|H|1\uparrow,2\uparrow,-3\uparrow\rangle\right|^2\cong 0,$$
$$W_{13}(\uparrow\uparrow)=2\pi\left|\langle 3\uparrow,4\uparrow,-2\uparrow|H|1\uparrow\rangle\right|^2\cong 0. \tag{2}$$

The interparticle interaction in polarized dipolar gases that the dipoles are polarized along the z-axis reads,

$$V_d(\boldsymbol{q}) = \frac{2I^2}{3\sqrt{2\pi}}\left(3\cos^2\theta_q - 1\right), \tag{3}$$

where $\boldsymbol{I}$ is magnetic dipole, and θ_q is the angle between the momentum $\boldsymbol{q}$ and the z-axis [21] . Choosing $\boldsymbol{p}_1$ along the z-axis, $\theta_{1i}\,(i = 2,3,4)$ is the angle between $\boldsymbol{p}_1$ and $\boldsymbol{p}_i$, and φ is the angle between the plane of $(\boldsymbol{p}_1, \boldsymbol{p}_2)$ with $(\boldsymbol{p}_3, \boldsymbol{p}_4)$. The relations $\cos\theta_{13}$ and $\cos\theta_{14}$ according to the momentum conservation are [22]

$$\cos\theta_{13} = \cos^2(\theta/2) + \sin^2(\theta/2)\cos\varphi, \tag{4}$$

$$\cos\theta_{14} = \cos^2(\theta/2) - \sin^2(\theta/2)\cos\varphi. \tag{5}$$

At low temperatures more particles are contributing in contacts about Fermi momentum [19] . By replacing Equations (3)-(5) into (2) and considering the fact that the size of momentums are equal to Fermi momentum and by neglecting the potential effects of trap in Fermi momentum, transition probability is obtained in the form below:

$$W(\theta,\varphi) = \frac{8I^4}{\pi}\cos^4(\theta/2)\sin^4(\theta/2)\cos^2\varphi. \tag{6}$$

3. COLLISION INTEGRAL

The Boltzmann transport equation reads

$$\frac{\partial n}{\partial t} + \frac{\partial n}{\partial \boldsymbol{r}}\cdot\frac{\partial \varepsilon}{\partial \boldsymbol{p}} - \frac{\partial n}{\partial \boldsymbol{p}}\cdot\frac{\partial \varepsilon}{\partial \boldsymbol{r}} = I(n), \tag{7}$$

where $I(n)$ is the collision integral. The full distribution function is given by $n(\boldsymbol{p},\boldsymbol{r}) = n_0(\boldsymbol{p}) + \delta n(\boldsymbol{p})$. where $n_0(\boldsymbol{p}) = \left(\exp(\varepsilon_0 - \mu)/k_B T + 1\right)^{-1}$

and $\delta n(\boldsymbol{p})$ is a small departure from equilibrium. We wrote the deviation from equilibrium as:

$$\delta n(\boldsymbol{p}) = -\frac{n_0(\boldsymbol{p})(1-n_0(\boldsymbol{p}))}{T}\chi(\boldsymbol{p},\boldsymbol{r}), \tag{8}$$

by using (8) in (7) we got [23]

$$-\frac{1}{2}\frac{\partial n_0}{\partial E_{p_0\uparrow}} p_i \frac{\partial E_{p_0\uparrow}}{\partial p_k}\left(\frac{\partial u_i}{\partial r_k}+\frac{\partial u_k}{\partial r_i}-\frac{2}{3}\delta_{ij}\nabla\cdot\boldsymbol{u}\right) = I(\delta n_\uparrow(p)). \tag{9}$$

The collision integral is form [23]

$$I(\boldsymbol{p}_1,\uparrow) = \frac{(m^*)^3 k_B T}{8\pi^4}\int \mathrm{d}x \int \mathrm{d}y \int \frac{\mathrm{d}\Omega}{2\pi} \times \int \frac{d\varphi_2}{2\pi}\frac{W(\theta,\varphi)}{\cos(\theta/2)} n_0(t) n_0(x+y-t)$$
$$[1-n_0(x)][1-n_0(y)][\chi_1+\chi_2-\chi_3-\chi_4], \tag{10}$$

here $x=(\varepsilon_3-\mu)/k_BT$, $y=(\varepsilon_4-\mu)/k_BT$, $t=(\varepsilon_1-\mu)/k_BT$.

We expressed $\chi(\boldsymbol{p})$ as a series of spherical harmonics in the polar angles Θ_1, Φ_1. Thus

$$\Psi(\boldsymbol{p}_1) = \Psi(p_1,\Theta_1,\Phi_1)\sum_{n=0}^{\infty}\sum_{m=-n}^{n}\Psi_n^m(t)P_n^{|m|}(\cos\theta_1)\mathrm{e}^{im\Phi}, \tag{11}$$

by inserting (11) in (10) and integrate over y we have

$$I_n^m(t) = \frac{(m^*)^3 k_B T}{8\pi^4}\int \mathrm{d}x \times \int \frac{\sin\theta \mathrm{d}\theta \mathrm{d}\varphi}{2\pi}\frac{W(\theta,\varphi)}{\cos(\theta/2)}\left(-\frac{\partial n_0}{\partial \varepsilon}\right)K(t,x)$$
$$\times\left[\Psi_n^m(t)+\Psi_n^m(-x)\times P_n(\cos\theta)-\Psi_n^m(x)\{P_n(\cos\theta_3)+P_n(\cos\theta_4)\}\right], \tag{12}$$

where

$$K(t,x) = \frac{\mathrm{e}^{-t}+1}{\mathrm{e}^{-x}+1}\frac{(x-t)}{\mathrm{e}^{(x-t)}-1}. \tag{13}$$

We divided $\Psi_n^m(t)$ in to even and odd functions of t, and in the same way separated $I_n^m(t)$. Since $K(t,x)=K(-t,-x)$, only the even part of $\Psi_n^m(t)$ contributes to the even part of I, and similarly for the odd part [24] .

$$\int \mathrm{d}x K(t,x)\left\{\Psi_{ns}^m(t)-\lambda_{ns}\Psi_{ns}^m(x)\right\} = \text{an even function of } t, \tag{14}$$

$$\int \mathrm{d}x K(t,x)\left\{\Psi_{ns}^m(t)-\lambda_{ns}\Psi_{ns}^m(x)\right\} = \text{an odd function of } t, \tag{15}$$

that an even function of t used for calculating of shear viscosity and odd function used for thermal conductivity [24] .

4. SHEAR VISCOSITY

The coefficients shear viscosity describes the response of the momentum current density to a transverse velocity field

$$\Pi_{ij} = -\sum_{ijkl}\eta_{ijkl}\left(\frac{\partial u_i}{\partial r_k}+\frac{\partial u_k}{\partial r_i}-\frac{2}{3}\delta_{ij}\nabla\cdot\boldsymbol{u}\right), \tag{16}$$

Π_{ij} is expressed in terms of the bogoliubov qusipartcle distribution function δn_p by

$$\Pi_{ij} = \sum_p p_i\left(n_p\right)_j \delta n_p', \tag{17}$$

to solve the linearized Boltzmann equation, it is suitable to define $q(t)$ as

$$\chi(t) = p_i\frac{\partial E_1}{\partial p_{1k}}q(t)\left[\frac{\partial u_i}{\partial r_k}+\frac{\partial u_k}{\partial r_i}-\frac{2}{3}\delta_{ij}\nabla\cdot\boldsymbol{u}\right]. \tag{18}$$

By substituting (18) in (9) and then in (17) and compared with (16), we got [24]

$$\eta_{ijkl} = -\frac{4p_f^5}{(2\pi)^3 m^*}\int d\Omega_p \hat{p}_i \hat{p}_j \hat{p}_k \hat{p}_l \left(\int dt \frac{\partial n_0}{\partial t} q_{2s\uparrow}(t)\right). \tag{19}$$

By solving the even part of t in (15) and with regard to the fact that system is polarized, we have [24] ,

$$\eta_{ijkl} = \frac{p_f^5}{\pi^5 m^*}\int d\Omega_p \hat{p}_i \hat{p}_j \hat{p}_k \hat{p}_l \left(\frac{B_\uparrow}{1-\lambda_{2s\uparrow}} C(\lambda_{2s\uparrow})\right), \tag{20}$$

in this equation $B_\uparrow$, $\lambda_{2s\uparrow}$ and $C(\lambda_{2s\uparrow})$ are

$$B_\uparrow = \frac{16\pi^5}{(m^*)^3 T^2}\left[\int \frac{\sin\theta d\theta d\varphi}{\cos(\theta/2)} W(\theta,\varphi)\right]^{-1}, \tag{21}$$

$$\lambda_{2s\uparrow} = \frac{\int \frac{\sin\theta d\theta d\varphi}{\cos(\theta/2)} W(\theta,\varphi)\left\{1-\frac{3}{4}(1-\cos\theta)^2 \sin^2\varphi\right\}}{\int \frac{\sin\theta d\theta d\varphi}{\cos(\theta/2)} W(\theta,\varphi)}, \tag{22}$$

$$\begin{aligned} C(\lambda_{2s\uparrow}) &= \frac{1-\lambda_{2s\uparrow}}{4}\sum_{n=0}^{\infty}\frac{4n+3}{(n+1)(2n+1)\{(n+1)(2n+1)-\lambda_{2s\uparrow}\}} \\ &= \frac{\lambda_{2s\uparrow}-1}{2\lambda_{2s\uparrow}}\left\{\gamma+\ln 2+\frac{1}{2}\psi_d(s_1)+\frac{1}{2}\psi_d(s_2)\right\}, \end{aligned} \tag{23}$$

and $\gamma = 0.577\cdots$ is Euler's constant, ψ_d is a digamma function and

$$s_1 = \frac{3}{4}+\frac{1}{4}\sqrt{8\lambda+1}, \quad s_2 = \frac{3}{4}-\frac{1}{4}\sqrt{8\lambda+1}. \tag{24}$$

$W(\theta,\varphi)$ only replaces with $\frac{1}{4}W_{22}(\uparrow\uparrow)$, because $W_{13}(\uparrow\uparrow)$ and $W_{31}(\uparrow\uparrow)$ are nearly zero.

For obtaining $\lambda_{2s\uparrow}$ and $B_{\uparrow}$, we used (6) and note that θ is small for superfluid case and its maximum value is $\pi T/\Delta(0)$ [20] , where maximum gap, $\Delta(0)$, due to strong coupling effects is equal to $1.76T_c$ [25] . Then, we obtained:

$$\lambda_{2s\uparrow} = 1 - \frac{9}{320}\theta_m^4 \tag{25}$$

$$B_{\uparrow} = \frac{192\pi^5}{\left(m^*\right)^3 k_B^2 T^2 I^4 \theta_m^6}, \tag{26}$$

$$C\left(\lambda_{2s\uparrow}\right) = \left[0.75 + 0.003675\theta_m^4\right]. \tag{27}$$

After substituting $B_{\uparrow}$, $\lambda_{2\uparrow}$, and $C\left(\lambda_{2\uparrow}\right)$ in (20) and taking the angular integrations, we have

$$\eta_{xx\uparrow} = \eta_{yy\uparrow} = 1943\cdot 237\frac{p_f^5}{\left(m^*\right)^4 k_B^2 I^4}\frac{T_c^4}{T^6}, \tag{28}$$

$$\eta_{zz\uparrow} = 156\cdot 62\frac{p_f^5}{\left(m^*\right)^4 k_B^2 I^4}\frac{T_c^8}{T^{10}}, \tag{29}$$

$$\eta_{xy\uparrow} = 3\cdot 503\times 10^{-6}\frac{p_f^5}{\left(m^*\right)^4 k_B^2 I^4}\frac{T_c^4}{T^6}, \tag{30}$$

$$\eta_{xz\uparrow} = 123\cdot 84\frac{p_f^5}{\left(m^*\right)^4 k_B^2 I^4}\frac{T_c^6}{T^8}, \tag{31}$$

$$\eta_{yz\uparrow} = 124 \cdot 63 \times 10^{-6} \frac{p_f^5}{\left(m^*\right)^4 k_B^2 I^4} \frac{T_c^6}{T^8}. \tag{32}$$

5. THERMAL CONDUCTIVITY

The diffusive thermal conductivity tensor K_{ij} is defined by

$$J_i = K_{ij} \nabla_j T, \tag{33}$$

where J is the diffusive heat current. It is noted that, in addition to heat transfer by a random diffusive process of thermal excitations described by Equations (33), there is a convective contribution to the heat current in a superfluid, even in the absence of mass flow, owing to the possibility of normal-superfluid counter flow.

At low temperatures, this contribution is negligible [24] -[26] and the measurement of the components of thermal conductivity at these temperatures is more meaningful than the measurement of the thermal conductivity components in the vicinity of T_c.

The diffusive heat current may be written in terms of the polarized quasi-particle distribution functions:

$$J_i = \int \mathrm{d}\tau E \frac{\partial E}{\partial p_i} \delta n_{p\uparrow}, \tag{34}$$

$\delta n_{p\uparrow}$ is defined at below[26]

$$\delta n_{p\uparrow} = \frac{-n_0(p)\left(1+n_0(p)\right)}{T} \left(\frac{\partial E_1}{\partial p_{1k}} \right) q(t) \left[\frac{\partial T}{\partial r_k} \right]. \tag{35}$$

By using (35) in (34) and compared with (33) and using odd function of t in (15) for polarized gas we got [24]

$$K_{ij} = -\frac{4p_f^3}{(m^*)^3(2\pi)^3} T\int d\Omega_p \hat{p}_i \hat{p}_j \left(\int dt \frac{\partial n_{0\uparrow}}{\partial t} q_{1a\uparrow}(t) t \right). \tag{36}$$

By using Sykes and Brooker procedure we defined K_{ij} as bellows

$$K_{ij} = \frac{8p_f^3 T}{(m^*)^3(2\pi)^3} T\int d\Omega_p \hat{p}_i \hat{p}_j \left(\frac{B_{1a\uparrow}}{3-\lambda_{1a\uparrow}} H(\lambda_{1a\uparrow}) \right), \tag{37}$$

$\lambda_{1a\uparrow}$ and $H(\lambda_{1a\uparrow})$ are introduced as below

$$\lambda_{1a\uparrow} = \frac{\int \frac{\sin\theta d\theta d\varphi}{\cos(\theta/2)} W(\theta,\varphi)(1+2\cos\theta)}{\int \frac{\sin\theta d\theta d\varphi}{\cos(\theta/2)} W(\theta,\varphi)}, \tag{38}$$

$$H(\lambda) = \frac{3-\lambda_{1a\uparrow}}{4} \sum_{n=0}^{\infty} \frac{(4n+5)}{(n+1)(2n+3)\{(n+1)(2n+3)-\lambda_{1a\uparrow}\}}$$
$$= \frac{\lambda_{1a\uparrow}-3}{2\lambda_{1a\uparrow}} \left\{ \gamma + \ln 2 - 1 - \frac{1}{2\lambda_{1a\uparrow}} + \frac{1}{2}\psi_d\left(s_1 - \frac{1}{2}\right) + \frac{1}{2}\psi_d\left(s_2 - \frac{1}{2}\right) \right\}, \tag{39}$$

where s_1 and s_2 are defined by (24).

By using (6) in (38), (39) we took the following explicit forms

$$\lambda_{1a\uparrow} = 1 - \frac{3}{4}\theta_m^2, \tag{40}$$

$$H(\lambda_{1a\uparrow}) = 0.42 - 0.595\theta_m^2. \tag{41}$$

By using the Sykes and Brooker procedure [24] , we calculated the diffiusive thermal conductivity tensor. After substituting $\lambda_{1a\uparrow}$, $B_\uparrow$, and $H(\lambda_{1a\uparrow})$ in (37), and taking the angular integrations, we got:

$$K_{xy\uparrow} = K_{xz\uparrow} = K_{yz\uparrow} = 0, \tag{42}$$

$$K_{xx\uparrow} = K_{yy\uparrow} = 97.78 \frac{p_f^3 T_c^4}{(m^*)^6 k_B^2 I^4 T^5}, \tag{43}$$

$$K_{zz\uparrow} = 122.88 \frac{p_f^3 T_c^6}{(m^*)^6 k_B^2 I^4 T^7}. \tag{44}$$

6. DISCUSSION AND CONCLUDING REMARKS

As a result, in a superfluid dipolar Fermi gases we calculated the transition probabilities for the cases where the Bogoliubov quasiparticles are presented in coalescence, decay and binary processes and obtained that binary processes are dominated in calculation of transition probabilities, and other processes are nearly zero. Then, by using collision integral in the Boltzmann equation contained binary processes, the components of shear viscosity and thermal conductivity were calculated. We obtained $\eta_{xx\uparrow} = \eta_{yy\uparrow}, \eta_{xy\uparrow}$; $\eta_{xz\uparrow} \cong \eta_{yz\uparrow}$; and $\eta_{zz\uparrow}$ with temperatures dependences T^{-6}; T^{-8}; and T^{-10} respectively (see Equations (28)-(32)); and we have found the components of the diffusive thermal conductivity, $K_{xx\uparrow} = K_{yy\uparrow}$ with temperature dependency T^{-5} and $K_{zz\uparrow}$ which is proportional to T^{-7} and other components which are zero. Temperature dependency of shear viscosity and thermal conductivity along the z-axis (T^{-10} and T^{-7} respectively) were showing that the polarization is along the z-axis.

Then, we had to compare our results in transport coefficients with [13] [14] . (They found the shear viscosities η_{xy}, η_{xz} and η_{zz} of A_1-phase of superfluid ^3He are proportional to $(T/T_c)^{-2}$.). Two major differences were shown: a) Transition probabilities in superfluid dipolar Fermi gas are the function of temperature, but in superfluid ^3He-A_1 are constant. b) ^3He-A_1 have two kind of the quasiparticles, one for Bogoliubov quasiparticle, and another for normal quasiparticle. The wave function of the system in

A_1 phases of ^3He is $|\uparrow\uparrow\rangle$ that interacts with Bogoliubov quasipaticle $(\uparrow)$ and normal quasiparticle $(\downarrow)$, but in superfluid dipolar Fermi gas only $|\uparrow\uparrow\rangle$ contributes to Bogoliubov quasiparticles $(\uparrow)$ in interaction.

It is noted that, in superfluid Fermi gas in p-wave state $W_{22}(\uparrow\uparrow)$, $W_{22}(\uparrow\downarrow)$ were contributed to calculations [16] [17] but in superfluid dipolar Fermi gas only $W_{22}(\uparrow\uparrow)$ with temperature dependency contributes to calculations.

Also, in [15] the components of thermal conductivity tensor of superfluid ^3He-A were calculated using approximate collision integrals at low temperatures, and the thermal coefficients $K_{\perp}$ and $K_{||}$ have been obtained with temperature dependences T^{-1} and T, respectively. This is due to the fact that λ_1^- $\left(\lambda_1^- = 1 + \left(2/\langle W\rangle\right)\langle W(\theta,\varphi)\cos\theta\rangle\right)$ was taken to be a constant parameter with respect to temperatures.

Maryam Khademi Dehkordi (18,39)">18] , $W_{22}(\uparrow\uparrow)$, $W_{22}(\downarrow\downarrow)$ and $W_{22}(\uparrow\downarrow)$ were contributed in calculations and the wave function $|\uparrow\uparrow\rangle$ interacted with Bogoliubov and normal quasiparticles, but in superfluid dipolar gas only $W_{22}(\uparrow\uparrow)$ contributed to calculation and only $|\uparrow\uparrow\rangle$ interacted with Bogoliubov quasiparticles. These differences in transition probabilities of $^3He - A_1$ and superfluid Fermi gas in p-wave state with superfluid dipolar Fermi gas, have shown themselves in transport coefficients.

REFERENCES

1. deMarco, B. and Jin, D.S. (1999) Onset of Fermi Degeneracy in a Trapped Atomic Gas. Science, 285, 1703-1706.
2. deMarco, B., Papp, S.B. and Jin, D.S. (2001) Pauli Blocking of Collisions in a Quantum Degenerate Atomic Fermi Gas. Physical Review Letters, 86, Article ID: 5409.

3. Truscott, A.G., Strecker, K.E., McAlexander, W.I., Partridge, G.B. and Hulet, R.G. (2001) Observation of Fermi Pressure in a Gas of Trapped Atoms. Science, 291, 2570-2572.
4. Schreck, F., Khaykovich, L., Corwin, K.L., Ferrari, G., Bourdwel, T., Cubizolles, J. and Salomon, C. (2001) Quasipure Bose-Einstein Condensate Immersed in a Fermi Sea. Physical Review Letters, 87, Article ID: 080403.
5. Leggett, A.J. (1980) Cooper Pairing in Spin-Polarized Fermi Systems. Journal de Physique Archives, 41, 19-26.
6. Nozières, P. and Schmitt-Rink, S. (1985) Bose Condensation in an Attractive Fermion Gas: From Weak to Strong Coupling Superconductivity. Journal of Low Temperature Physics, 59, 195-211.
7. Jochim, S., Bartenstein, M., Altmeyer, A., Hendl, G., Riedl, S., Chin, C., Denschlag, J.H. and Grimm, R. (2003) BoseEinstein Condensation of Molecules. Science, 302, 2101-2103.
8. Bourdel, T., Khaykovich, L., Cubizolles, J., Zhang, J., Chevy, F., Teichmann, M., Tarruell, L., Kokkelmans, S.J.J.M.F. and Salomon, C. (2004) Experimental Study of the BEC-BCS Crossover Region in Lithium 6. Physical Review Letters, 93, Article ID: 050401.
9. Braby, M., Jingyi, C. and Schäfer, T. (2010) Thermal Conductivity and Sound Attenuation in Dilute Atomic Fermi Gases. Physical Review A, 82, Article ID: 033619.
10. Shahzamanian, M.A. and Yavary, H. (2002) The Shear Viscosity of Superfluid ^{6}Li. Physica B, 321, 385-387.
11. Shahzamanian, M.A. and Yavary, H. (2002) The Second Viscosity of Superfluid ^{6}Li. International Journal of Modern Physics B, 16, 3797-3802.
12. Stoof, H.T.C., Houbiers, M., Sacket, C.A. and Hulet, R.G. (1996) Superfluidity of Spin-Polarized ^{6}Li. Physical Review Letters, 76, Article ID: 10.
13. Shahzamanian, M.A. and Afzali, R. (2004) Shear Viscosity of Superfluid ^{3}He-A$_1$ at Low Temperature. Annals of Physics, 309, 281-294.
14. Shahzamanian, M.A. and Afzali R. (2004) Quasiparticle Relaxation Rate and Shear Viscosity of Superfluid ^{3}He-A$_1$ at Low Temperatures. Physica B, 348, 108-114.
15. Shahzamanian, M.A. (1989) Diffusive Thermal Conductivity of Superfluid ^{3}He-A1 at Low Temperatures. Journal of Physics: Condensed Matter, 1, 1965-1970.
16. Nasirimoghadam, S., Nabipoor, F., Khademi Dehkordi, M. and Shahzamanian, M.A. (2012) Transport Coefficients of the Superfluid Fermi gas in P-Wave State at Low Temperature. Physica C: Superconductivity, 483, 109-112.
17. Khademi Dehkordi, M., Nasirimoghadam, S., Nabipoor, F. and Shahzamanian, M.A. (2012) Diffusive Thermal Conductivity of p-Wave State at Low Temperatures. Communications in Theoretical Physics, 58, 891-894.
18. Afzali, R. and Ebrahimian, N. (2005) Diffusive Thermal Conductivity of the A$_1$-Phase of Superfluid ^{3}He at Low Temperatures. Journal of Physics: Condensed Matter, 17, 4441.
19. Abrikosov, A.A. and Khalatnikov, I.M. (1959) The Theory of a Fermi Liquid. Reports on Progress in Physics, 22, 329.
20. Takagi, S. (1975) On the NMR Properties near the ^{3}He A-Normal Transition. Journal of Low Temperature Physics, 18, 309-314.
21. Baranov, M.A. (2008) Theoretical Progress in Many-Body Physics with Ultracold Dipolar Gases. Physics Reports, 464, 71-111.
22. Pethick, C.J., Smith, H. and Bhattacharyya, P. (1977) Transport Processes in Superfluid ^{3}He-B at Low Temperatures. Physical Review B, 15, 3384.
23. Volhardt, D. and Wolfle, P. (1990) The Superfluid Phases of Helium 3. Taylor & Francis, London.
24. Sykes, J. and Brooker, G.A. (1970) The Transport Coefficients of a Fermi Liquid. Annals of Physics, 56, 1-39.
25. Pethick, C.J. and Smith, H. (2002) Bose-Einestein Condensation in Dilute Gases. Cambridge University Press, Cambridge.
26. Wolfle, P. and Einzel, D. (1978) Transport and Relaxation Properties of Superfluid ^{3}He. Journal of Low Temperature Physics, 32, 39-56.

CHAPTER 4

Percolation Transitions of the Ideal Gas and Supercritical Mesophase

Leslie V. Woodcock

Department of Physics, University of Algarve, Faro, Portugal

ABSTRACT

High-temperature and pressure boundaries of the liquid and gaseous states have not been defined thermodynamically. Standard liquid-state physics texts use either critical isotherms or isobars as ad hoc boundaries in phase diagrams. Here we report that percolation transition loci can define liquid and gas states, extending from super-critical temperatures or pressures to "ideal gas" states. Using computational methodology described previously we present results for the thermodynamic states at which clusters of excluded volume (V_E) and pockets of available volume (V_A), for a spherical molecule diameter σ, percolate the whole volume ($V = V_E + V_A$) of the ideal gas. The molecular-reduced temperature (T)/pressure (p) ratios ($T^* = k_BT/p\sigma^3$) for the percolation transitions are T^*_{PE} = 1.495 ± 0.01 and T^*_{PA} = 1.100 ± 0.01. Further MD computations of percolation loci for the Widom-Rowlinson (W-R) model of a partially miscible binary liquid (A-B) show the connection between the ideal gas percolation transitions and the 1st-order phase-separation transition. A phase diagram for the penetrable cohesive sphere (PCS) model of a one-component liquid-gas is then obtained by analytic transcription of the W-R model thermodynamic properties. The PCS percolation loci extend from a critical coexistence of gas plus liquid to the low-density limit ideal gas. Extended percolation loci for argon, determined from literature equation-of-state measurements exhibit similar phenomena. When percolation loci define phase bounds, the liquid phase spans the whole density range, whereas the gas phase is confined by its percolation boundary within an area of low T and p on the density surface. This is contrary to a general perception, and reopens a debate of "what is liquid". We append this contribution to the science of liquid-gas criticality and liquid-state bounds with further open debate.

Keywords: Percolation Transition, Mesophase, Ideal Gas, Criticality, Liquid State

1. INTRODUCTION

Almost 40 years ago, in their classic review on the status of liquid state theory [1] , Barker and Henderson began with the words "Liquids exist in a relatively small part of the enormous range of temperatures and pressures existing in the universe". The tiny liquid area, in the T-p projection of Gibbs density surface, was defined within either a critical isotherm, or isobar, and a triple point. Above a critical temperature (or pressure) and below the triple-point, the liquid state did not exist. Not everyone agreed with these ad hoc bounds, however. Recent research on percolation transition loci on Gibbs thermodynamic surfaces [2] shows that J. D. Bernal, for example, suggested the liquid phase extended to a metastable random close-packed state at sub-triple-point temperatures. Besides noting that the liquid state, albeit metastable, should extend down to absolute zero, Bernal also argued that there may be no supercritical continuity of liquid and gas, where it is bounded from the gas phase by a "hypercritical" line of discontinuity. A review of recent computer studies of RCP, and also percolation loci in model and real fluids, show that Bernal may have been closer to the scientific truth than many of his peers [2] .

Here, we report results for ideal gas properties, which, alongside real experimental p-V-T properties of a typical real fluid (argon), comprise compelling evidence that the liquid state is not bounded, by either the critical isotherm or isobar. Liquid and gas phases are terminated by percolation loci along any isotherm. Moreover, we find that percolation loci extend all the way from critical coexistence to low density states with ideal gas properties.

The equation-of-state of a real gas with finite molecular size (diameter σ) behaving ideally within a low-density limit, is simply

$$p^* = \rho^* \tag{1}$$

where p* is a molecular-reduced pressure ($p\sigma^3/k_BT$), T is temperature (K), k_B is Boltzmann's constant ρ* is a reduced density ($\rho = N\sigma^d/L^d$), where L^d is length (L, d = 1), area (A, d = 2) or volume (V, d = 3). Equation (1) has an abiding role in the description of thermodynamic properties of real molecular fluids. Pressure is everywhere continuous; second and all higher derivatives of p(ρ) are zero. Because of this simplicity, all state functions are exactly known for any d. Equation (1) is a universal scaling law that spans the dimensions.

Within the ideal gas limit of obedience to Equation (1), real fluids with finite size, i.e. σ > 0, however, exhibit various properties that cannot scale with d, linear transport coefficients, for example. Percolation transitions, not unrelated to the transport coefficients, are also strongly dimension dependent in form, and are known to determine thermodynamic phase changes in model lattice gases [3] . Percolation transitions of the available volume (V_A) and excluded volume (V_E) for the insertion of one more molecule of a finite diameter are properties relating to Gibbs energies that effect phase transitions.

For hard-core fluids,

$$V = V_A + V_E \tag{2}$$

then, the ensemble averages $\langle V_A \rangle$ and $\langle V_E \rangle$ equate with chemical potential (μ_i) of species i

$$\mu_i = -k_BT \log_e \left(\langle V_A \rangle / V \right) \tag{3}$$

Equation (3), with Equation (2), defines $\langle V_A \rangle$ and $\langle V_E \rangle$ for real fluids.

For the ideal gas, percolation of V_E is defined as a density above, or temperature below which, the overlapping exclusion spheres of radius σ from a point in a uniformly random distribution of N points, form clusters that can span the whole of V. V_A comprises a distribution in configuration space of accessible pockets in which there are no ideal gas point molecules within one sphere diameter anywhere in the pocket. The percolation transition for V_A is the density above, or temperature below which, the empty pockets coalesce to span the system. For temperatures above percolation, V_A comprises a network of connecting pathways to the whole system accessible to a diffusing sphere in the static ideal gas equilibrium configuration.

2. DETERMINATION OF PERCOLATION TRANSITIONS

We designate the percolation transition reduced temperatures as T^*_{PE} and T^*_{PA} respectively. Relationships between dimensionality and percolation transitions can be summarized:

(note: for an ideal gas $T^* = 1/p^* = V^* = 1/\rho^*$)

d = 1 no percolation

d = 2 PE and PA coincide $T^*_{PE} = T^*_{PA}$ and $\rho^*_{PE} = \rho^*_{PA}$

d = 3 there is an inequality $T^*_{PE} > T^*_{PA}$ and $\rho^*_{PE} < \rho^*_{PA}$

There is a fundamental difference between 2 and 3 dimensions. For d = 2, there are two regions, "gas-like" $T^* > T^*_{PE}$ and "liquid-like" $T^* < T^*_{PE}$, whereas for d = 3 there are three regions, gas-like $T^* > T_{PA}$, liquid-like $T^* < T_{PA}$ and a mesophase, $T^*_{PA} > T^* > T^*_{PE}$. In the mesophase, both the pockets of availability and clusters of exclusion sites percolate the system. The mesophase is both gas-like and liquid-like.

PE both for d = 2 and 3 has been investigated by a number of authors for the present and related systems [4] - [11] Rough estimates of PE for d = 2 and 3 ideal gases can be gleaned from figure 3 and figure 4 of the paper by Bug et al. [4] . Extrapolating their data points, when the attraction parameter $\varepsilon = 0$, to zero density one obtains for d = $2 \Phi_{PE} \sim 1.2$ and for d = $3 \Phi_{PE} \sim 0.35$, where Φ is the excluded volume fraction ($=\pi\rho\sigma^3/6$). Heyes and coworkers [5] - [8] report related investigations of PE for various models by MC and MD simulations. From an interpolation to zero density of the hard-sphere fluid variable-exclusion shell percolation threshold Heyes et al. [7] obtain $\Phi_{PE} = 0.346$ (d = 3). An extensive study of ideal-gas PE states for exclusion squares, cubes, and other geometric shapes has been reported [9] .

Estimates of $\rho^*_{PE/PA}$ for d = 2 can easily be obtained pictorially in just a few minutes using an EXCEL spreadsheet. Figure 1 shows a typical configuration, 2000 random numbers from a uniform distribution 0 - 1, i.e. an ideal gas (N = 1000) both above and below, and also in the vicinity of, the percolation transition. Disc diameters do not vary when the square area is expanded or contracted on the display. Fixing the diameter at 5 mm, percolation occurs at L~14.0 cm hence $\rho^*_{PE/PA} = 1.28$ or $T^*_{PE/PA} = 0.77$. Computation of PE and PA for the ideal gas d = 3 is not so easy.

There are no reports of PA (d = 3) having been previously investigated or determined for the ideal gas, although the transition density ρ^*_{PA} is known to be 0.537 ± 0.05 for the hard-sphere fluid [10] . Here, we use the same methods, and criteria for percolation, as described previously for the hard-sphere fluid. Both T^*_{PE} and T^*_{PA} have been computed for a range of finite size systems; the results are summarized in Figure 2(a) and Figure 2(b) respectively. Thermodynamic limiting values (N ® ∞) are obtained from the linear trendlines.

Every configuration either has a percolating cluster or it does not. Clearly, for small finite systems, there will be configurations that percolate, and some that do not, in the vicinity of PE. The percolation threshold in the computations of Heyes et al. [5] - [8] was defined when 50% of configurations have a percolating cluster, with details described by Seaton et al. [11] . Here, we define PE using an ensemble average definition of a percolation density [10] ; i.e. ρ^*_{PE} is the saddle-point density above which the cluster size probability distribution P(n) is bimodal. This is the normalized probability of a site belonging to a cluster of size n. Above T^*_{PE}, P(n) is a monotonic gas-like distribution, for all PE below T^*_{PE} it is bimodal. Plotting the saddle-point definition of $T^*_{PE}(N)$ against $1/N^{1/3}$ (Figure 2(a)) gives a linear trendline that interpolates to the result $T^*_{PE}(N \rightarrow \infty) = 1.495 \pm 0.01$ ($\rho^*_{PE} = 0.668$).

Our method for determining T^*_{PA} is essentially that described previously for hard spheres [10] , except that it is easier for the ideal gas. Here we use N-V-T MD for non-additive binary spheres that can simulate the Widom-

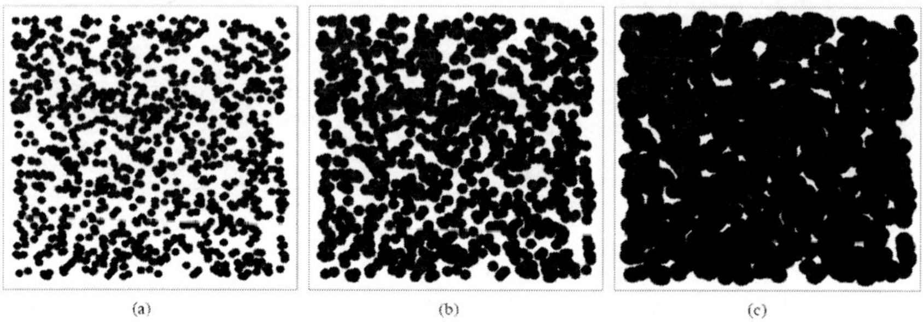

Figure 1. Excluded and accessible areas (black and white respectively) for a configuration of a two-dimensional ideal gas: (a) "Gas-like" density below the percolation transition; (b) Close to the percolation transition; and (c) "Liquid-like" density above the percolation transition: for d = 2 both V_E (black) and V_A (white) percolate at the same density.

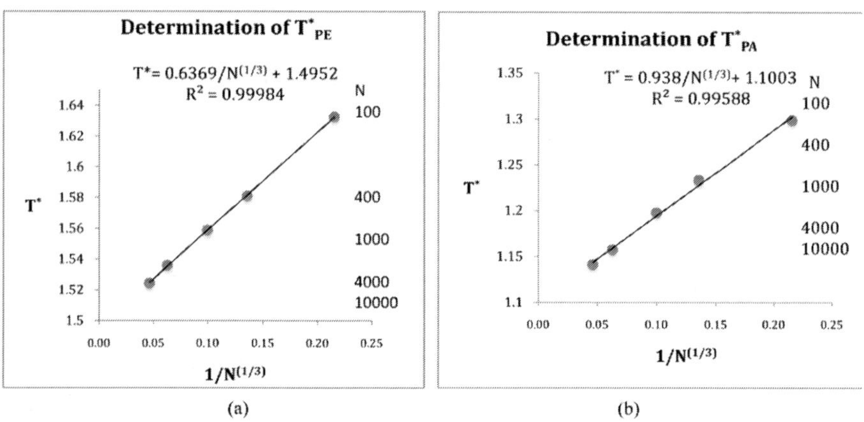

Figure 2. (a) Ideal gas PE transition (T^*_{PE}) from the mean cluster size distribution saddle-point method [10] for a range of finite systems; (b) Ideal gas PA transition (T^*_{PA}) from zero-diffusivity limit method [10] .

Rowlinson (W-R) model fluid [12] [13] . This belongs to the general class of symmetric binary non-additive hard-sphere fluid mixtures defined by collision diameters

$$\sigma_{AA} = \sigma_{BB} = (1+\delta)\sigma_{AB}$$

where δ is a dimensionless non-additivity, that varies from −1, for the W-R penetrable-sphere model binary fluid, via zero for one-component hard spheres, to infinity. Positive δ relates to ionic liquids and ionic crystal structures when mole fraction $X_B = 0.5$.

The MD program solves equations of motion of a binary mixture $N_A + N_B$. The results for the PA values in Figure 2(b) are obtained by the mean-squared displacements of B average over many frozen random configurations of ideal gas A. As the B particles do not interact with themselves, we average over any N_B in the same MD simulation run. All the values in Figure 2(b) were obtained for equimolar systems. Plotting the point of zero diffusivity, $D_i(\rho, N)$ ® 0, against N (N_A in MD run in Figure 2(b)), gives a linear trendline with the result $T^*_{PA}(N \to \infty) = 1.100 \pm 0.01$ ($\rho^*_{PA} = 0.908$).

3. PARTIALLY MISCIBLE BINARY LIQUID: W-R MODEL

We have also determined T^*_{PE} along isopleths of the binary W-R model fluid; T^* is defined as $T^* = 1/p^*$ and $p^* = p\sigma^3/k_BT$. MD simulations have some advantages over Grand Canonical Monte Carlo [13] (GCMC). Not least is the direct extraction of transport properties for determination of T^*_{PA} loci. These are obtainable by "freezing" component A whilst allowing B to diffuse. The cluster distributions that determine T^*_{PE} also yield accurate values for coexisting X_B by integrating the solute cluster probability distribution P(n) which decreases monotonically, from a maximum at n = 1, to zero for clusters of B in solution of A, or vice-verser. Accurate MD pressures are calculated from A-B collision frequencies.

What is the effect on the percolation transitions of increasing the mole fraction of B from the ideal gas limit ($X_B = 0$)? For the isopleth at $X_B = 0.1$, and for N = 10,000, the reduced pressures at which the two transitions occur, i.e. p^*_{PE} and p^*_{PA} are 0.715 and 0.923 respectively. We find, up to $X_B = 0.1$ and beyond, both percolation pressures increase with X_B, PE more so than PA, roughly according to

$$p^*_{PE}\left(\rho^*_B\right) \sim p^*_{PE}(0)\left[1+\rho^*_B\right]$$

$$p^*_{PA}\left(\rho^*_B\right) \sim p^*_{PA}(0)\left[1+\rho^*_B/2\right]$$

where $\rho^*_B = \rho^* X_B$. The percolation transition pressures increase with ρ^*_B because, as B are added at constant T, the system expands with both V_A and V_E increasing, but V_E increases more than V_A; adding B causes A-sites to cluster more, whilst creating more spherical B-pockets. The pressures of percolation transitions for finite X_B appear to be coincident with higher-order discontinuities (Figure 3) in the supercritical region. Weak thermodynamic discontinuities have been both predicted theoretically for real systems [14] , and reportedly seen experimentally [15] . Changes in pressure slopes are evident from the MD excess pressures defined relative to the ideal gas.

$$p^*_{ex} = p^* - \rho^* \tag{4}$$

Figure 3(a) shows the isopleth $X_B = 0.1$ has four distinct regions. At high density, in the two-phase region, the MD pressures averaged over 100 million A-B collisions still show fairly large uncertainties. The maximum pressure along any isopleth coincides with the first-order mixing-demixing transition. This reflects the thermodynamic equilibrium condition of minimal Gibbs energy (G) (since $dG_T = \rho^{-1}dp$) for equilibrium on either side of the transition. At the mole fraction $X_B = 0.1$ in the mesophase region pressure increases linearly with density. In the one-phase region, the MD data is sufficient to observe that the percolation loci appear to be associated with changes in slope that could reflect higher-order thermodynamic phase transitions, but presently not sufficiently accurate to establish the order or strength of discontinuities.

The vertical dashed lines in Figure 3 correspond to the percolation transition densities computed explicitly by the methods described in the text and referenced [5] : they coincide with changes in the slope of the excess pressure. The MD results for $X_B = 0.5$ show three regions; there is no PA, just the PE transition at the density 0.65. The change in the slope of p^*_{ex}, and hence also p^*, is more pronounced. The rigidity function $(dp/d\rho)_T$ is again constant in the mesophase, with a very slight slope.

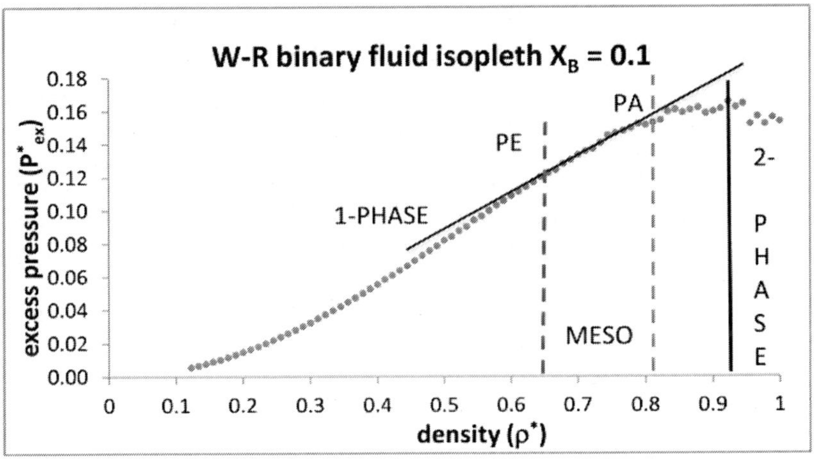

(a)

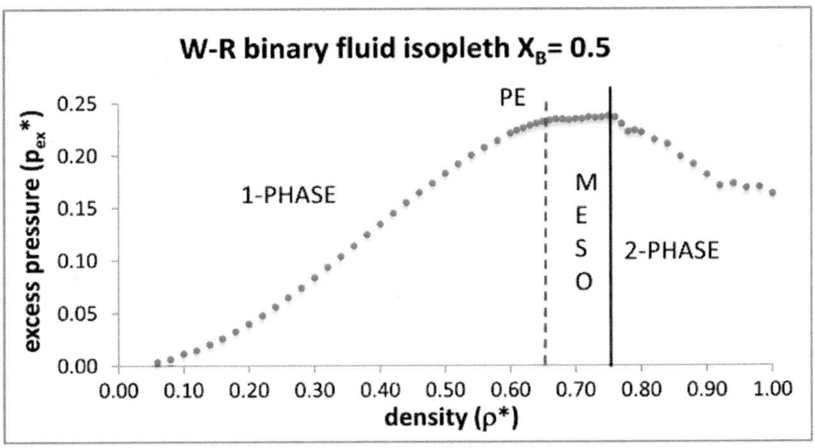

(b)

Figure 3. Excess pressures of the Widom-Rowlinson binary fluid mixture: (a) along the isopleth X_B = 0.1 (N = 10,000): (b) along the isopleths X_B = 0.5 (N = 8000).

From the coexistence pressures, and direct computations of PA and PE, we are able to construct a phase diagram for the W-R binary fluid. The temperature loci of the percolation transitions fit the trendlines (dashed lines in Figure 4)

$$T^*_{PE} = 1.495 - 0.750 X_B$$

$$T^*_{PA} = 1.100 + 0.045 X_B - 1.004 X_B^2$$

showing the connection between percolation and the demixing phase transition. Note the perfect symmetry about the isopleth $X_B = 0.5$.

The percolation loci on the T^*-X_B surface (Figure 4), to within the uncertainty of the data, decrease with X_B an intersection occurs when $T^* = 1.00 \pm 0.01$ with the uncertainty estimated from the combined regressions. At this intersection the fluid phase separates as solutions of A in B and B in A have the same T, p, and μ of both species (chemical potential). Solving for X_B when $T^*_{PE} = T^*_{PA}$ gives $X_B = 0.339$ for the critical coexistence mole fraction.

The present results show a coexistence line at T^*_e rather than an Ising-like singular critical point. The evidence for higher-order discontinuities at PE and PA loci may invalidate some mean-field theoretical approaches. The W-R binary fluid is essentially a simple model of partially miscible liquids, e.g. cyclohexane and methanol. The high-density fluid states at low temperatures are immiscible ideal gases. Here we have an example of what could arguably be described as "liquid states" of the ideal gas.

4. ONE COMPONENT LIQUID-GAS: PCS MODEL

Probably the simplest 3D model Hamiltonian of a molecular fluid, which is continuous in phase space, and exhibits liquid-gas criticality and two-phase gas-liquid coexistence, is the penetrable cohesive sphere (PCS) fluid [12] [13] . The internal energy (U) is simply

$$U = 3Nk_BT/2 + \varepsilon\left(\langle V_E/v_0\rangle - N\right) \tag{5}$$

where k_B is Boltzmann's constant and T is temperature (K); the angular brackets denote a configurational aver age. Equation (5) defines an attractive molecular energy complementary to the volume of overlapping clusters, i.e. V_E as defined above for an ideal gas, of a configuration of N penetrable spheres, and v_0 ($=4\pi\sigma^3/3$) is the volume of a sphere. At low temperatures, this model exhibits the exact properties of an ideal gas in both the low-density (gas phase) and high-density (liquid-phase) limits. Here again, there is a liquid-like state with the properties of the ideal gas. Both W-R and PCS models, therefore raise a curious conundrum. Could a dense fluid with the properties of an ideal gas be described as "liquid"?

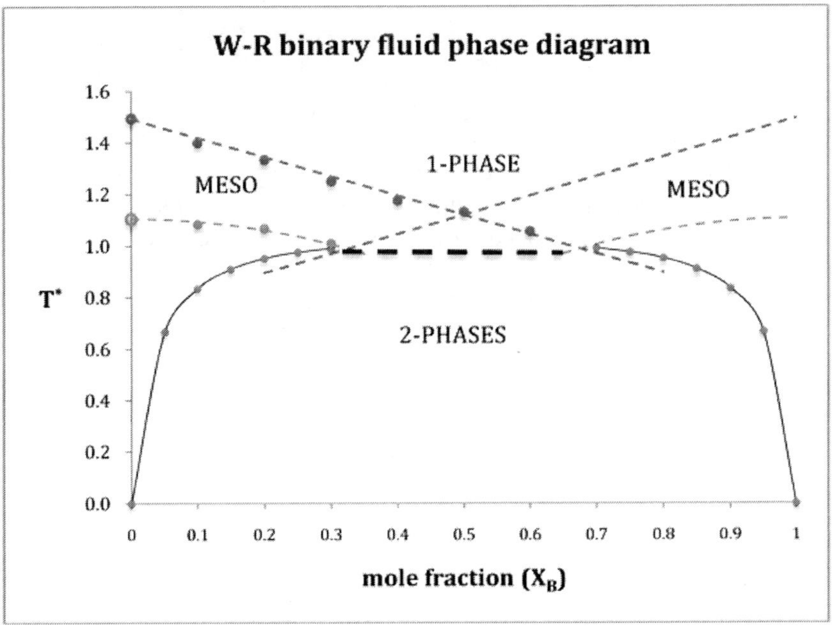

Figure 4. Phase diagram of the Widom-Rowlinson binary fluid; T^*_{PE} and T^*_{PA} loci blue and green respectively; percolation transitions of the ideal gas open circles at $X_B = 0$; computed 2-phase coexistence state points are the red circles.

Every state of the PCS fluid corresponds to a transcribed state of the W-R binary model fluid. The equations for the transcription from the W-R binary percolation and coexistence pressures (Figure 5(a)) to the PCS one-component gas-liquid pressure can be derived, for example, from an analysis of the respective grand partition functions as described by de Miguel et al. [13] . The transcription equations we need here are

pressure [PCS] $p\sigma^3/\varepsilon$ = [W-R] $\left(p^* - Z^*_A/v_o\right)k_B T/\varepsilon$ (6)

density [PCS] $\rho\sigma^3$ = [W-R] ρ^*_B (7)

where Z^*_A is the thermodynamic activity of a component defined as $\ln Z^* = \mu/k_B T$ and μ is Gibbs chemical potential relative to the ideal gas at the same T, p. G change, hence Z^*, can be obtained by integrating the excess pressure loci at constant T, with respect to density.

$$\mu/T=\int(p-p_{id})\,\mathrm{d}\log_e(\rho_B) \tag{8}$$

Figure 5(b) shows that the simplest imaginable continuous Hamiltonian model to exhibit liquid-gas coexistence and a critical temperature has a coexistence line at the intersection percolation loci as observed for square-well and Lennard-Jones model fluids and many real fluids [2] [18] - [23] . There is a liquid-gas critical coexistence line, and a supercritical mesophase. The liquid state extends to an ideal gas zero density and pressure limit. This raises the question: could there exist a high temperature limit of the percolation transition loci in real fluids in the ideal gas limit?

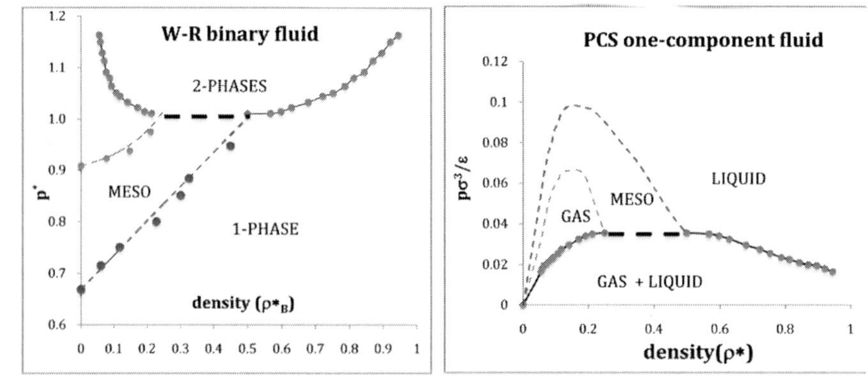

Figure 5. (a) Phase diagram of the W-R binary fluid in the pressure-density projection: p^*_{PE} and p^*_{PA} loci are dashed blue and green respectively; coexistence data points are red circles; (b) Phase diagram of the PCS liquid-gas system obtained by transcription of W-R pressures in (a).

The evidence suggests this is indeed the case, provided we re-interpret the experimental thermodynamic properties of real fluids in the light of our knowledge of percolation transitions. For 80 important gases or liquids, including the simplest real liquid argon, in the NIST "Thermophysical Properties of Fluid Systems" data bank [16] [17] , equation-of-state p-V-T data with 6-figure accuracies are obtainable. These $p(\rho)_T$ supercritical isotherms have been formulated, however, using a large number of parameters and an assumption of a supercritical continuity of liquid and gas. If there were to be no continuity of liquid and gas, there would need to exist three different equations-of-state to describe the gas and liquid phases, bounded by percolation loci, and the mesophase region in between. Present findings suggest theory-based equations-of-state with far fewer parameters, and with scientifically correct functional forms should eventually replace the NIST equations. Present observations indicate a virial expansion for the gas phase, perhaps a free-volume expansion for the liquid, and a linear combination for the mesophase [18] - [21] .

5. FLUID PHASES OF ARGON

Although lacking a molecular-level definition, for any real fluid, for which an exact Hamiltonian is generally unknown, the percolation loci can be defined and obtained phenomenologically along any thermodynamic equilibrium isotherm by the rigidity inequalities [18] [19] . In the case of real fluids with attractive potentials, the percolation transition bounding the gas phase, i.e. the counterpart of PE for impenetrable spheres, has been designated PB, as it is a percolation of bonded clusters.

Rigidity is the work required to increase the density of a fluid; with dimensions of molar energy, it relates directly to the change in Gibbs energy (G) with density at constant T according to

$$\omega_T = \left(\mathrm{d}p/\mathrm{d}\rho\right)_T = \rho\left(\mathrm{d}G/\mathrm{d}\rho\right)_T \quad (9)$$

The inequalities that distinguish gas from liquid are:

gas $\rho < \rho_{PB}$ $\left(\mathrm{d}\omega/\mathrm{d}\rho\right)_T < 0$ (10)

meso $\rho_{PB} > \rho > \rho_{PA}$ $\left(\mathrm{d}\omega/\mathrm{d}\rho\right)_T = 0$ (11)

liquid $\rho > \rho_{PA}$ $\left(\mathrm{d}\omega/\mathrm{d}\rho\right)_T > 0$ (12)

It is clear from Equation (9) that ω ³ 0, i.e. rigidity must always be positive: Gibbs energy cannot decrease with pressure when T is constant. By these definitions, moreover, not only can there be no "continuity" of gas and liquid, but the gas and liquid states are fundamentally different in their thermodynamic description. Rigidity is determined by number density fluctuations at the molecular level, which fundamentally different in each phase. There are many small clusters in a gas with one large void, and many vacant pockets in the liquid with one large cluster.

The isotherms [16] [17] of argon from the critical temperature (T_c = 151 K) to 500 K are plotted in Figure 6. All the isotherms below a temperature of 450K show that there is a flat meso region, i.e. $\left(\mathrm{d}\omega/\mathrm{d}\rho\right)_T = 0$. In this narrow range, for each supercritical isotherm, to within the uncertainty of

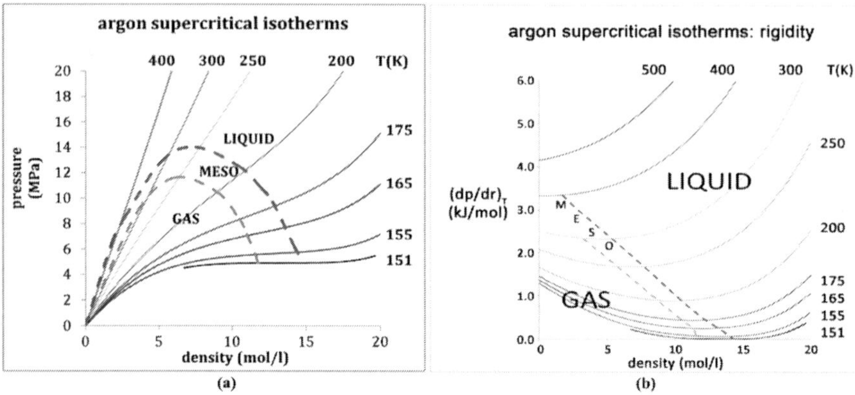

Figure 6. Supercritical isotherms of fluid argon from NIST thermophysical tables [16] [17] : loci of PB and PA are green and blue respectively (a) pressure isotherms; (b) rigidity isotherms.

the original experimental data, the rigidity is constant. A value can be obtained by a linear fit over a finite density range with a linear trendline regression between 0.999999 and 1.0 for all the supercritical isotherms in Figure 6. An accurate equation-of-state for the meso region is thus obtained

$$p(\text{meso}) = \omega\rho + p_0 \tag{13}$$

where ω along any isotherm T(K) and p_0 is a pressure constant. The percolation loci densities can be estimated by observing the differences p(gas) ? p(meso), and p(liquid) ? p(meso), both decrease quadratically with density, and interpolate to zero at the percolation loci densities ρ_{PB} and ρ_{PA}.

6. CONCLUSIONS

The results presented here for the percolation transition loci comparing both real and model fluids reaffirm previous observations [2] [18] - [23] that there is no critical point singularity on Gibbs density surface for gas-liquid condensation. Rather, there is a coexistence boundary line at the critical temperature, above which there exists a mesohase between the percolation loci that bound the liquid and gas phases. From the present W-R results it can be seen to further infer that partially miscible liquid-liquid mixtures will also show an upper critical consolute temperature with a dividing line separating one- and two-phase regions. Above this critical divide, there will be a mesophase bounded by the percolation loci that may

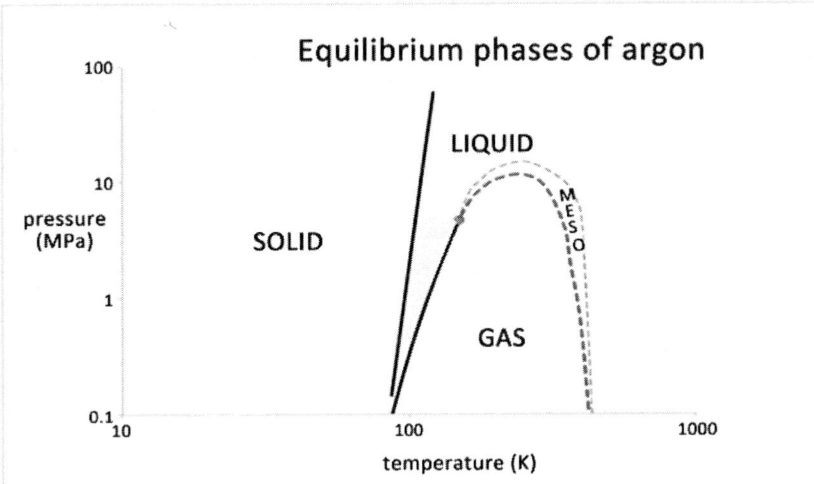

Figure 7. A putative phase diagram for argon showing the percolation loci and mesophase that bound the liquid and gas states: the tiny shaded area is often referred to as "the liquid state" [1] .

extend to the ideal dilute solution limits. Experimental studies of liquid-liquid UCST phenomena, however, are generally at ambient pressures and terminate at the boiling temperatures.

Returning to the question about universality and dimensionality dependence of the description of criticality, we observe that for d = 2, since PE = PA for all densities (or concentration X_B), the phase behavior and criticality will be quite different; there can be no mesophase in 2-dimensions. We conjecture, therefore, that the percolation locus intersects the equimolar isopleth with a critical singularity at $X_B = 0.5$ for the d = 2 W-R model. Another consequence of the absence of a d = 2 mesophase would be no metastability beyond the first-order phase boundaries, and, unlike d = 3, no metastability and hence no spinodals within the subcritical bimodals. The existence of a mesophase is a property only of d = 3 systems. This difference in the description of liquid-gas criticality between 2 and 3 dimensions vitiates the hypothetical "universality" concept as applied to liquid-gas and binary-liquid, criticality.

Finally, we conjecture an answer to the question of Barker and Henderson "What is liquid?" [1] . The present results for various percolation loci suggest that all real atomic and molecular fluids will have a liquid state that is bounded only by a percolation transition and its equilibrium freezing transition along any isotherm. The boundary may be defined phenomenologically by inequality Equation (12). The locus of this liquid-state boundary is seen to extend all the way from the liquid critical temperature and pressure, to an ideal gas at limiting low density. On the basis of these observations, it is the "gas phase", i.e. defined by the inequality Equation (10) that exists in a limited area of the universal T-p plane. We note, however, that there can be no zero of density/pressure for a real fluid, as they become logarithmic to high vacuum levels. The truly ideal gas is a ficticious concept; the sign of the second virial coefficient determines the designation gas or liquid in this limit. The liquid area of existence extends to infinite pressure and temperature; whereas the gas phase extends to infinite vacuum, but only below a certain temperature. It appears that the liquid phase will extend upwards in temperature, for all pressures to perhaps continuously become plasma (Figure 7). This is contrary to what is hitherto generally believed to be a "liquid" [1] and hence opens further debate on the subject (Appendices 1 & 2).

ACKNOWLEDGEMENTS

We acknowledge correspondence regarding the scientific content of this paper courtesy of Prof. G. Jackson Editor of Molecular Physics (Appendix 1) and Prof. R. Piazza Dep. Editor IoP J. Physics (C) (Appendix 2). These discussions add to the previous debates on a new science of liquid-gas criticality published in Natural Science [22] - [24] .

CITE THIS PAPER

Leslie V. Woodcock (2016) Percolation Transitions of the Ideal Gas and Supercritical Mesophase. *Open Access Library Journal*,**03**,1-19.

REFERENCES

1. Barker, J.A. and Henderson, D. (1976) What Is Liquid? Understanding the States of Matter. Reviews of Modern Physics, 48, 587-671.
2. Finney, J.L. and Woodcock, L.V. (2014) Renaissance of Bernal's Random Close Packing and Hypercritical Line in the Theory of Liquids. Journal of Physics: Condensed Matter, 26, Article ID: 463102.
3. Stauffer, D. (1985) Introduction to Percolation Theory. Taylor and Francis, London.
4. Bug, A.L.R., Safran, S.A., Grest, G.S. and Webman, I. (1985) Do Interactions Raise or Lower a Percolation Threshold? Physical Review Letters, 55, 1896-1899.
5. Heyes, D.M. and Melrose, J.R. (1988) Percolation Thresholds of Simple Fluids. Journal of Physics A: Mathematical and General, 21, 4075.

6. Heyes, D.M. (1990) Monte Carlo Simulation of Continuum Percolation of 3D Well Fluids. Journal of Physics: Condensed Matter, 2, 2241-2249.
7. Heyes, D.M., Cass, M. and Branka, A.C. (2006) Percolation Threshold of Hard-Sphere Fluids in between Soft-Core and Hard-Core Limits. Molecular Physics, 104, 3137-3146.
8. Heyes, D.M. and Okumura, H. (2006) Equation of State and Structural Properties of WCA Fluids. Journal of Chemical Physics, 124, Article ID: 164507.
9. Baker, D.R., Paul, G., Sreenivisen, S. and Stanley, H.E. (2002) Continuum Percolation Thresholds for Interpenetrating Squares and Cubes. Physical Review E, 66, Article ID: 046136.
10. Woodcock, L.V. (2011) Percolation Transitions in the Hard-Sphere Fluid. AIChE Journal, 58, 1610-1618.
11. Seaton, N.A. and Glandt, E.D. (1987) Aggregation and Percolation in a System of Adhesive Spheres. Journal of Chemical Physics, 86, 4668-4677.
12. Widom, B. and Rowlinson, J.S. (1970) New Model for the Study of Liquid-Vapor Phase Transitions. Journal of Chemical Physics, 52, 1670-1684.
13. de Miguel, E., Almarza, N.G. and Jackson, G. (2007) Surface Tension of the Widom-Rowlinson Model. Journal of Chemical Physics, 127, Article ID: 034707.
14. Ma, T. and Wang, S. (2011) Third-Order Gas-Liquid Phase Transition and the Nature of Andrews Critical Point. AIP Advances, 1, Article ID: 042101.
15. Nishikawa, K. and Morita, T. (1998) Fluid Behavior at Supercritical States Studied by Small Angle X-Ray Scattering. Journal of Supercritical Fluids, 13, 143-148.
16. Stewart, R.B. and Jacobsen, R.T. (1989) Thermodynamic Properties of Argon from the Triple Point to 1200 K at Pressures to 1000 MPa. Journal of Physical and Chemical Reference Data, 18, 639-798.
17. Tegeler, C., Span, R. and Wagner, W. (1999) A New Equation of State for Argon Covering the Fluid Region for Temperatures from the Melting Line to 700 K at Pressures up to 1000 MPa. Journal of Physical and Chemical Reference Data, 28, 779-850.
18. Woodcock, L.V. (2014) Gibbs Density Surface of Fluid Argon: Revised Critical Parameters. International Journal of Thermophysics, 35, 1770-1784.
19. Woodcock, L.V. (2016) Thermodynamics of Gas-Liquid Criticality: Rigidity Symmetry on Gibbs Density Surface. International Journal of Thermophysics, 37, 24.
20. Woodcock, L.V. (2013) Observation of a Liquid-Gas Critical Coexistence Line and Supercritical Phase Bounds from Percolation Loci. Fluid Phase Equilibria, 351, 25-33.
21. Heyes, D.M. and Woodcock, L.V. (2013) Critical and Supercritical Properties of Lennard-Jones Fluids. Fluid Phase Equilibria, 356, 301-308.
22. Woodcock, L.V. (2013) Fluid Phases of Argon: A Debate on the Absence of van der Waals' Critical Point. Natural Science, 5, 194-206.
23. Woodcock, L.V. (2014) Gibbs Density Surface of Water and Steam: 2nd Debate on the Absence of van der Waals "Critical Point". Natural Science, 6, 411-432.
24. Magnier, H.J., Curtis, R.C. and Woodcock, L.V. (2014) Nature of the Supercritical Mesophase. Natural Science, 6, 797-807.

CHAPTER 5

Temperature Effects on Gas Sensing Properties of Electrodeposited Chlorine Doped and Undoped n-Type Cuprous Oxide Thin Films

Nayana Bandara[1,2], Charith Jayathilaka[1,3], Dhammika Dissanayaka[4], Sumedha Jayanetti[1]

[1]*Department of Physics, University of Colombo, Colombo, Sri Lanka*

[2]*Department of Physics, Open University of Sri Lanka, Nawala, Sri Lanka*

[3]*Department of Physics, University of Kelaniya, Kelaniya, Sri Lanka*

[4]*Department of Chemistry, University of Colombo, Colombo, Sri Lanka*

ABSTRACT

As one of the most widely used domestic fuels, the detection of possible leakages of Liquefied Petroleum (LP) gas from production plants, from cylinders during their storage, transport and usage is of utmost importance. This article discusses a study of the response of undoped and chlorine doped electrodeposited n-type Cuprous Oxide (Cu_2O) films to of LP gas. Undoped n-type Cu_2O films were fabricated in an electrolyte bath containing a solution of sodium acetate and cupric acetate whereas n-type chlorine doped Cu_2O thin films were prepared by adding a 0.02 M cuprous chloride ($CuCl_2$) into an electrolyte solution containing lactic acid, cupric sulfate and sodium hydroxide. The n-type conductivity of the deposited films was determined using spectral response measurements. The structural and morphological properties of the fabricated films were monitored using X-ray diffraction (XRD) and Scanning Electron Microscopy (SEM). Due to doping, the overall conductivity of the chlorine doped n-type Cu_2O films increased by several orders of magnitude. The temperature dependent gas responses of both the undoped and chlorine doped n-type Cu_2O thin films to the LP gas was monitored by measuring the electrical resistance (R), and using the contact probe method at a constant gas flow rate of 0.005 ml/s. Upon exposure to gases, both doped and undoped films showed a good response to the gas by increasing/decreasing the electrical resistance by ΔR. The undoped n-type Cu_2O thin films showed a negative response ($\Delta R < 0$) at all temperatures resulting in a maximum response around 85°C. However, the chlorine doped n-type Cu_2O thin films initially showed a positive response ($\Delta R > 0$) to the LP gas which then reversed its sign to give a negative response which peaked at 52°C. The positive response shown by the chlorine doped Cu_2O films vanished completely at 42°C.

Keywords: Liquefied Petroleum Gas, Electrodeposition, Cu_2O Thin Films, Chlorine Doped, Undoped, Gas Response

1. INTRODUCTION

Environmentally hazardous gases are being released continuously to the atmosphere due to industrialization, increased human activities and the natural processes that take place as a result of drastic changes in the environment. Therefore, monitoring of environment has been of extreme importance for the safety and well-being of human and animal life and nature in general. Consequently, gas sensing has become an important area of research that leads to the development of highly responsive gas sensing devices capable of detecting minute amounts of gases of different types. One such gas that requires monitoring in the context of developing countries is the highly inflammable Liquefied Petroleum (LP) gas which is being used very widely as a domestic fuel. At most times leakage of LP gas from production plants or from cylinders occur during their storage, transport and usage. These leakages at most times can be found from the odor of the gas, however, by then a significant amount of gases may have leaked out to the surroundings. Gas sensor devices enable the early detection of such leakages thus preventing accidents and wastage while helping maintain a safer and cleaner environment.

With the pioneering work reported in 1962 by Seiyema et al. [1] , much technological effort has been made in the field of gas sensing aiming towards improvement of the gas response, selectivity, stability and feasibility for practical use. A principal mechanism employed in gas sensing is the monitoring of the electrical resistance of the sensor material upon its exposure to a particular gas of low concentration. Among the many sensor materials available, use of polycrystalline oxide semiconductor thin films as sensing materials is rapidly expanding. Extensive studies have been made on oxide semiconductors such as SnO_2, ZnO and WO_3, for their gas sensing applications for various types of gases [2] [3] . Recently, there have been a few studies on the use of p-type Cu_2O thin films as a potential gas sensing material, Alahapitiya et al. [4] have used thermally oxidized p-type Cu_2O for methane sensing. Shishyano et al. [5] have used electrodeposited p-type Cu_2O for NO_2 gas sensing for LP gas sensing. Dhawale et al. [6] have explored the possibility of using electron beam irradiated chemically deposited TiO_2 thin films. Shukla et al. [7] have reported thesynthesis of tin oxide thick films and their appli- cations in LP gas sening at room temperature.

It has been known that the electrodeposition can be used to fabricate Cu_2O thin films with both n-type and p-type conductivities [8] . Thus the ability to control the conductivity type, the surface morphology etc. coupled with the ease and associated low cost of fabrication make electrodeposited Cu_2O thin films a suitable candidate for gas sensing applications. The inherent high resistivity associated with the elctrodeposited Cu_2O thin films has been considered a drawback that can be addressed using suitable doping methods for related device applica- tions. For example, chlorine doping during the elctrodeposition has lowered the resistivity of Cu_2O thin films significantly [9] . This paper reports the use of electrodeposited chlorine doped n-type Cu_2O thin films for monitoring LP gas while making a comparison with the previously reported LP gas sensing behavior of undoped Cu_2O thin films [10] . The temperature dependent variation of the gas sensing sensitivity is discussed including the reaction mechanisms that cause the observed behaviour.

2. EXPERIMENTAL

Both undoped and chlorine doped n-type Cu_2O thin films were deposited on Ti substrates, a process that has been well established in the research group [9] . Prior to the deposition, the substrates were cleaned thoroughly with detergent, dilute nitric acid, in an acetone bath, and lastly with distilled water. The undoped Cu_2O thin films were potentiostatically electrodeposited in a three-electrode electrochemical cell that contained aqueous solutions of 0.1 M sodium acetate and 0.01 M cupric acetate and hence termed as an acetate bath. Deposition was carried out for duration of 45 minutes at 60°C underpotentiostatic condition of a −200 mV [9] .

However, difficulties were encountered when chlorine doping was attempted using an acetate bath due to precipitation. Doping was therefore accomplished in an electrolytic bath containing aqueous solutions of 3 M lactic acid, 0.45 M cupric sulfate, and 4 M sodium hydroxide. Sodium hydroxide was used to adjust the pH of the electrolyte around 9.5. A solution of $CuCl_2$ was used as the chlorine precursor by varying its concentration. Resistivity measurements of the resulting Cu_2O thin films showed that initially the resistance to drop drastically and then gradually beyond a concentration of 0.02 M $CuCl_2$. Thus for gas sensing measurements, Cu_2O films chlorine doped using a 0.02 M $CuCl_2$ solution was used. The deposition was carried out for a period of 45 minutes, at a constant temperature of 60°C, under potentiostatic conditions of −275 mV. The n-type conductivity of the deposited films was verified using the spectral response measurements [9] . The surface morphological and structural characterization of the films was determined using SEM (Philips XL40) and XRD (SHIMADZU SSX-550) analysis.

A film sample was then enclosed in a gas sensing chamber made of stainless steel. Chamber contained two compartments; the top through which the gas was flown and the bottom where the heating element was housed. In order to measure the electrical resistance, contact probes were fixed on to the surface of the film sample which was placed on an asbestos heating platform. Externally, the probes were connected to a multimeter, which in turn was connected to a computer data logger (Figure 1).

All the measurements were made under atmospheric conditions by using a flow through technique. While maintaining a constant gas flow rate of 0.005 ml/s, films were exposed to LP gas for approximately 30 - 40 s. and then the gas flow was stopped. The sensing temperature was varied between 30°C and 100°C while monitoring the temperature with a thermocouple (type K) which was in contact with surface area of the substrate which was not covered by the Cu_2O film. The temperature was controlled using a thermostat with a temperature controller. The electrical resistivity measurements were made using computer interfaced a Keithley 2100 digital multimeter. The measurements were taken over a period of approximately 120 s after the gas was sent in to the chamber and repeated when the temperatures were brought down to the room temperature. This procedure was repeated at different temperatures in order to determine the temperature that corresponds to the maximum gas sensitivity.

3. RESULTS AND DISCUSSION

3.1. Structure and Morphology

3.1.1. X-Ray Diffraction (XRD)

Figure 2 shows the XRD patterns of the undoped and chlorine doped Cu_2O thin films. It can be seen that both doped and undoped Cu_2O thin films show single phase, polycrystalline characteristics. It can be seen from spectrum B in Figure 2, that the undoped Cu_2O films show a preferred orientation resulting in a stronger (220) reflection. However, doping has caused the preferred orientation to change to produce a stronger (200) reflection in the chlorine doped Cu_2O film samples.

3.1.2. Scanning Electron Microscopy

Figure 3(a) and Figure 3(b) show the SEM pictures of electrodeposited undoped and chlorine doped Cu_2O samples indicating that both films have uniform polycrystalline coverage. However, effect of doping has caused the average polycrystalline grain size to reduce towards nanoscale compared to that of the undoped Cu_2O thin films in which the average grain size is in the micro scale. Through SEM measurements, it was found that the average

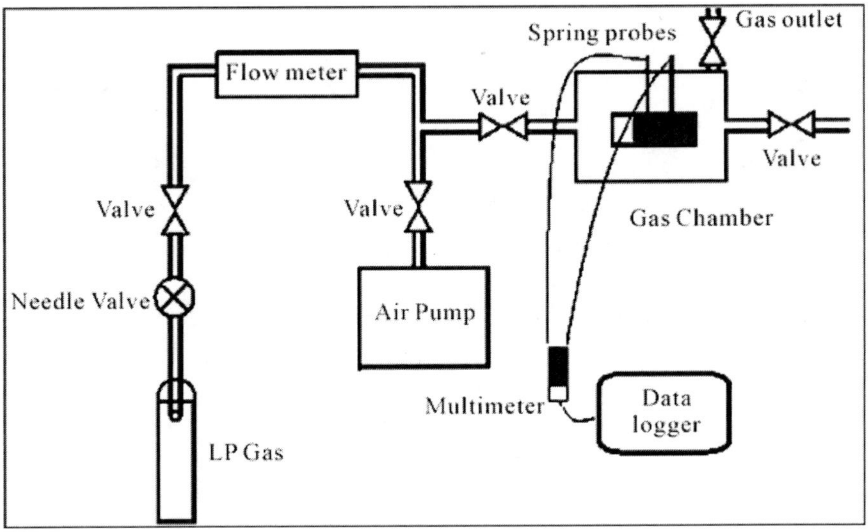

Figure 1. Block diagram of gas sensing setup.

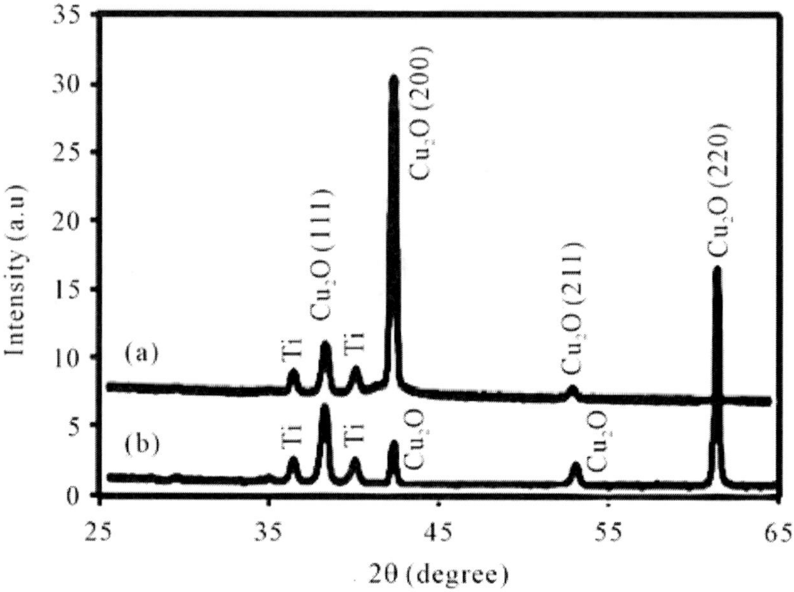

Figure 2. X-ray diffraction spectra of (a) chlorine doped Cu_2O thin film and (b) undoped Cu_2O thin film.

Figure 3. SEM pictures of (a) undoped Cu_2O thin film deposited on Ti substrate and (b) chlorine doped Cu_2O thin film deposited on Ti substrate with a 0.02 M $CuCl_2$ added to the electrolyte bath.

polycrystalline grain size of chlorine doped Cu_2O thin films showed a concentration dependence on $CuCl_2$. It was observed that the increased $CuCl_2$concentration caused the average grain size to reduce gradually providing a larger effective surface area to interact with the gas molecules [11] .

3.1.3. Resistivity Measurements

Figure 4 shows the resistivity measurements taken of the chlorine doped Cu_2O samples in which the $CuCl_2$ precursor concentration in the electrolyte was varied during the deposition of the films. It can be seen that initially, the resistivity of the films reduces from MΩ cm range down to kΩ cm range when the $CuCl_2$ precursor concentration in the electrolyte was 0.02 M. Further increase in the concentration causes the resistivity to reduce gradually through the kΩ cm range.

This decrease in the resistivity was quite significant when compared to the resistivity of undoped Cu_2O thin films fabricated electrochemically using the acetate bath which produced Cu_2O thin films in which the resistivity was larger by over 5 orders of magnitude than the chlorine doped Cu_2O thin films obtained with the 0.02 M $CuCl_2$ precursor concentration in the electrolyte in the lactate bath.

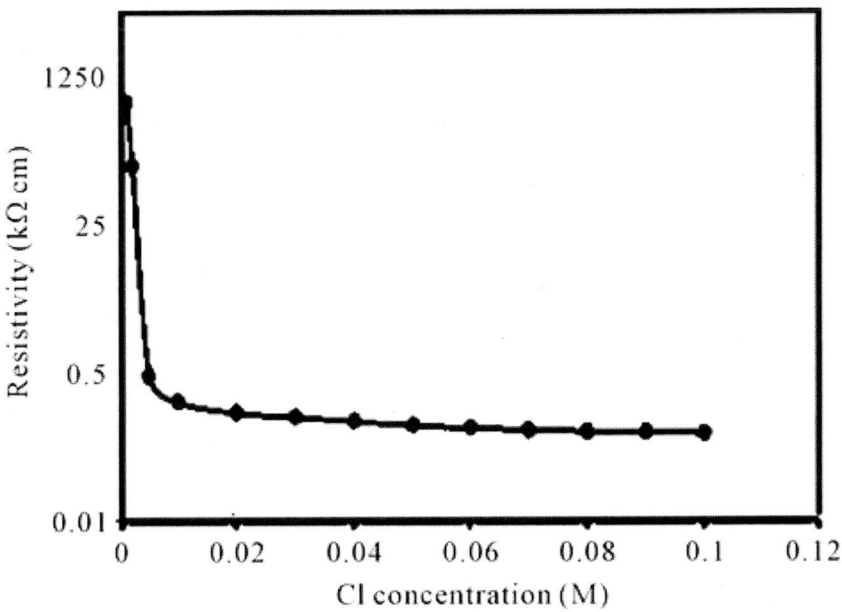

Figure 4. Variation of resistivity of chlorine doped Cu_2O thin films with $CuCl_2$ concentration in the electrolyte solution maintained at pH 9.5 and temperature 60°C.

3.2. Gas Sensing Properties

As mentioned above, the gas sensing properties were measured by monitoring the resistance of the Cu_2O thin films upon exposure to LP gas and its stoppage after a certain period. The response to the gases are represented as resistance vs. time curves and the magnitude of sensitivity vs. time curves where the magnitude of sensitivity is termed as

$$\text{Sensitivity, } S = \left| \frac{R_{LPG} - R_{air}}{R_{air}} \right| \quad (1)$$

Here R_{LPG} is the resistance of the film upon exposure to LP gas and R_{air} is the resistance of the film when it is under normal atmospheric conditions.

Response at Low Temperatures

Figure 5(a) shows the variation of resistance of undoped Cu_2O films as a function of time around room temperature. When the undoped Cu_2O films were exposed to the LP gas, films showed a negative response, i.e. the resistance of the film decreased with the exposure time and gradually recovered to its initial value under the normal atmospheric conditions when the gas flow was stopped [10] . In a magnitude of sensitivity vs. time graph, this behavior is depicted by a single response peak as shown in the Figure 5(b).

This behavior can be well understood in terms of the mechanism given by Shukla [7] on LP gas sensing behavior of SnO_2 thin films. Under atmospheric conditions, the oxygen adsorbed on the surface of the n-type Cu_2O thin film extracts electrons from its conduction band to form O_2^- on the film surface causing the film to be at a higher equilibrium resistance initially. Once the LP gas is sent into the chamber, it reacts with the chemisorbed oxygen removing them from the surface giving the electrons back to the conduction band of the Cu_2O film as shown by the Equation (2).

$$2C_nH_{2n+2}(ad) + O_2^-(ad) \rightarrow 2C_nH_{2n+1} - O + H_2O + 2e^- \quad (2)$$

This reduces the resistance of the film until the LP gas flow in to the chamber is stopped. The stoppage of the gas flow causes atmospheric oxygen to gradually adsorb on the Cu_2O film once again allowing the resistance to recover back to its value under atmospheric conditions.

In contrast, upon exposure to LP gas, the resistance of the chlorine doped Cu_2O films increased ($R_{LPG} > R_{air}$) initially showing a positive response and then showed a negative response ($R_{LPG} < R_{air}$) as in the case of undoped

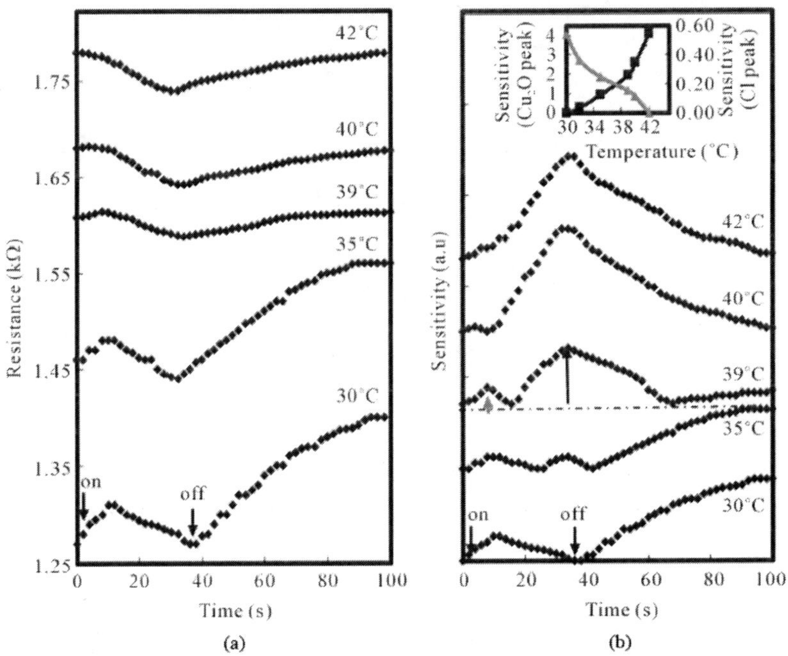

Figure 5. Variation of (a) resistance and, (b) sensitivity of undoped Cu_2O thin films due to the exposures of films to LP gas at a flow rate 0.005 ml/s at different temperatures.

Cu_2O films yielding two peaks in the magnitude of sensitivity vs. time graph. Figure 6(a) and Figure 6(b) show the resistance and the magnitude of sensitivity variations at temperatures of 30°C, 35°C, 39°C, 40°C and 42°C obtained for chlorine doped Cu_2O thin films upon exposure to the LP gas. In contrast to the response given by undoped Cu_2O samples in which the variation of resistance gives rise to only a single peak, it can be seen that at temperatures of 30°C, 35°C and 39°C, there are two responses, with the first peak appearing around 10 s and the second peak appearing around 30 s, before the gas is made to stop flowing. The second peak in the sensitivity vs. time graph in the Figure 6(b) is comparable to the peaks seen when undoped Cu_2O was used as shown in Figure 5(b).

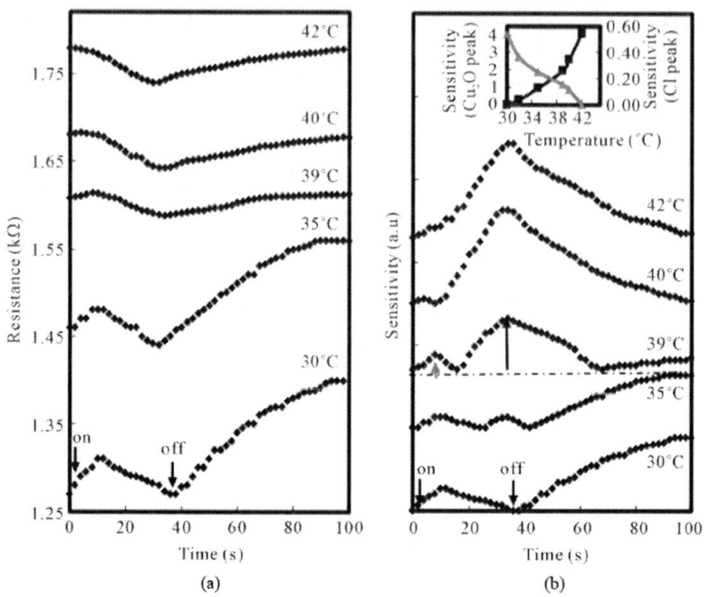

Figure 6. Temperature dependent variation of (a) resistance and (b) sensitivity of chlorine doped Cu_2O thin films deposited in electrolyte solution with 0.02 M $CuCl_2$ when films were exposed to LP gas at a flow rate of 0.005 ml/s. (Inset: ▲ variation of maximum sensitivity in the chlorine peak, and ■ variation of maximum sensitivity in the Cu_2O peak of the chlorine doped Cu_2O thin films with the temperature).

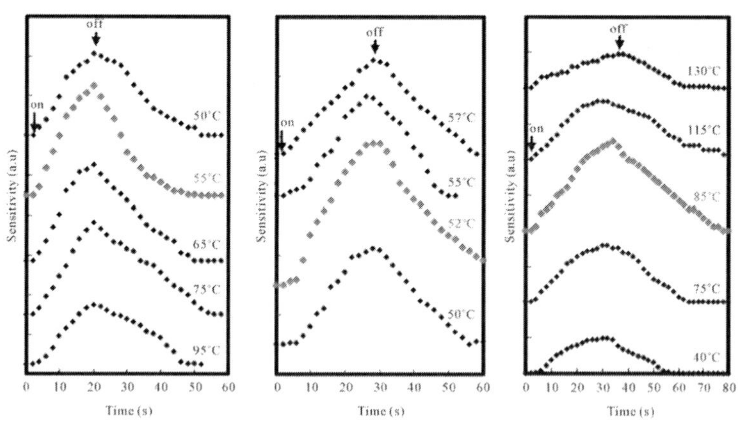

Figure 7. Variation of sensitivity of (a) chlorine doped Cu_2O thin films at the high temperatures, (b) chlorine doped Cu_2O thin films at temperatures around 55°C and, (c) undoped Cu_2O thin films at the different temperatures due to the exposure of films to LP gas. (Highlighted in red are the variations of sensitivity that yields the maximum peak sensitivity.)

Therefore, the first peak appears to have arisen due to the presence of chloride ions in the doped Cu_2O films. Thus for the sake of clarity, the first peak is termed as the chlorine peak whereas the second peak is termed as the Cu_2O peak. Furthermore, it can be seen that the chlorine peak reduces its intensity with increasing temperature disappearing completely at 42°C whereas the intensity of the Cu_2O peak continues to increase.

This behavior can be understood in the following manner. Overall reduction of the resistance of the chlorine doped thin films measured under atmospheric conditions is attributed to the increase of electron density in the conduction band of the Cu_2O thin film which occurs due to the electrons coming from the donor impurity levels. Thus the doping concentration controls the overall resistivity of the Cu_2O film. However, when the films are exposed to the LP gas, due to the formation of electrostatic interactions between induced dipoles of LP gas molecules with the chloride ions in the film, the electron pumping process to the conduction band is retarded causing the resistance to increase initially. When the interaction with chloride ions reaches its saturation, the adsorbed oxygen in the film starts to react with the LP gas molecules as stated above according to the Equation (2) causing the resistance to decrease similar to the undoped Cu_2O films until the gas flow is stopped. It can also be seen that compared to the undoped Cu_2O films, the initial response to the LP gas arising from the chlorine doped Cu_2O films is quicker.

When the chlorine doped thin films are exposed to LP gas at temperatures > 42°C, the absence of the chlorine peak indicates that the gas is not responsive to chloride ions present in the film sample due to the weakening of the dipole interactions between the LP gas molecules and the chlorides at higher temperature. Thus the contribution to the sensing behavior arises only from the adsorbed O_2^- on the surface of the film. Therefore, the chlorine doped Cu_2O films show a response similar to the undoped Cu_2O thin film samples however, showing contrasting differences in the peak sensitivities measured as a function of temperature. Figure 7(a) and Figure 7(b) show the sensitivity measurements made of the chlorine doped samples for corresponding resistance variation when the LP gas was sent changing the film temperature in the range from 50°C - 95°C. Measurements showed that the intensity of the Cu_2O sensitivity peak of the chlorine doped sample increased with temperature up to about 55°C and decreased with further increase in temperature. By varying the temperature around 55°C, it was found that the maximum sensitivity was resulted in at 52°C. In contrast, the undoped Cu_2O films showed the maximum sensitivity around 85°C as shown by the Figure 7(c).

Above behavior was repeatedly observed when both doped and undoped Cu_2O thin film samples were exposed to the LP gas indicating that doping affects the overall gas sensing behavior of the films also.

4. CONCLUSION

In conclusion, it can be clearly seen that the chlorine doping has altered the gas sensing behavior of n-type Cu_2O thin films significantly. While chlorine doping causes the resistance of Cu_2O films to go down drastically, the presence of chloride ions in the film has resulted in an additional sensitivity peak that exists closer to the room temperature. It can be interpreted that the doping has caused LP gas molecules to interact initially with more active chloride ions providing a faster positive response, i.e. $\Delta R(=R_{LPG}-R_{air})>0$ and then with the O_2^- ions on the Cu_2O film providing a negative response, i.e. $\Delta R < 0$ as in the case of undoped Cu_2O films. It can also be seen that the maximum sensitivity of chlorine doped thin films to the LP gas occurs at a lower temperature compared to the undoped Cu_2O films.

ACKNOWLEDGEMENTS

Nayana Bandara, Charith Jayathilaka,, Dhammika Dissanayaka, Sumedha Jayanetti University Grant Commission of Sri Lanka is gratefully acknowledged for financial assistance provided through the research grant UGC/ICD/RG 2011. KNDB is thankful to the Open University of Sri Lanka, for granting leave to carry out this research study.

REFERENCES

1. Seiyama, T., Kato, A., Fujiishi, K. and Nagatani, M. (1962) A New Detector for Gaseous Components Using Semiconductive Thin Films. Analytical Chemistry, 34, 1502-1503.
2. Parmar, M. and Rajanna, K. (2011) Copper (II) Oxide Thin Film for Methanol and Ethanol Sensing. International Journal on Smart Sensing and Intelligent Systems, 4, 710-725.
3. Yamazoe, N., Sakai, G. and Shimanoe, K. (2003) Oxide Semiconductor Gas Sensors. Catalysis Surveys from Asia, 7, 63-73.
4. Ahalapitiya, H.J., Samarasekara, P. and Kun, G. (2009) Methane Gas Sensors Application of Cuprous Oxide Synthesized by Thermal Oxidation. Physica Status Solidi (a), 206, 332-337.
5. Shishiyanu, T.S. and Lupan, O.I. (2006) Novel NO_2 Gas Sensor Based on Cu_2O Thin Films. Sensors and Actuators B: Chemical, 113, 468-476.
6. Dhawle, D.S., et al. (2009) Liqueified Petoleum Gas (LPG) Sensing Perfomance of Electron Beam Irradiated Chemically Deposited Ti_2O Thin Film. Sensors and Actuators B: Chemical, 141, 58-64.
7. Shukla, T. (2012) Synthesis of Tin Oxide Thick Film and Its Investigation as a LPG Sensor at Room Temperature. Journal of Sensor Technology, 2, 102-108.

8. Siripala, W. and Kumara, K.P. (1989) A Photo Electrochemical Investigation of n- and p-Type Semiconducting Behavior of Copper Oxide Films. Semiconductor Science and Technology, 4, 465.
9. Jayathilaka, K.M.D.C., Siripala, W. and Jayanetti, J.K.D.S. (2013) Efficiency Improvement of Cuprous Oxide Based Cl Doping. Proceedings of the International Conference (Solar Asia), Kuala Lumpur, 91-96.
10. Bandara, K.N.D., Jayathilaka, K.M.D.C., Siripala, W. and Jayanetti, J.K.D.S. (2012) Electrodeposited Nano Crystalline Cu_2O Thin Films for Gas Sensing Applications. Proceedings of the 68th Technical Sessions, Sri Lanka Association for the Advancement of Science (SLAAS), 56.
11. Han, X.F. (2011) Electrochemical n-Type Doping in Metal Oxides and Its Application in Photovoltaic. Ph.D. Thesis, University of Texas at Arlington, Arlington, 32-33.

CHAPTER 6

Spin Polarization of the Uniform Three-DimensionalElectron Gas

H. Van Cong

Groupe de Physique des Semi-Conducteurs, Département de Physique, Université de Perpignan via Domitia, LAMPS, Perpignan, France

ABSTRACT

A simple-and-analytic form for total energy (or ground-state energy) in the uniform three-dimensional electron gas, expressed as a function of any Wigner-Seitz radius $r_s$and relative spin polarization ζ is obtained with a very good accuracy of 0.036% from the Stoner model and our interpolation between high-and-low density limits with use of a two-point approach for the correlation energy and spin stiffness at $r_s = 1$ and 70. This suggests a satisfactory desciption of some physical properties such as: paramagnetic-ferromagnetic phase transition and thermodynamic-and-optical phenomena.

Keywords: Electron Gas, Correlation Energy; Spin Stiffness, Total Energy, Spin Susceptibility, Paramagnetic-and-Ferromagnetic Phase Transitions, Wigner States

1. INTRODUCTION

First of all, in the uniform three-dimensional electron gas (U3DEG) at zero temperature, it should be noted that the state of this system is entirely specified by the WignerSeitz radius rs being related to the total electron density n by: $n(r_s) \equiv 3/(4\pi r_s^3 a_B^3)$, where a_B = 0.53 × 10^{-8} cm is the Bohr radius, and the relative spin polarization ζ defined by: $-1 \le \zeta \equiv (n^{\uparrow} - n^{\downarrow})/n \le 1$. Here, $n \equiv n^{\uparrow} + n^{\downarrow}$, and $n^{\uparrow}$ and $n^{\downarrow}$ mean the up- and down-spin electron densities, respectively, noting that the two particular values of ζ (= 0 and ±1) denote the paramagnetic (P) and ferromagnetic, (F) states, respectively. Furthermore, all the energies given in the following are calculated in Rydbergs (Ry). Such an U3DEG is one of the basic models of many-body physics, and has been investigated for over 73 years [1-14]. In particular, for ζ = 0, some physical properties at any $r_s$such as: Ferrell's condition for the stability of the total energy [7], electronic pressure and compressibility [7], virial theorem [12], asymptotic forms for plasmon-dispersion. Coefficients obtained in low-and-high plasmon-energy limits [10], and spin susceptibility [8,9,11], were expressed in terms of the total energy and correlation energy, and their respective derivatives, being denoted in general forms for a presentation simplicity as: $\frac{\partial^n}{\partial x^n} f(x,y) \equiv [f(x,y)]_x^{(n)}$, $\frac{\partial^n}{\partial y^n} f(x,y) \equiv [f(x,y)]_y^{(n)}$, for n = 1, 2, ...

The aim of the present paper is to investigate a simpleand-analytic form for $E(r_s, \zeta)$ in the U3DEG, obtained with a very good accuracy of 0.036% from the Stoner model, an interpolation between high-and-low density limits (HLDL), and a two-point approach for correlation energy $E_c(r_s, \zeta)$ and spin stiffness $\alpha_c(r_s)$ at r_s = 1 and 70, giving rise to a satisfactory description of some physical properties such as: PF-phase transition and thermodynamic-and-optical phenomena.

2. ACCURATE AND SIMPLE ANALYTIC FORM FOR E(Rs, Z) AT ANY Rs AND Z

Accounting for a neutralizing positive background and denoting $\alpha(4/9\pi)^{1/3}$, the total energy per particle is given by:

$$E(r_s,\zeta) = E_{HE}^{stoner}(r_s,\zeta) + E_C(r_s,\zeta),$$
$$E_{HE}^{stoner}(r_s,\zeta) \equiv \left[K_0(r_s,\zeta) + \Delta K_0(\zeta;g)\right] + E_x(r_s,\zeta), \quad (1)$$

where $E_x + (r_s,\zeta) = 3h(\zeta)/(2\pi\alpha r_s)$ is the exchange energy, $K_0(r_s,\zeta) = 3k(\zeta)/5(\alpha r_s)^2$ is the kinetic energy at Fermi surface, $\Delta K_0(\zeta,g) = 3g(1-\zeta^2)/10\alpha^2$ is the kinetic energy correction calculated within the mean-field Hartree-Fock approximation using the noninteracting wave function (i.e., a Stoner model [10]), and g is an empirical parameter chosen so that the relative errors of $E(r_s, \zeta)$ for any r_s and are minimized (its value is given in **Table 1**). Moreover, we use in this work a general spin-interpolation formula for $E_c(r_s, \zeta)$ proposed by Vosko et al. [5]

$$E_c(r_s,\zeta) = E_c(r_s,0) + \alpha_c(r_s)\left[f(\zeta)/[f(\zeta)]^{(2)}_{\zeta(=0)}\right]$$
$$(1-\zeta^4) + \left[E_c(r_s,1) - E_c(r_s,0)\right] f(\zeta)\zeta^4. \quad (2)$$

In Equations (1) and (2), the three functions $k(\zeta)$, $h(\zeta)$ and $f(\zeta)$ are defined by:

$2k(\zeta) \equiv (1+\zeta)^{5/3} + (1-\zeta)^{5/3}$, $2h(\zeta) \equiv (1+\zeta)^{4/3} + (1-\zeta)^{4/3}$, $\left[2^{1/3}-1\right] f(\zeta) \equiv h(\zeta) - 1$ that gives: $[f(\zeta)]^{(2)}_{\zeta} \equiv 2\left[(1+\zeta)^{-2/3} + (1-\zeta)^{-2/3}\right]/9(2^{1/3} - 1) = 1.709921$ for $\zeta = 0$.

From Equations (1) and (2) one notes that: $E(r_s, \zeta) = E_c(r_s, -\zeta)$ and $E_c(r_s, \zeta) = E_c(r_s, -\zeta)$, reducing to the interval: $0 \le \zeta \le 1$, and that if the three functions: $E_c(r_s, 0)$, $E_c(r_s, 1)$ and $\alpha_c(r_s)$, represented by $X(r_s; a)$, are known, then $E_c(r_s, \zeta)$ and $E(r_s, \zeta)$ are therefore determined. Here, the function $X(r_s; a)$ can be expressed in terms of its two exact asymptotic results given in HLDL, $G(r_s)$ and $W(r_s, a)$,

$$X(r_s;a) \equiv W(r_s;a) + \left[G(r_s) - W(r_s;a)\right]/\left[1+\lambda r_s^{\gamma}\right] \quad ,(3)$$

reducing in the HDL to: $\lim_{r_s \to 0} X(r_s;a) = G(r_s)$ and in the LDL to: $\lim_{r_s \to \infty} X(r_s;a) = W(r_s;a)$. Thus, the present method used to obtain our result (3) for $X(r_s; a)$ is the interpolation between HLDL, being similar to those also used in [8,9,11]. It should be noted that this is the unique method used to obtain the analytic expression for $X(r_s; a)$ at any r_s.

In the HDL, in which the random-phase approxima

Table 1. Parameters of the best fit to Equations (4) and (7). Energies in mRy.

For	$E_{cP}(r_s)$	$E_{cF}(r_s)$	$\alpha_c(r_s)$
X_1 [9]	−119.571	−63.185	80.64
X_{70}	−8.68307 [10]	−5.533 [10]	5.3205 [9]
x [9]	875.53	637.356	−407.26
y [9]	62.18	31.09	−33.77
Z [9]	93.288	51.198	−70.95
a	0.0908	0.085	0
	$g = 7.7 \times 10^{-3}$ from Stoner model		
λ	0.03848	0.02287	0.02968
γ	1.67379	1.63995	2.01735

tion (RPA) may be used, the function $G(r_s)$ can be written under a condensed Gell-Mann-and-Bruecker's as [2,9]:

$$G(r_s) = y\ln(r_s) - z, \tag{4}$$

where the values of y and z [13] are given in **Table 1**.

In the LDL (or beyond RPA), the function $W(r_s)$ can be written under a condensed Wigner's form [1,9] as:

$$W(r_s) = -x/(a + r_s), \tag{5}$$

where the values of the empirical parameter a are chosen so that the relative errors of $E(r_s, \zeta)$ for any r_s and ζ are minimized. The values of a and x [9] are reported in **Table 1**.

Going back to Equation (3), in order to determine the constants λ and γ, we now use a two-point approach by denoting: $X(r_s = 1) \equiv X_1$ [9] and $X(r_s = 70) \equiv X_{70}$ [9,10], being reported in **Table 1**. Thus, those constants λ and γ are determined by:

$$\lambda = [G(r_s) - X_1]/[X_1 - W(r_s)] \text{ for } r_s = 1, \quad (6)$$

and

$$\gamma = [1/\ln(r_s)] \times \ln([G(r_s) - X_{70}]/\lambda[X_{70} - W(r_s)]) \text{ for } r_s = 70 \quad (7)$$

In summary, from Equations (3)-(7) and **Table 1**, the correlation energy $E_c(r_s, \zeta)$ given in Equation (2) and total energy $E(r_s, \zeta)$ in Equation (1) are thus determined. Then, in order to calculate the relative errors (RE) of $E(r_s, \zeta)$ from the present result (1) and those from other approximate results evaluated from [4,8,9,11], neglecting the Stoner model, we will use the accurate results obtained from their diffusion Monte Carlo (DMC) method [4] by using backflow wave functions and twist averaged boundary conditions to obtain the accurate values of $E(r_s, \zeta)$ at low densities $40 \le r_s \le 100$ and for $0 \le \zeta \le 1$ with standard errors in units of 10^{-8} Ry. This formula of RE is defined by: 1-(Approximate results/ZLC's results). So, it is found that our present results of $E(r_s, \zeta)$ are accurate to within 0.036% while other approximate ones [4, 8,9,11] only give: 0.24%, 0.29%, 0.42% and 0.471%, respectively. In **Table 2**, for a comparison, we only report our results of RE and those obtained from [4,11].

Furthermore, if neglecting the Stoner model [i.e., g = 0], then the maximal RE in absolute value of the present results is now found to be equal to 0.0625% (>0.036%), suggesting that such a Stoner model used is needed to obtain a better accuracy.

Now, it is interesting to extend our numerical results of $E(r_s, \zeta)$ for $0 \le \zeta \le 1$ at higher density range: $r_s \le 30$ $0.1 \le$, being reported in **Table 3**, in which for a comparison we also include some CA's results [4].

Some physical properties derived from our accurateand-simple forms for $E(r_s, \zeta)$ and $E_c(r_s, \zeta)$ given in Equa tions (1,2) are investigated as follows.

Table 2. REs (%) of the present results of $E(r_s, \zeta)$ (Present) are compared with those of CA [4] and SPS [11].

r_s	ζ	0	0.185	0.333	0.519	0.667	0.852	1
40	Present	−0.036		−0.006		0.036		−0.032
50	CA	0.200						0.240
	Present	−0.004	0.002	0.017	0.024	0.029	0.027	−0.035
60	Present	−0.004	−0.008	0.001	0.024	0.002	−0.001	−0.019
70	Present	0.001	−0.002	0.006	0.004	−0.006	−0.011	−0.015
75	SPS	0.456	0.459	0.468		0.471		0.444
	Present	0.001	0.001	0.005		−0.011		−0.006
85	Present	0.008		0.003		−0.021		0.004
100	CA	0.210						0.110
	Present	0.008		−0.010		−0.035		0.015

Table 3. Present results of total energy for $0 \leq \zeta \leq 1$ and in $0.1 \leq r_s \leq 30$ compared with CA [4].

r_s/ζ	0	0.185	0.333	0.519	0.667	0.852	1	
0.1								
Present	211.587	215.727	225.039	224.470	266.323	302.237	339.130	
1 CA	1.17401							
Present	1.17401	1.21047	1.29243	1.46323	1.65502	1.92929	2.29031	
2 CA	0.00410						0.25170	
Present	0.00383	0.01196	0.03021	0.06820	0.11082	0.18078	0.25174	
3								
Present	−0.13519	−0.13189	−0.12449	−0.10909	−0.09184	−0.06359	−0.03517	
4								
Present	−0.15628	−0.15455	−0.15067	−0.14260	−0.13359	−0.11891	−0.10436	
5 CA	−0.15120						−0.12140	
Present	−0.15291	−0.15187	−0.14950	−0.14462	−0.13918	−0.13040	−0.12186	
6								
Present	−0.14373	−0.14303	−0.14146	−0.13822	−0.13463	−0.12888	−0.12341	
10 CA	−0.10675						−0.10130	
Present	−0.10775	−0.10753	−0.10704	−0.10603	−0.10492	−0.10321	−0.10173	
15								
Present	−0.08020	−0.08012	−0.07993	−0.07955	−0.07913	−0.07851	−0.07804	
20 CA	−0.06329							
Present	−0.06379	−0.06371	−0.06362	−0.06344	−0.06325	−0.06296	−0.06276	
30								−0.06251
Present	−0.04531	−0.04527	−0.04527	−0.04522	−0.04516	−0.04507	−0.04503	

3. PHASE TRANSITION

3.1. First-Order Phase Transition from the TotalEnergy

From the three states such as: the P-state represented by $E_P(r_s)$, the F-state by $E_F(r_s)$, and the Fermi Wigner crystal (W) state by [13]:

$$E_w(r_s) = -\frac{1.79186}{r_s} + \frac{2.65279}{r_s^{3/2}} - \frac{0.73}{r_s^2} + 0\left(r_s^{-5/2}\right)$$

being obtained at low electron densities and beyond RPA, we can evaluate the total-energy differences: $\left[E_P(r_s)-E_F(r_s)\right] D_{FP}(r_s) \equiv$ and $D_{WF}(r_s) \equiv \left[E_F(r_s)-E_W(r_s)\right]$.

1) The $D_{FP}(r_s)$ is negative for $r_s < 78.147$ (the P-phase) and positive for $r_s > 78.147$, giving rise to a first-order PE-phase transition which occurs at a critical value: $r_{sC} = 78.147$, in accordance with CA's data [4]: $r_{sC} = 75 \pm 5$ and with ZLC's data [10]: $75 < r_{sC} < 85$. One notes that this critical value: $r_{sC} = 78.147$ corresponds to a critical density: $n_C = 3.37 \times 10^{18}$ cm^3 being found to be equivalent to the metal-insulator transition in the n-type excited intrinsic Si occurring at 2.71×10^{18} cm^3 [13] and in the P-Si at 3.52×10^{18} cm^3 [14].

2) The $D_{WF}(r_s)$ is negative for $r_s < 115.49$ (the F-phase) and positive for $r_s \geq 115.49$, giving rise to the WFphase transition which occurs at a critical value: $r_{sC} = 115.49$, in good agreement with CA's data [4]: $r_{sC} = 100 \pm 20$.

We futher evaluate the total-energy difference times as: $\left[E(r_s,\zeta)-E(r_s,0)\right]\times r_s^{3/2}$ (Ry) and plot versus spin polarization ζ for different values of r_s in the following **Figure 1**.

Some concluding remarks are given below.

1) At $r_s \leq 63$, the system is P, with the unpolarized phase stable.

2) As the electron density decreases, at $r_s \approx 64.4$, the system becomes unstable with respect to spin fluctuations while ZLC [10] obtained the corresponding instable point at $r_s \approx 50$.

3) The partially polarized states become stable at $r_s \approx 65.6$.

4) As the electron density continues to decrease, the fully polarized state has a lower energy with respect to unpolarized state at $r_s \geq 78.147$ while a corresponding ZLC's estimation [10] is $r_s \geq 80$. We also find that the partially polarized state has an even lower energy.

3.2. Second-Order Phase Transition from the Spin Susceptibility

First of all, if denoting the unit vector, Bohr magneton, magnetic field and spin susceptibility, respectively, by: $\vec{u} \equiv \vec{M}/M$, $\mu_B = 9.274096 \times 10^{-21}(erg/G)$, B(G) and χ (erg/cm³·G²) the magnetization vector is given by: $\vec{M} \equiv (n^{\uparrow} - n^{\downarrow})\mu_B \vec{u} = (n\mu_B)(\zeta\vec{u}) = x(B\vec{u})$, leading to: B

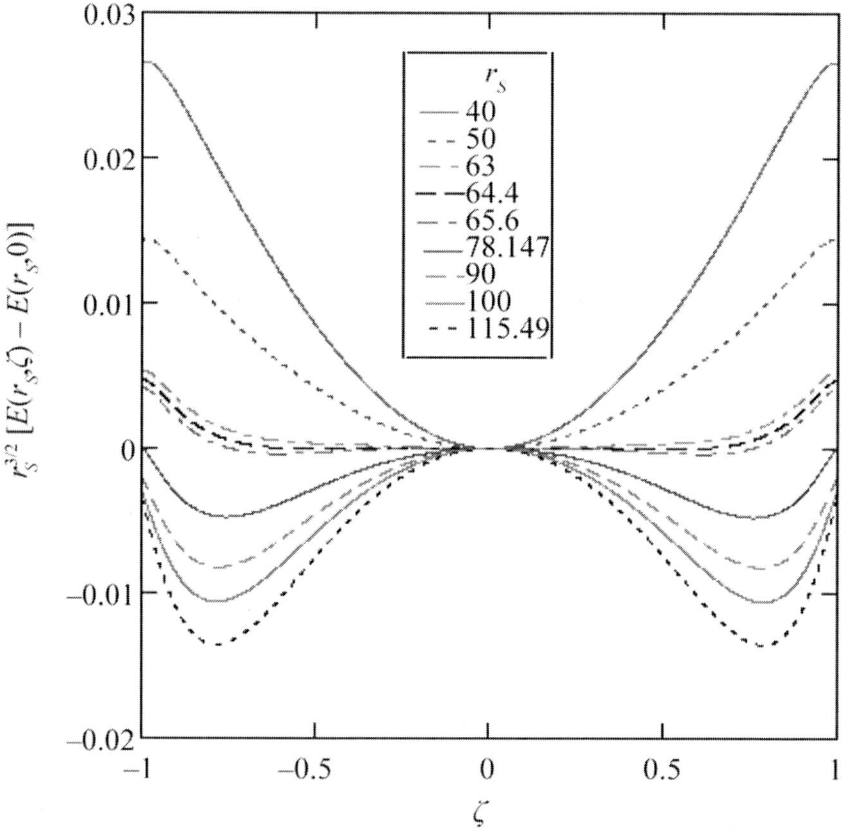

Figure 1. $r_s^{3/2}\left[E(r_s,\zeta)-E(r_s,0)\right]$ vs ζ for different r_S.

B/χ. Then, the interaction-energy density U required to build up this magnetization is given by = ζnμ

$$U = \int_0^M B\mathrm{d}M' = \int_0^M (M'/\chi)\mathrm{d}M' = (1/2\chi)M^2$$
$$= (1/2\chi)(\zeta n \mu_B)^2$$
.(8)

Now, from Equation (1), by a variable change: x = cζ, c being a length unit, x thus varies from –c to c and the U3DEG induces from the Hook's law an elastic (or conservative) force for $|\zeta| \ll 1$: $= -k_1(r_s)x\vec{u}$ $\boldsymbol{F}(x) = -\mathrm{d}E(r_s, x/c)/\mathrm{d}x\vec{u}$, where the spring constant (or force constant) is defined by: $\equiv (\mathrm{d}^2E(r_s,\zeta)/c^2\mathrm{d}\zeta^2)_{\zeta=0} = [k(r_s)/c^2]$ $k_1(r_s) = (\mathrm{d}^2E(r_s, x/c)/\mathrm{d}x^2)_{x=0} \equiv$, that gives:

$k(r_s) = (2/3r_s^2\alpha^2) - (3g/5\alpha^2) - (2/3\pi r_s\alpha) + \alpha_c(r_s)$ being a simple-and-analytic form and used to compute following physical quantities.

Then, in such an elastic-force model, the work required to build up the magnetization is

$$\Delta E\left(r_s, \frac{x}{c} = \zeta\right) \equiv E\left(r_s, \frac{x}{c} = \zeta\right) - E(r_s, 0) = \int_0^x k_1 x' \mathrm{d}x'$$
$$= (1/2)k_1x^2 = (1/2)k\zeta^2,$$

being used to define the interaction-energy density by: $U = n\Delta E(r_s, \zeta)$, identical to the result (8), that leads to a simple form for the spin susceptibility χ(erg/cm³·G²) as:

$$\chi(r_s) \equiv N(r_s)\mu_B^2/k(r_s) \quad (9)$$

We now evaluate our present result (9) and plot versus r_s in the following **Figure 2**, in which we also report other results from [8,9], indicating that our present result of χ shows a singularity at a critical value: 65.6, defining thus the second-order PF-phase transition, which can be compared with ZLC's result [10]: 50.

4. SOME THERMODYNAMIC-OPTICAL PHENOMENA

1) We first investigate the Ferrell's condition for the stability of the total energy E(r_s, ζ) defined by [7]:

$$\left[r_s^2 E(r_s,\zeta)\right]_{r_s}^{(2)} = 2E(r_s,\zeta) + 4r_s\left[E(r_s,\zeta)\right]_{r_s}^{(1)} + r_s^2\left[E(r_s,\zeta)\right]_{r_s}^{(2)} \leq 0. \tag{10}$$

Our numerical calculation indicates that it is valid up to a superior value of r_s found to be given by: r_{sS} =1200, 1400, 2200, 2300, 2600, 2700, 16000 and +∞ for ζ = 0, 0.5, 0.78, 0.79, 0.83, 0.84, 0.99 and 1, respectively, suggesting that r_{sS} increases with increasing ζ.

2) The electronic pressure is defined [7] and related to the generalized virial theorem given in Equation (A4) of

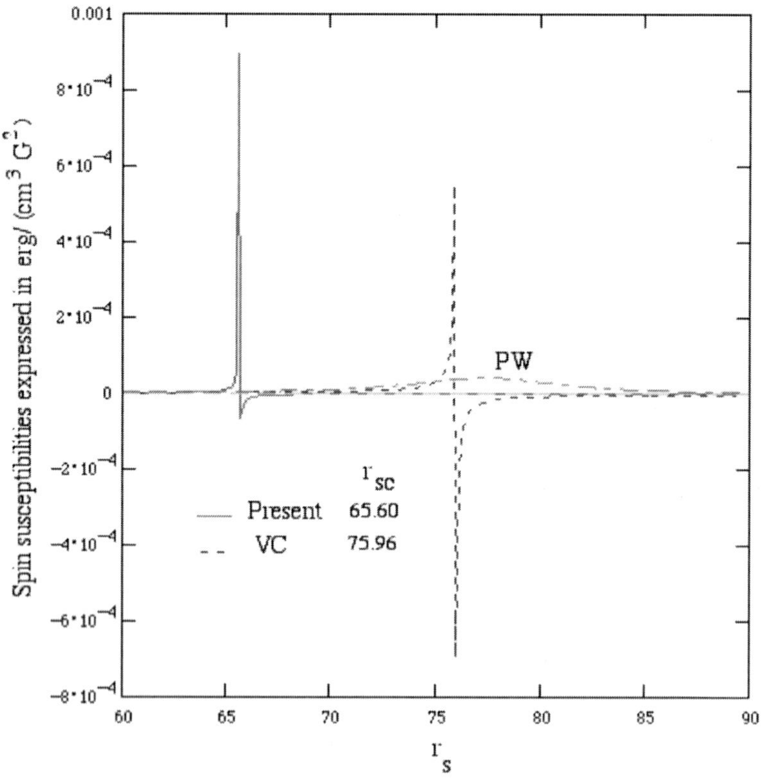

Figure 2. Spin susceptibilities vs r_s.

the Appendix A by:

$$P(r_s,\zeta) = -nr_s \times \left[E(r_s,\zeta)\right]_{r_s}^{(1)} / 3$$
$$= (n/3)\left[2T(r_s,\zeta) + V(r_s,\zeta)\right], \quad (11)$$

which becomes negative for $r_s \geq 4.1908$ at $\zeta = 0$ (in good accordance with the Ichimaru's result [7], $r_s > 4.2$), for $r_s \geq 4.569$ at $\zeta = 0.5$, and for $r_s \geq 5.653$ at $\zeta = 1$. This means that the equilibrium densities of the total energy $E(r_s, \zeta)$ occur at 4.1908, 4.569 and 5.653 for $\zeta = 0$, 0.5 and 1, respectively, suggesting that those values increase with increasing ζ.

3) The compressibility of the non-interacting U3DEG is defined by: $\kappa_o(r_s,\zeta) = 3/\left[2nE_F(r_s,\zeta)\right]$, where $E_F(r_s, \zeta)$ is the Fermi energy, and via the compressibility sum rule, the inverse reduced compressibility can thus be evaluated by:

$$\kappa_o(r_s,\zeta)/\kappa(r_s,\zeta) = 1 - \left[4\alpha r_s \gamma_o(r_s,\zeta)\right]/\pi \quad ,(12)$$

where $\gamma_o(r_s, \zeta)$ is determined in Equation (B3) of the the Appendix B.

Our numerical calculation indicates that it becomes negative for $r_s \geq 5.2597$ at $\zeta = 0$ (in good accordance with the Ichimaru's result [7]: $r_s > 5.3$), for $r_s \geq 5.08$ at $\zeta = 0.5$, and for $r_s \geq 4.55$ at $\zeta = 1$, noting also that this negative compressibility does not imply an instability of the system when a rigid background of compensating charge is assumed.

4) Finally, in order to compare the two functions γ_∞ and γ_o determined in Equations (B2) and (B3) of the Appendix B, we can study the asymptotic forms for plasmon dispersion coefficient β as follows.

For sufficiently low plasmon energies ($\hbar\omega_P$) and if one disregards the relaxation effect in the short-time domain, β is thus reduced to [7]: $\beta_o(r_s,\zeta) \approx \beta_{RPA}(r_s,\zeta) - (\omega_P/4\omega_F)\gamma_o(r_s,\zeta)$, where $\omega_F \equiv E_F/\hbar$ and in the RPA: $\beta_{RPA} = 3\omega_F/5\omega_p$, and for a sufficiently high plasmon energies, one obtains [7]: $\beta_\infty(r_s,\zeta) \approx \beta_{RPA}(r_s,\zeta) - (\omega_P/4\omega_F)\gamma_\infty(r_s,\zeta)$. Therefore, one gets:

$$R(r_s,\zeta) \approx \frac{\beta_o(r_s,\zeta) - \beta_{RPA}(r_s,\zeta)}{\beta_\infty(r_s,\zeta) - \beta_\infty(r_s,\zeta)} = \frac{\gamma_o(r_s,\zeta)}{\gamma_\infty(r_s,\zeta)} \quad (13)$$

giving for any r_s and ζ: $1.3 \le R(r_s, \zeta) \le 1.7$, meaning that $(r_s, \zeta) > 1$ or $\beta_o(r_s, \zeta) > \beta_\infty(r_s, \zeta)$.

5. CONCLUDING REMARKS

In summary, we have developed simple analytic forms for $E(r_s, \zeta)$ and $E_c(r_s, \zeta)$, by basing on the Stoner model and interpolation between correct HLDL with the use of a two-point approach for correlation energy and spin stiffness at $r_s = 1$ and 70, giving rise to:

1) A best precision of the order of 0.036% for our present form for total energy, being found to be more accurate compared with other works [4,8,9,11], as given in **Table 2**, and 2) A satisfactory description of some physical properties such as: PF-phase transition and thermodynamic and optical phenomena, as given in above Sections 3 and 4.

6. ACKNOWLEDGEMENTS

H. Van Cong (his) helpful remarks and suggestions, which have greatly improved the presentation of this paper, and also Dr. Chesneau Xavier for his interest in this work.

REFERENCES

1. E. P. Wigner, "On the Interaction of Electrons in Metal," Physical Review, Vol. 46, No. 11, 1934, pp. 1002-1011.
2. M. Gell-Mann and K. A. Bruecker, "Correlation Energy of an Electron Gas at High Density," Physical Review, Vol. 106, No. 2, 1957, pp. 364-368.
3. R. A. Coldwell-Horsfall and A. A. Maradudin, "ZeroPoint Energy of an Electron Lattice," Journal of Mathematical Physics, Vol. 1, No.5, 1960, pp. 395-405.
4. D. M. Ceperley and B. J. Alder, "Ground State of the Electron Gas by a Stochastic Method," Physical Review Letters, Vol. 45, No. 7, 1980, pp. 566-569.
5. S. H. Vosko, L. Wilk and M. Nusair, "Accurate SpinDependent Electron Liquid Correlation Energies for Local Spin Density Calculations: a Critical Analysis," Canadian Journal of Physics, Vol. 58, 1980, pp. 1200- 1211.
6. J. P. Perdew and A. Zunger, "Self-Interaction Correction to Density-Functional Approximations for Many-Electron Systems," Physical Review B, Vol. 23, No. 10, 1981, pp. 5048-5079.
7. S. Ichimaru, "Strongly Coupled Plasmas: High-Density Classical Plasmas and Degenerate Electron Liquids," Review of Modern Physics, Vol. 54, No. 4, 1982, pp. 1017- 1059.[Citation Time(s):12]

8. J. P. Perdew and Y. Wang, "Accurate and Simple Analytic Representation of the Electron-Gas Correlation Energy," Physical Review B, Vol. 45, No. 23, 1992, pp. 13244-13249.
9. H. Van Cong, "About a Simple Expression for the Homogeneous Electron Gas Correlation Enery, Accurate at Any Electron Density and Relative Spin Polarization, and Its Applications," Physica Status Solidi B, Vol. 205, 1998, pp. 543-552.
10. F. H. Zong, C. Lin and D. M. Ceperley, "Spin Polarization of the Low-Density Three-Dimensional Electron Gas," Physical Review E, Vol. 66, No. 3, 2002, Article ID: 036703.
11. J. Sun, J. P. Perdew and M. Seidl, "Correlation Energy of the Uniform Electron Gas from an Interpolation between Highand Low-Density Limits," Physical Review B, Vol. 81, No. 8, 2010, Article ID: 085123.
12. N. H. March, "Kinetic and Potential Energies of an Electron Gas," Physical Review, Vol. 110, No. 3, 1958, pp. 604-605.
13. H. Van Cong, "Bandgap Changes in Excited Intrinsic (Heavily Doped) Si and Ge," Physica B, Vol. 405, No. 4, 2010, pp. 1139-1149.
14. H. Stupp, M. Hornung, M. Lakner, O. Madel and H. V. Löhneysen, "Possible Solution of the Conductibility Exponent Puzzle for the Metal-Insulator Transition in Heavily Doped Uncompensated Semiconductors," Physical Review Letters, Vol. 71, No. 16, 1993, pp. 2634-2637.

CHAPTER 7

Natural Gas Flaring—Alternative Solutions

Azeez G. Aregbe

Department of Chemical and Petroleum Engineering, University of Lagos, Lagos, Nigeria

ABSTRACT

Gas flaring is one of the major problems in the world. It consumes useful natural resources and produces harmful wastes, which have negative impacts on the society. It is one of the most tedious energy and environmental problems facing the world today. It is a multi-billion dollar waste, a local environmental catastrophe and environmental problem which has persisted for decades. From the year 1996-2010, in Nigeria, 12,602,480.25 million ft^3 of natural gas was flared (NNPC). This is equivalent to losing about $12,967.952 \times 10^{12}$ Btu of energy that would have been used to generate power or converted to other forms of energy. In 2015, the World Bank estimated that 140 billion cubic meters of natural gas produced with oil is flared annually, mostly in developing countries without gas processing infrastructures, or other means of utilizing the produced gas. It is widely known that flaring or even, venting of gas contributes significantly to greenhouse gas emissions, with negative impacts on the environment. Thus, alternative solutions to reduce or utilize the quantity of gas flared are crucial issues. Therefore, the need to study and provide detailed understanding of these alternative solutions to gas flaring is important. This paperoutlined the harmful effects of gas flaring and the different possible alternatives to gas flaring. The proposed alternative solutions are gas for secondary oil recovery, feedstock for petrochemical plants, domestic uses, LNG & CNG, as well as energy conservation by storing as gas hydrate for future use or other purposes. Gas hydrate is stable above the freezing point of water and sufficiently high pressure. It is relatively stable under its saturation temperature and pressure and also much denser than normal ice. This property of gas hydrate can be experimentally investigated and capitalized on, to effectively store natural gas as hydrate for energy conservation instead of flaring the gas wastefully. The alternative solutions will convincingly reduce and in the nearest future stop gas flaring globally.

Keywords: Natural Gas, Gas Flaring, Harmful Effects, LNG, CNG, Gas Hydrate

1. INTRODUCTION

Natural gas is a naturally occurring hydrocarbon gas mixture consisting primarily of methane, but also includes varying amounts of other higher alkanes- ethane, propane, butane, etc., and sometimes a small percentage of carbon dioxide, nitrogen, and/or hydrogen sulfide. It is formed when layers of decomposing plant and animal matter are exposed to intense heat and pressure over thousands of years. The energy that the plants originally obtained from the sun is stored in the form of chemical bonds in the gas. Natural gas is often informally referred to simply as "gas", especially when compared to other energy sources such as oil or coal [1] .

It is a fossil fuel used as a source of energy for heating, cooking, and electricity generation. It is also used as fuel for vehicles and as a chemical feedstock in the manufacture of plastics and other commercially important organic chemicals. It is one of the cleanest, safest, and most useful forms of energy in our day-to-day lives [2] .

Natural gas can be measured in a variety of ways, although the most common unit of measurement is the Gigajoule (GJ), which signifies one billion joules, the metric measure for heat or energy. A gigajoule (GJ) is the energy equivalent of about 30 liters of gasoline. (Alberta Energy Company). It can also be measured in Mcf (thousand cubic feet), Btu (British thermal unit) as well as Standard cubic metre (scm).

Natural gas can be "associated" (found in oil fields), or "non-associated" (isolated in natural gas fields), and is also found in coal beds (as coalbed methane). The natural gas industry is extracting an increasing quantity of gas from challenging resource types: sour gas, tight gas, shale gas, and coalbed methane. When the gas is produced with petroleum, it is often vented or flared.

In 2015, the World Bank estimated that 140 billion cubic meters of natural gas produced with oil is flared at thousands of oil fields worldwide, mostly in developing countries that do not have gas processing infrastructure. The result: emissions equivalent to that of 77 million automobiles, or translated into power generation, more electricity than the entire continent of Africa currently consumes.

Physical and Chemical Properties of Natural Gas

Natural gas is a naturally occurring gas mixture, consisting mainly of methane and other minor components. Table 1 outlines the typical components of natural gas on the Union Gas system and the typical ranges for these values (allowing for the different sources). Union Gas is an Energy company that gets most of their gas supply from western Canada; some gas is supplied from other sources, including the United States and Ontario producers.

Since the different gas supplies enter the Union Gas system at different locations, the exact value of each component at any site will vary among the different regions. The system average heating value will depend on the mix of gas supplies and therefore can vary from the values listed below.

Sulphur: In the Union Gas system, the typical sulphur content is 5.5 mg/m^3.

Table 1. Compositions of natural gas (union gas system).

S/N	Component	Typical analysis (mole %)	Range (mole %)
1	Methane	95.0	87.0 - 97.0
2	Ethane	3.2	1.5 - 7.0
3	Propane	0.2	0.1 - 1.5
4	Butanes (iso/normal)	0.03	0.01 - 0.3
5	Pentanes (iso/normal)	0.01	Trace - 0.04
6	Hexanes plus	0.01	Trace - 0.06
7	Nitrogen	1.0	0.2 - 5.5
8	Carbon Dioxide	0.5	0.1 - 1.0
9	Oxygen	0.02	0.01 - 0.1
10	Hydrogen	Trace	Trace - 0.02
11	Specific gravity	0.58	0.57 - 0.62
12	Gross heating value (MJ/m^3), dry basis*	38.0	36.0 - 40.2

*The gross heating value is the total heat obtained by complete combustion at constant pressure of a unit volume of gas in air, including the heat released by condensing the water vapour in the combustion products (gas, air, and combustion products taken at standard conditions).

This includes the 4.9 mg/m^3 of sulphur in the odorant (mercaptan) added to gas for safety reasons.

Water: The water vapour content of natural gas in the Union Gas system is less than 80 mg/m^3, and is typically 16 to 32 mg/m^3.

In its natural state, it is colorless, odorless and tasteless but in order to warn off its presence in case of leaks, an odorant (mercaptan) is added to it, giving it its characteristic smell of rotten eggs. Its relative density makes it lighter than air, so that leaks or emissions quickly dissipate into the upper layers of the atmosphere, making it less likely to form explosive mixtures in air. It is also non corrosive in nature [3] . Some of the physical and chemical properties of Methane (Major component of Natural gas) are shown in Table 2.

2. NATURAL GAS FLARING

Gas flaring is a process whereby natural gas associated with crude oil during production of crude oil is been burnt. Flaring is a major concern in petroleum- producing areas where there exists insufficient infrastructure to utilize the produced natural gas. It serves as a way of disposing the gas produced in those areas. This as simple as it may sound, creates series of negative impacts on the people in those arrears as well as the environment in general [4] .

In the early days of petroleum exploration, natural gas was not considered a useful product because of the difficulties in transporting it to where it can be utilized or the problems associated with it storage. As a result, gas was simply burned off at the well or vented into the atmosphere, to create rooms for other operations and supposedly to save the whole system from being burnt down by gas explosion. Table 3 shows the top ten gas flaring countries in the world as at 2012.

Table 2. Physical and chemical properties of natural gas.

S/N	Property	Value
1	Molecular weight of mixture	18.2
2	Boiling point at 1 atmosphere	−160.0°C
3	Melting point	−180.0°C
4	Vapor density (air = 1) at 15.5	0.61
5	Liquid density (water = 1) at 0°/4°C	0.554
6	Water solubility at 20°C	Slightly soluble (0.1% - 1.0%)

Table 3. Top ten gas flaring countries in the world (world bank).

S/N	Country	Gas flared (10^6 cm^3)	Rank
1	Russia	35,000	1st
2	Nigeria	17,000	2nd
3	Iran	10,000	3rd
4	Iraq	10,000	4th
5	United States	5000	5th
6	Algeria	4800	6th
7	Kazakhstan	4700	7th
8	Venezuela	4300	8th
9	Saudi Arabia	4100	9th
10	Angola	4000	10th

Even today, flaring and venting continue in locations where local markets and gas transportation infrastructures are not available, or where the gas itself is contaminated with other incombustible gases [5] . In 2015, the World Bank estimated that 140 billion cubic meters of natural gas produced with oil is flared at thousands of oil fields worldwide, mostly in developing countries that do not have gas processing infrastructure.

Table 4 shows the total volumes of gas produced and flared in Nigeria-one of the major oil and gas producing countries in the world, from 1996-2010 [1] . Out of the total gas production of 846,655.4 million cubic metres, a total volume of 356,862.5 million cubic metres of gas was flared (approximately 42.15% of total gas production) during the 15 years period.

The Independent Statistic & Analysis Report from Energy Information Administration (EIA) in 2016 stipulated that the power (kWh) generated from a given fuel is a function of the heat rate of the power plant used and the heat content of the given fuel. Thus, to generate a kilowatt-hour (kWh) of electricity, the amount of fuel required can be calculated as;

$$A_{fuel}\left(\frac{\text{physicalunit}}{\text{kWh}}\right)=\frac{\text{Heatrate}(\text{Btu/kWh})}{\text{Fuelheatcontent}(\text{Btu/massorvolume}))} \tag{1}$$

Table 5 shows the findings and reports during the analysis, (EIA, 2016), with three different types of fuels.

Table 4. The statistic of gas flaring in Nigeria, 1996-2010.

Year	Gas produced (10^6 m^3)	Gas flared (10^6 m^3)	% of gas flared
1996	35,450.00	26,590.00	75.01
1997	37,150.00	24,234.00	65.23
1998	37,039.00	23,632.00	63.80
1999	43,636.00	22,362.00	51.25
2000	42,732.00	24,255.00	56.76
2001	52,453.00	26,759.00	51.02
2002	48,192.45	24,835.58	51.53
2003	51,766.03	23,943.03	46.25
2004	58,963.61	25,090.91	42.55
2005	59,284.97	23,002.71	38.80
2006	82,036.86	28,584.39	34.84
2007	84,707.34	27,307.13	32.24
2008	80,603.61	21,811.00	27.06
2009	64,882.86	17,987.59	27.72
2010	67,757.65	16,468.18	24.30

Table 5. Fuel heat rates and heat contents of fossil fuels.

S/N	Fuel	Plant heat RATES (Btu/kWh)	Fuel heat contents
1	Coal	10,080	9710 (Btu/lbs)
2	Petroleum	10,156	139,713 (Btu/US gal)
3	Natural gas	10,408	1029 (Btu/ft^3)

Note: Heat contents of coal vary widely by types of coal.

In Table 5, using a plant heat rate of 10,408 Btu/kWh, one ft^3 of natural gas has a fuel heat content of about 1029 Btu. Therefore, from the year 1996-2010, in Nigeria, 12,602,480.25 million ft^3 of natural gas was flared. This is equivalent to losing about $12{,}967.952 \times 10^{12}$ Btu of energy that would have been used for something profitable.

2.1. Flare Stack Configuration

A gas flare, which can also be referred to as a flare stack, is a gas combustion device used in industrial plants such as petroleum refineries, chemical plants, and natural gas processing plants as well as at oil or gas production sites having oil wells, gas wells, offshore oil and gas rigs and landfills.

In industrial plants, flare stacks are primarily used for burning off flammable gas released by pressure relief valves during unplanned over-pressuring of plant equipment. During plant or partial plant startups and shutdowns, flare stacks are also often used for the planned combustion of gases over relatively short periods. When industrial plant equipment items are over-pressured, the pressure relief valve is an essential safety device that automatically releases gases and sometimes liquids. Those pressure relief valves are required by industrial design codes and standards for safety purposes [6] .

The released gases and liquids are routed through large piping systems called flare headers to a vertical elevated flare. The released gases are burned as they exit the flare stacks. The size and brightness of the resulting flame is a function of the flammable material's flow rate in joules per hour (or Btu/hour).

Steam is very often injected into the flame to reduce the formation of black smoke. When too much steam is added, a condition known as "over steaming" can occur resulting in reduced combustion efficiency and higher emissions. To keep the flare system functional, a small amount of gas is continuously burned, like a pilot light, so that the system is always ready for its primary purpose as an over-pressure safety system.

The flow diagram in Figure 1 depicts the typical components of an overall industrial flare stack system and the major components of a flare stack are:

- Knockout drums to remove any oil and/or water from the relieved gases.
- A water seal drum to prevent any flashback of the flame from the top of the flare stack.
- An alternative gas recovery system for use during partial plant startups and/ or shutdowns as well as other times when required. The recovered gas is routed into the fuel gas system of the overall industrial plant.
- A steam injection system to provide an external momentum force used for efficient mixing of air with the relieved gas, which promotes smokeless burning.

- A pilot flame (with its ignition system) that burns all the time so that it is available to ignite relieved gases when needed.
- The flare stack, including a flashback prevention section at the upper part of the stack.

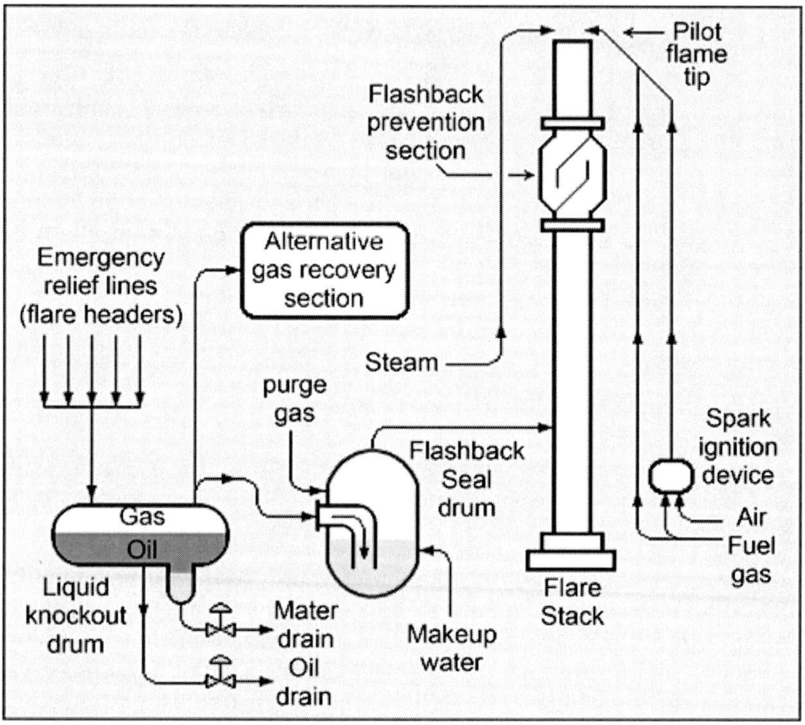

Figure 1. Schematic flow diagrams of an overall vertical, elevated flare stack system.

There is also a safe method to divert the flare gas which is insertion of Liquid U seal with Liquid Hold up vessel. The Liquid U seal is designed to take pressure up to permitted back pressure of the system. This helps to divert the flare gas to recovery system. In case of plant upset, pressure rises and liquid in the U seal will move into Liquid Hold up vessel. On normalization, the Liquid U seal will start diverting the gas again.

2.2. Effects of Natural Gas Flaring

Gas flaring is one of the most challenging energy and environmental problems facing the world today. Environmental consequences associated with gas flaring have a considerable impact on local populations, often resulting in severe health issues. Generally, gas flaring is normally visible and emitted both noise and heat [7] . The effects of gas flaring are:

2.2.1. Climate Change

Gas flaring has serious implications on climate change. The burning of fossil fuel, mainly coal, oil and gas has led to warming up the earth and is projected to get worse according to the inter-governmental panel on climate change (IPCC). Climate change is particularly serious for developing countries, and Africa as a continent is regarded as highly vulnerable with limited ability to adapt. Gas flaring contributes to climate change by emission of carbon dioxide, the main greenhouse gas. Venting of the gas without burning, a practice for which flaring seems often to be treated as a synonymy, releases methane, the second main greenhouse gas.

2.2.2. Acid Rain

The primary causes of acid rain are emissions of sulphur dioxide (SO_2) and nitrogen oxides (NO) which combine with atmospheric moisture to form sulfuric acid and nitric acid respectively. These acidify lakes and streams and damage vegetation. In addition, acid rain accelerates the decay of building materials and paints. Prior to falling to the earth, SO_2 and NO_2 gases and their particulate matter derivatives, sulfates and nitrates, contribute to visibility degradation and harm public health.

2.2.3. Effects on Agriculture

The flares give rise to atmospheric contaminants. These include oxides of Nitrogen, Carbon and Sulphur, (NO_2, CO_2, CO, SO_2) particulate matter and hydrogen sulphide (H_2S). These contaminants acidify the soil, hence depleting soil nutrient. In some cases, there is no vegetation in the areas surrounding the flare due to the tremendous heat that is produced and acid nature of soil pH. The effects of the changes in temperature on crops include stunted growth, scotched plants and such other effects as withered young crops.

2.2.4. Pollution

Gas flaring leads to the emission of pollutants which are harmful to both humans and the society at large. Incomplete combustion of gas leads to the production of carbon monoxide which is one of the major pollutants with adverse effects on human health and the society at large. The economic and environmental ramifications of this high level of gas flaring are serious because this process is a significant waste of potential fuel which is simultaneously polluting water, air, and soil

2.2.5. Health Implications

The implication of gas flaring on human health are all related to the exposure of those hazardous air pollutants emitted during incomplete combustion of gas flare. These pollutants are associated with a variety of adverse health impacts, including cancer, neurological, reproductive and developmental effects. Deformities in children, lung damage and skin problems have also been reported. Hydrocarbon compounds are known to cause some adverse changes in hematological parameters. These changes affect blood and blood-forming cells negatively. Aside the health and environmental consequences of gas flaring, there are also loses of billions of dollars' worth of gas which is literally burnt off daily in the atmosphere. Much of this can be used for secondary oil recovery as well as to power he oil platforms through gas turbines and other means. Some can be converted for domestic use, LNG, CNG gas-to-liquid and for electricity generation [8] .

3. ALTERNATIVE SOLUTIONS TO GAS FLARING

Flaring of natural gas doesn't solve any problem on the other hand, it creates more problems. It affects lives of people and pollutes the environment. There are so many alternative solutions to gas flaring. The different alternative solutions are as follows;

3.1. Reinjection for Secondary Oil Recovery

Natural gas produced in association with crude oil can be used for either gas injection or gas lift which is more profitable and economical compared to gas flaring. Gas injection is when gas is introduced into a depleted oil reservoir in order to increase the pressure within the reservoir and thereby increase the production of crude oil from the reservoir. This is not to be confused with gas lift, where gas is injected into the annulus of the well rather than the reservoir. After the crude has been pumped out, the natural gas is once again recovered. Since many of the wells found around the world contain heavy crude, natural gas can be used to increase the production of this heavy crude.

The process of injecting natural serves as a way of increasing the pressure in the oil well, thus causing more gas molecules to dissolve in the oil lowering its viscosity and thereby increasing the well's output. Air is not suitable for repressuring wells because it tends to cause deterioration of the oil, thus carbon dioxide or natural gas is used for this process.

3.2. Source of Energy and Feedstock for Petrochemical Plants

Natural gas is most times referred to as "clean burning" because it produces fewer undesirable by-products per unit energy than coal or oil. Like other fossil fuels, its combustion emits carbon dioxide, but at about half the rate of coal per kilowatt hour of electricity generated. It is also more energy efficient [9] .

Natural gas is a major source of electricity generation through the use of cogeneration, gas turbines and steam turbines. It is also well suited for a combined use in association with renewable energy sources such as wind or solar. Particularly high efficiencies can be achieved through combining gas turbines with a steam turbine in combined cycle mode. Natural gas burns more cleanly than other hydrocarbon fuels, such as oil and coal, and produces less carbon dioxide per unit of energy released. For an equivalent amount of heat, burning natural gas produces about 45 percent less carbon dioxide than burning coal for power. The US Energy Information Administration reports the following emissions in million metric tons of carbon dioxide in the world for 2012:

1) Natural gas: 6799.

2) Petroleum: 11,695.

3) Coal: 13,787.

Natural gas is also useful for domestic purposes. If dispensed in a residential setting, can generate temperatures in excess of 1100°C (2000°F) making it a powerful domestic cooking and heating fuel. In much of the developed world it is supplied through pipes to homes, where it is used for many purposes including ranges and ovens, gas-heated clothes dryers, heating/cooling, and central heating. Heaters in homes and other buildings may include boilers, furnaces, and water heaters.

Natural gas is a major feedstock for the petrochemical plants. It is used in the production of ammonia, via the Haber process, for making fertilizers. It is also used to produce hydrogen, with one common method being the hydrogen reformer. Hydrogen has many applications: it is a primary feedstock for the chemical industry, a hydrogenating agent, an important commodity for oil refineries, and the fuel source in hydrogen vehicles. Natural gas is also used in the manufacture of fabrics, glass, steel, plastics, paint, and other products.

3.3. Liquefied Natural Gas (LNG)

Another alternative solution to gas flaring is to liquefy the gas and stored in vessels or bottles as liquid natural gas. This is relatively safer and more economical compared to flaring of gases. The natural gas produced from either a gas well or in association with crude oil production can be liquefied in series of processes. And in most cases, start with the removal of impurities like water, acid gases among others produced with the gas. Solid particles may also be present in the produced gas and must be removed before processing to maximize the mechanical efficiency the LNG equipment. The major composition of natural gas is methane and for liquefaction to occur i.e. to liquefy methane gas, the temperature of the system must be brought down to −160°C. This of course is also a function of other factors such as the components part of the gas to be liquefied. The stored LNG can be used domestically and also for industrial purposes. This will also create jobs and reduce the amount of pollutants released into the environment due to gas flaring. Figure 2shows the major breakdown/subsystems of the LNG system.

3.4. Compressed Natural Gas (CNG)

Compressed natural gas refers to methane stored at high pressure. It is made by compressing natural gas to less than 1% of the volume it occupies at standard atmospheric pressure. It is stored and distributed in hard containers at a pressure ranging from 20 - 25 MPa, usually in cylindrical or spherical shapes. In transportation, natural gas is used either as CNG or as liquefied natural gas (LNG) [10] . Because natural gas has a lower energy density than liquid petroleum fuels, it is either compressed or liquefied to store more energy. Therefore, to use natural gas, vehicles must have a CNG- or LNG-specific fuel storage and delivery system installed. The cost and placement of fuel storage tanks is the major barrier to wider/quicker adoption of CNG as a fuel.

CNG is also used in internal combustion engine automobiles that have been modified or in vehicles which were designed for CNG use, either alone or with a segregated gasoline system to extend range or in conjunction with another fuel such as diesel. Natural gas vehicles are increasingly used in different countries like Iran, Pakistan, and some parts in South America, Europe and North America because of rising gasoline prices and availability of natural gas. On CNG vehicles, natural gas must be compressed before being stored on the vehicle. This compression takes place at the fueling site, using specialized equipment.

In Figure 3, the costs of CNG are compared with those of Gasoline and Diesel fuels in United State, from 2006-2012. CNG vehicles can generally be operated longer between oil changes because natural gas produces less soot and combustion by-products. Gaseous fuels also do not "wash" the lubrication from the cylinder walls, so cylinder and ring wear are significantly reduced. This increases the engine overhaul interval-some CNG users have reported engine overhaul intervals two to three times greater than gasoline vehicles running similar duty cycles. The United States Environmental Protection Agency regulates emissions for all vehicles in the United States. In recent years, the emissions benefits of natural gas fueling have been reduced because all vehicles are held to higher standards [11] . Using CNG reduces some regulated emissions, including hydrocarbons, oxides of nitrogen, and carbon monoxide. These reductions differ based on the type of vehicle and its duty cycle.

3.5. Natural Gas in Transportation

Natural gas can serve as an oil replacement in transportation markets in three ways. Firstly, natural gas can be converted to methanol—an alcohol with similar properties to ethanol—that can be burned in internal combustion engines with slight vehicle modifications.

Figure 2. A schematic of liquefied natural gas (LNG) trade chain.

Secondly, light- and medium-duty vehicles using existing engine technologies can also burn compressed natural gas (CNG). Here the natural gas is stored at pressure, typically around 3000 psi. Because of the pressure, the CNG storage tanks are larger than existing gasoline storage tanks, so vehicles often have less trunk space and can cover less distance than conventional gasoline cars without refueling [12] . The Honda Civic GX, currently sold in the United States, for example, has a CNG capacity equivalent to eight gallons of gasoline. A number of CNG vehicles sold in Europe are bi-fuel vehicles capable of burning both CNG and gasoline in their engines, When the CNG tank empties, and the engine shift to the gasoline tank for fuel. Bi-fuel vehicles will frequently use gasoline first because the cold-start properties of gasoline are better than CNG.

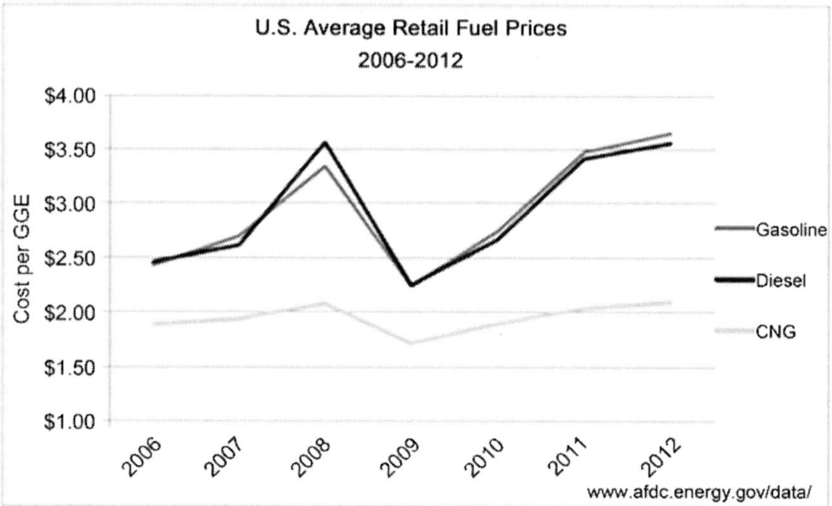

Figure 3. Costs of CNG compared to gasoline and diesel fuels in united state.

Thirdly, medium- and heavy-duty vehicles can run off of either CNG or liquefied natural gas (LNG), which is stored at very low temperatures (−160°C). The advantage of LNG over CNG is that it requires 30 percent less space allowing for longer driving distances. One disadvantage of LNG is that storing it for long periods is expensive, therefore LNG is often considered as a replacement fuel for vehicles that are in continuous use (e.g., heavy duty). Most industry followers envision LNG technologies as the likely replacement for diesel in the largest classes of heavy-duty vehicles [13] .

3.6. Gas Hydrate

Methane hydrate, commonly known as Gas hydrate is an ice-like crystalline compound formed when natural gas and water molecule exist together at relatively low temperature and high pressure. It is an assemblage of methane molecule (CH_4) that is bounded within the cage-like structure (crystal lattice) formed by water molecule (H_2O). Gas hydrate is relatively stable above the freezing point of water and sufficiently high pressure. It is relatively stable under its saturation temperature and pressure and also much denser than normal ice.

In the sea bed, once the methane is saturated in ocean with the appropriate temperature and pressure, ice crystal will start to grow into a gas-like lattice that traps the dissolved methane. There is actually no major chemical bond between the water molecule and methane other than Van der Waals forces. But the major influence is the presence of a guest molecule inside the ice crystals that increases the melting point of ice to several degrees above 0°C and makes the structure more stable. This property of gas hydrate can be experimentally investigated and capitalized on, to effectively store natural gas as hydrate for energy conservation instead of flaring the gas wastefully. Figure 4 shows the crystalline cage-like structure formed when water molecules formed crystal lattice around methane gas.

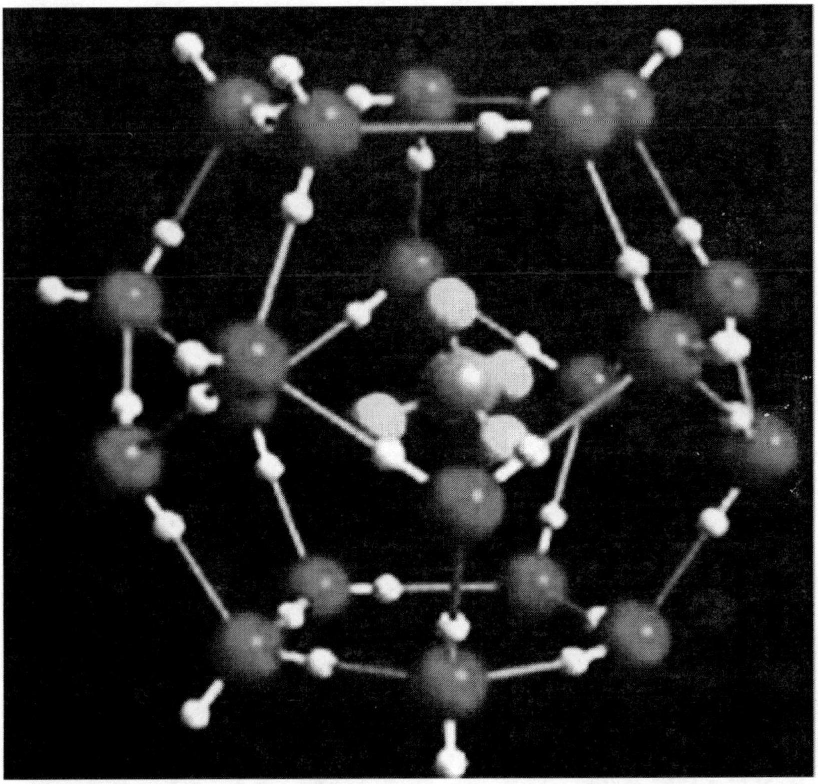

Figure 4. Gas hydrate cage-like structure (green: methane, red: water).

The research work conducted by Belosludov et al., showed that the thermal expansion of methane hydrate is considerably larger than that of ice of relatively same quantity. For practical application, clathrate hydrates can be used as storage materials but it is vital to know the different temperatures and pressure where the compound is stable [14] . It was discovered that methane hydrate is thermodynamically stable under overheating. This was due to the fact that the thermal expansion of the hydrate is much larger than that of ice. The thermal expansion of methane hydrate increases with that of ice because it can stay in a region of stability on methane hydrate phase diagram.

Gas hydrate has been one of the major problems of multiphase flow in regions where low temperature exists and different approaches are been conducted to prevent it formation. But in recent years, it was discovered that gas hydrate deposits hold enormous amount of energy which when trapped can solve the problem of energy generation in the world. This has led to the exploration of gas hydrate deposits. A sample of crystalline methane hydrate is shown in Figure 5.

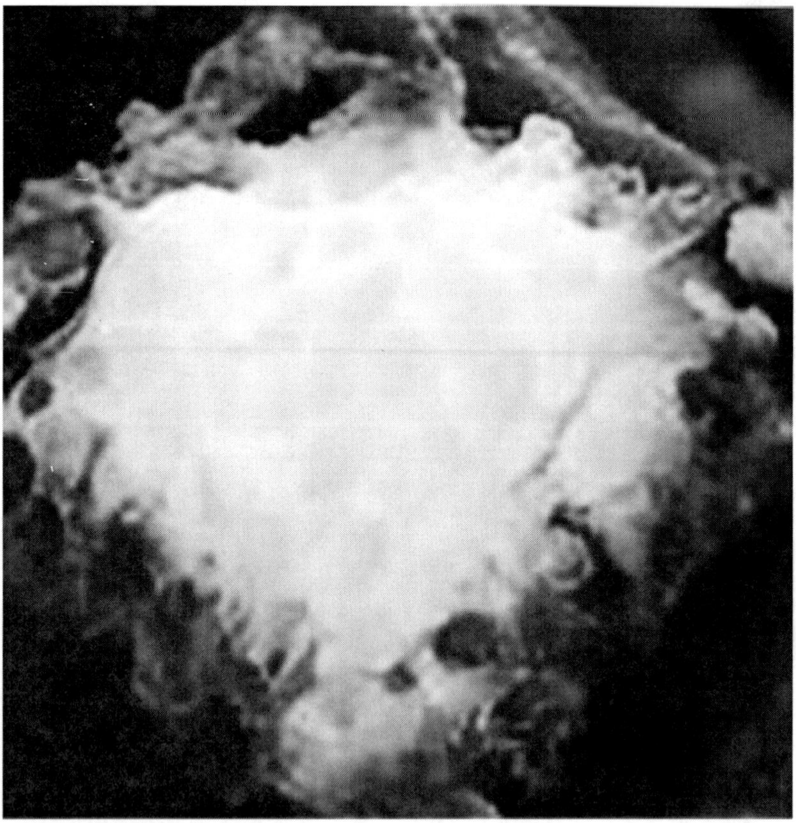

Figure 5. Sample of methane gas hydrate crystals.

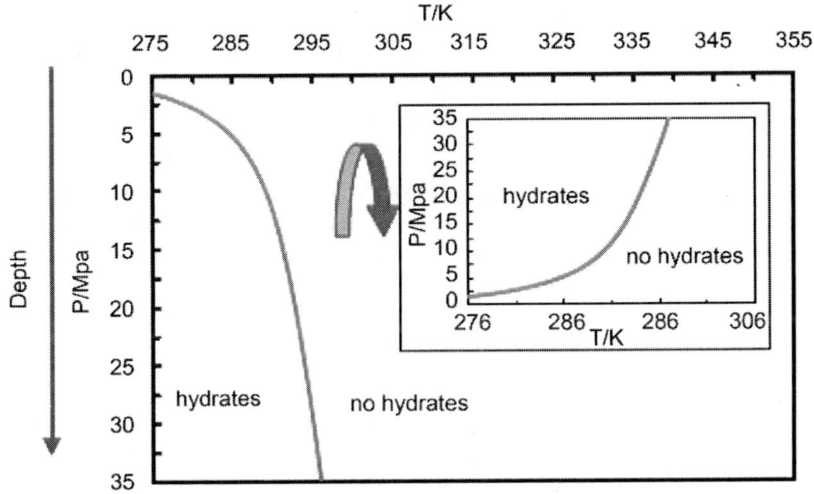

Figure 6. Gas hydrate phase diagram showing its stable conditions.

The experimental work conducted by Ruffine et al. showed that when the pressure and temperature in the gas hydrate system are increase, and if the ice structure surrounding the hydrate is stable to some extent, the methane hydrate will have larger pressure than the ice phase [15] . Their experiment depicted the function of the hydrate skin, somewhat, as a barrier, and the effect of breaking that barrier on the kinetics of the hydrate-ice structure in a closed system. This property needs to be taken into consideration when developing any kind of hydrate-based technology for natural gas transportation and storage. This also occurs during hydrate formation within sediments and is one of the reasons free gas exists within the gas hydrate stability zone.

In Figure 6, the stability of methane hydrate at different temperature and pressure is depicted. The gas hydrate is stable at relatively low temperature and high pressure but when the temperature increases and the pressure reduces, the hydrate dissociate into gas and water.

4. CONCLUSIONS

Gas flaring has more negative impacts than the positive impacts. It pollutes the environment and also affects the health of people in the areas in which gas flaring is practiced. It can be convincingly said that gas flaring is an intentional wasting of abundant resources which can meet the different challenges facing the society at large, in terms of power generation, energy conservation, global warming of the earth among others. From the year 1996-2010, in Nigeria, 12,602,480.25 million ft^3 of natural gas was flared (NNPC). This is equivalent to losing about $12{,}967.952 \times 10^{12}$ Btu of energy that would have been used for something profitable.

There are different ways the produced gas can be used instead of discarding it by flaring. Utilizing the gas in a safer way can solve the problems of energy generation, greenhouse effects and also create a safer environment. The produced gas can serve different purposes instead of being flared. It can be as simple as using the gas to increase oil production, channeling the produced gas to power the oil and gas platform, and manufacturing of petrochemical products, as well as electricity generation and also fueling of automobiles.

The proposed alternative solutions are gas for secondary oil recovery, feedstock for petrochemical plants, domestic uses, LNG & CNG, as well as energy conservation by storing as gas hydrate for future use or other purposes. Gas hydrate is relatively stable under its saturation temperature and pressure and also much denser than normal ice. This property of gas hydrate can be used to effectively store the produced gas as hydrate for energy conservation instead of flaring. The alternative solutions will convincingly reduce gas flaring and create rooms for innovation and development that will ultimately stop gas flaring.

CITE THIS PAPER

Aregbe, A.G. (2017) Natural Gas Flaring—Alternative So- lutions. World Journal of Engineering and Technology, 5, 139-153.

REFERENCES

1. Ismail, O.S. and Umukoro, G.E. (2012) Global Impact of Gas Flaring. Energy and Power Engineering, 4, 290-302.
2. Sonibare, A. and Akeredolu, A. (2006) Natural Gas Domestic Market Development for Total Elimination of Routine Flares in Nigeria's Upstream Petroleum Operations. Energy Policy, 34, 743-753.
3. Emam, E.A. (2015) Gas Flaring in Industry: An Overview. Petroleum & Coal, 57, 532-555.
4. Orimoogunje, O.I., Ayanlade, A., Akinkuolie, T.A. and Odiong, A.U. (2010) Perception on the Effect of Gas Flaring on the Environment. Research Journal of Environmental and Earth Sciences, 2, 188-193.
5. Zheng, P., Rebennack, S., Iliadis, N. A. and Pardalos, P.M. (2010) Optimization Models in The Natural Gas Industry. Department of Industrial & Systems Engineering, Center for Applied Optimization, University of Florida, Gainesville.

6. Madueme, S. (2010) Gas Flaring Activities of Major Oil Companies in Nigeria: An Economic Investigation. International Journal of Engineering and Technology, 2, 610-617.
7. Adewale O.O. and Mustapha, U. (2015) The Impact of Gas Flaring in Nigeria. International Journal of Science, Technology and Society, 3, 40-50.
8. Oghifo, O.T. (2011) Gas Flaring/Power Plants in Nigeria: Socio-Economic and Environmental Impact on the People of Niger Delta. BE306E Master's Thesis in Environmental Management, Bodø, 1-93.
9. Gervet, B. (2007) Gas Flaring Emission Contributes to Global Warming. Renewable Energy Research Group, Division of Architecture and Infrastructure, Luleå University of Technology, Luleå, 2-14.
10. Hassan, A. and Konhy, R. (2013) Gas Flaring in Nigeria: Analysis of Changes in Its Consequent Carbon Emission and Reporting. Accounting Forum, 37, 124-134.
11. Davies, P. (2001) The New Challenge of Natural Gas' Paper Presented at OPEC and the Global Energy Balance—Towards a Sustainable Future, Vienna, September 2001, 5-15.
12. Mercado, R.Z. and Sanchez, C.B. (2014) Optimization Problems in Natural Gas Transportation Systems: A State-of-the-Art Review. Universitaria San Nicolás de los Garza, NL 66450, San Nicolás de los Garza, 13-37.
13. Bassey, M. (2008) Gas Flaring: Assaulting Communities, Jeopardizing the World. National Environmental Consultation Hosted by the Environmental Rights Action in Conjunction with the Federal Ministry of Environment at Reiz Hotel, Abuja, 1-11.
14. Belosludov, V.R., Subbotin, O.S., Krupskii, D.S., Belosludov, R.V., Kawazoe, Y. and Kudoh, J. (2007) Physical and Chemical Properties of Gas Hydrates: Theoretical Aspects of Energy Storage Application. Society of Nano Science and Technology. Materials Transactions, 48, 704-710.
15. Ruffine, L., Donvala, J.P., Charloua, J.L., Cremièrea, A. and Zehnderb, B.H. (2010) Experimental Study of Gas Hydrate Formation and Destabilisation Using a Novel High-Pressure Apparatus. Marine and Petroleum Geology, 27, 1157-1165.

CHAPTER 8

Advanced Catalytic Membrane Characterisation and Gas Permeation Properties for Enhanced Ethyl Lactate Conversion

Edidiong Okon, Habiba Shehu, Ifeyinwa Orakwe, Edward Gobina*

Center for Process Integration and Membrane Technology, School of Engineering, the Robert Gordon University, Aberdeen, UK

ABSTRACT

In this work, membrane evaluation, gas permeation properties and characterisation have been presented. A silica composite membrane was prepared, characterized and used for the permeation tests with four carrier gases to determine the most suitable carrier gas for enhancing the analysis of esterification product with gas chromatograph. The carrier gases used for the permeation tests were carbon dioxide (CO_2), argon (Ar), helium (He) and nitrogen (N_2). The permeation analysis was carried out between the gauge pressure range of 0.10 - 1.00 bar and temperature of 60°C. The gas flow rate was found to increase with respect to gauge pressure. The order of the gas flow rate with respect to the gauge pressure was Ar > CO_2 > He > N_2. The surface morphology and elemental composition of the membrane were analysed using scanning electron microscopy coupled with energy dispersive analysis of x-ray (the Zeiss EVO LS10). The SEM results exhibited a defect-free surface while the EDAX results identified different elements on the spectra including titanium (Ti), silicon (Si) and oxygen (O). Liquid nitrogen adsorption method (Quantachrome 2013 model) was used for the surface area and pore size distribution analysis. The Brunauer-Emmette-Teller (BET) surface area results of the 5th and 6th dip-coated membranes were 1.497 and 0.253 m^2/g respectively, while the Barrette-Joyner-Halender (BJH) curves gave a pore size of 4.184 and 4.180 nm respectively for the 5th and 6th dip-coated membranes indicating a mesoporous structure. The BET curve exhibited a type IV isotherm. The BJH curve of the 6th dip-coated membrane showed a significant reduction in flow rate after the modification process. The membrane recorded a permeance in the range of 1.03×10^{-6} to 9.32×10^{-7} $mol\cdot m^{-2}\cdot s^{-1}\cdot Pa^{-1}$. The permeance relationship with the inverse square root of the gas molecular weight showed a linear proportionality with the flow of carrier gases confirming Knudsen flow mechanism of gas transport.

Keywords: Permeation, Characterisation, Inorganic Membrane, Carrier Gas Transport, Esterification and Adsorption Isotherm

1. INTRODUCTION

Membrane technologies have found application in diverse industrial applications including biotechnology, food and pharmaceutical industries. They can also be used for the treatment of industrial effluents. Membranes are also used in different processes such as microfiltration, nanofiltration, and ultrafiltration offering a range of separations which may be used to recover, purify and concentrate valuable solvents in industry [1] . Although membranes possess numerous advantages, they also exhibit a major drawback such as fouling. Fouling is the major problem in membrane process as it reduces the membrane performance resulting in high running costs, which are largely the biggest costs in the membrane industrial plant [1] . In order to reduce the influence of fouling on membranes, different methods including mechanical and chemical cleaning process have been designed. When characterising the membrane, the pore size distribution are normally obtained using liquid nitrogen adsorption/desorption analysis while the morphological characteristics could be obtained using scanning electron microscopy method [1] [2] . Other information could also be obtained from the gas permeability measurement. Membranes may be classified into organic and inorganic categories and the different configurations include: Dense, porous and composite membranes [3] . Currently, ceramic porous inorganic membranes have been widely employed in different fields such as chemical and petrochemical, bioengineering and environmental engineering. Compared to other membranes, ceramic membranes have higher thermal and chemical stability [4] .

Membranes of different shapes and characteristics are generally made from a wide variety of chemically and thermally stable precursors. Other materials including alumina (Al_2O_3), titania (TiO_2), zirconia (ZrO_2) and silica (SiO_2) are also being used [5] . Silica membranes have attracted a lot of attention in the gas separation industries because they are thermally and chemically stable in contrast to polymeric membranes. They can offer higher permeability and selectivity for small gas molecules including He and H_2 over larger molecules like CO_2, N_2 and C_3H_8 gases. Different modification methods can be employed for the preparation of membranes including sol-gel dip-coating, chemical vapour deposition and sintering. The sol-gel method involves the formation and deposition of a suitable sol, like colloidal boehmite onto a porous support followed by drying and thermal treatment [6] . Sol-gel dip-coating method has been widely employed for the formation of silica and alumina-based inorganic membranes [7] . Membranes prepared using this method usually exhibit higher permeance in contrast to those prepared with the chemical vapour deposition methods. Inorganic membrane may be classified based on their pore size as macrospores (>500 Å), micropores (20 Å) and mesopores (20 - 500 Å) [8] . Gas transport in inorganic membranes may be explained using various mechanism of transport through the pore. The different transport mechanisms include Knudsen, viscous, capillary condensation, molecular sieving and surface diffusion mechanisms. Knudsen transport takes place when the mean free path of the gas molecules is greater than the pore size. In this case, there are more collisions between the molecules than the pore walls. However, separation selectivities in this mechanism are proportional to the ratio of the inverse square root of the gas molecular weight. Knudsen flow mechanism is usually predominant in mesoporous and macroporous membranes [6] [8] .

The expression for Knudsen flow in porous membrane can be written using the following equation [8] :

$$D = \frac{\varepsilon \delta \rho}{3\tau}\left(\frac{8RT}{\pi M}\right)^{1/2} \tag{1}$$

where D is the Knudsen diffusion (m^2/s), ε = porosity of the membrane (%), R = gas molar constant ($J \cdot mol^{-1} \cdot K^{-1}$), M = the molar weight of the diffusing gas (g/mol), dp = pore diameter (m), τ = tortuosity (m) and T = temperature (K), π = 2.131.

The mean free path of the gas molecule is the average distance travelled in collisions and can be explained using the equation [9] .

$$\lambda = \frac{kT}{\sqrt{2}\pi d^2 P} \tag{2}$$

where λ is the mean free path (m), k is the Boltzmann constant ($J \cdot K^{-1}$), T is temperature (K), d is the collision diameter (m) and P is the pressure (Pa) [9] .

Viscous flow mechanism takes place if the pore radius of the membrane is greater than the mean free path of the permeating gas molecule [10] [11] . In this case more collisions will take place between the permeating gas molecules (molecule-molecule collision) than between the molecule and the pore wall of the membrane and viscous flow will dominate [10] .

The viscous flow mechanism of gas transport can then be expressed as:

$$F_{viscous} = \frac{r_p^2 (P_1 - P_2)}{16 L \mu R T} \left(\frac{mol}{m^2 \cdot s} \right) \tag{3}$$

where P_1 = absolute pressure (Pa), P_2 = atmospheric pressure (Pa), μ = gas viscosity ($Pa \cdot s^{-1}$), L = membrane wall thickness (m), r = membrane pore size (m), R = gas molar constant ($J \cdot mol^{-1} \cdot K^{-1}$), T = atmospheric temperature (K) [10] .

Gas separation by molecular sieving mechanism takes place when the pore diameter of the inorganic ceramic membrane is roughly the same as those of the permeating gas molecules [12] [13] . However, as the pores become smaller than approximately 0.5 nm, separation factors greater than 10 are usually achievable. If the pore size of the membrane is between the diameters of the larger and smaller gas molecules, then only the smaller gas molecule can permeate through the membrane leading to a more efficient separation [10] [14] . In the capillary condensation mechanism, separation takes place in the pores of the membrane with mesoporous layer in the presence of condensable gas specie such as water vapour [11] . The condensed feed component diffuses through the pores of the membrane blocking other components resulting in separation [15] . Surface diffusion mechanism of transport occur when the gas molecule exhibit a strong affinity for the membrane surface and adsorption along the pore walls. However, separation in this mechanism takes place due to differences in the amount of adsorption of the permeating molecules.

According to IUPAC (International Union of Pure and Applied Chemistry), the major physisorption isotherms can be classified into six different types [16] . However, the most important isotherm for a porous membrane material is the type 1 isotherm which relates to the microporous solids and type IV and V, which corresponds to mesoporous solid (most especially ceramic materials which undergo hysteresis and capillary condensation during desorption. The use of adsorption/desorption isotherms to characterise porous ceramic membrane materials possesses major drawback despite its widespread application in various analytical techniques. The different types of physisorption isotherms and hysteresis are explained as shown in Figure 1(a) and Figure 1(b) respectively [2] .

In this study, the characterisation and evaluation of a silica membrane with 15 nm pore size was investigated to determine the permeation properties of single gases Ar, He, N_2 and CO_2 with inorganic composite membrane for the esterification reactions.

2. EXPERIMENTAL

2.1. Membrane Preparation

The membrane preparation was carried out based on the dip-coating process patented by Gobina [17] . Prior to the permeation tests, the support was weighed to determine the actual weight before and after modification. 545 mL of iso- pentane (Sigma Aldrich ≥ 99%), was measured into 1000 mL glass cylinder and 50 mL of silicon elastomer was added to the solution

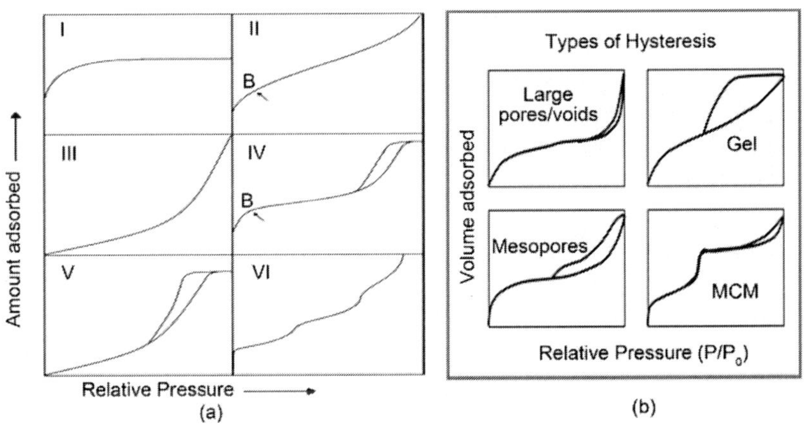

Figure 1. Types of physisorption isotherms (a) and types of hysteresis (b).

together with 5 mL of the curing agent. A magnetic stirrer was used to mix the three solutions and the mixture was allowed to stir for 30 minutes. After this period the support was withdrawn the solution and allowed in the solution for 30 minutes. After 30 minutes, the support was taken out from the solution and dried in the oven for 2 hrs at a constant temperature of 65°C in order to obtain an ultra-thin silica layer on the support. The support modification process was carried out based on the patented innovation by Gobina (2006) [17] [18] . Figure 2 shows the schematic diagram of the dip-coating process.

The membrane thickness after modification was calculated using the following formula [19] .

$$L = \frac{W_2 - W_1}{A\rho(1-\varepsilon)} \tag{4}$$

where L = membrane thickness (m), A = membrane area (m^2), ρ = theoretical density of alumina (3.95×10^{-3} kg$\cdot$m^{-3}), W_1 = initial weight of the alumina support (g), ε = membrane porosity (%), W_2 = total weight of the support and membrane (g).

2.2. Permeation Cell

The permeation cell consisted of a thermally resistant stainless steel shell. The membrane was centralized in the tube using a graphite seal each at either end. Through the use of various connections and valves the cell permits the measurement of the gas flux through the membrane at various feed pressures. A heating tape was wrapped over the stainless steel reactor which enables high- temperature studies to be carried out. Prior to permeation experiments, a leak test was carried out by monitoring the downstream pressure variation while the system remained totally closed and pressurized.

2.3. Permeation Tests

The four gases used for the permeation analysis include nitrogen (N_2), argon (Ar), helium (He) and carbon dioxide (CO_2). The gases were supplied by BOC, UK. The permeation analysis was carried out at a feed pressure drop of between 0.10 - 1.00 bar and at 60°C. This gauge pressure scale and the temperature were chosen to correspond with the esterification

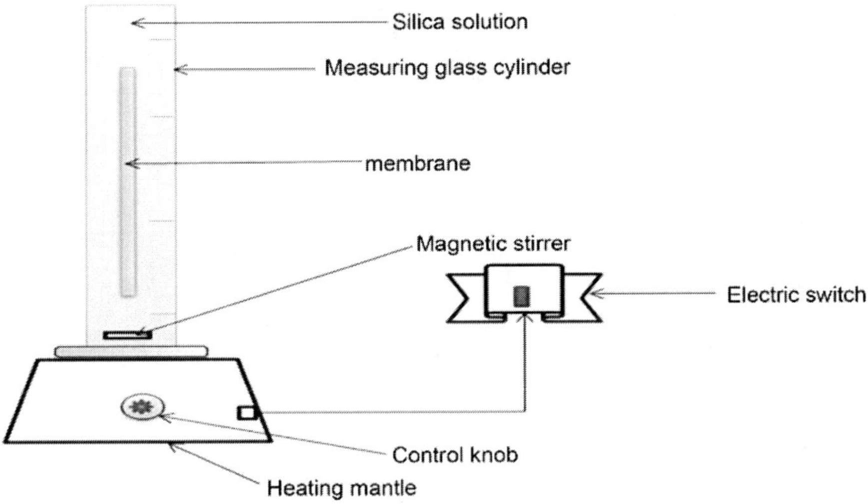

Figure 2. Schematic diagram of gas permeation setup.

reaction parameters for the analysis of the ester product. The total length of the membrane was 36.6 cm, while the inner and outer radius of the membrane was 7 and 10 mm respectively. Figure 3 shows the single gas permeation setup.

2.4. Liquid Nitrogen Adsorption

The surface area analysis of the support and coated support was examined using liquid nitrogen adsorption instrument shown in Figure 4. Prior to the analysis, a small fragment of the silica membrane was crushed and used for the liquid nitrogen analysis. The actual weight of the 5th and 6th dipping membrane samples was 4.3 g and 4.2 g respectively. The sample cell weight for the 5th and 6th dipping was 16.8 g and 27.7 g respectively. The degassing of the system was carried out using a flow of dry helium gas with heating in order to remove moisture from the sample. A similar procedure to that of Vospernik et al. [20] was used for the liquid nitrogen analysis with some modification in the temperature. The liquid nitrogen temperature was 77 K. The actual weights of the silica membranes, the cell weight and the weight of sample + cell before and after the degassing process are presented in Table 1.

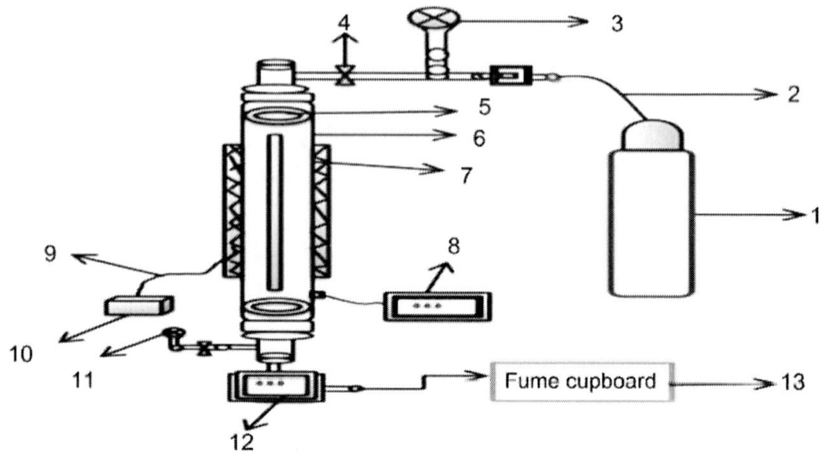

Figure 3. Schematic diagram of gas permeation setup which consists of; carrier gas cylinder (1), gas feed inlet (2), permeate pressure gauge (3), control valve (4), O-ring graphite seal (5), reactor (6), heating tape (7), temperature regulator (8), thermocouple (9), thermocouple box (10), retentate pressure gauge (11), flow meter (12) and fume cupboard (13).

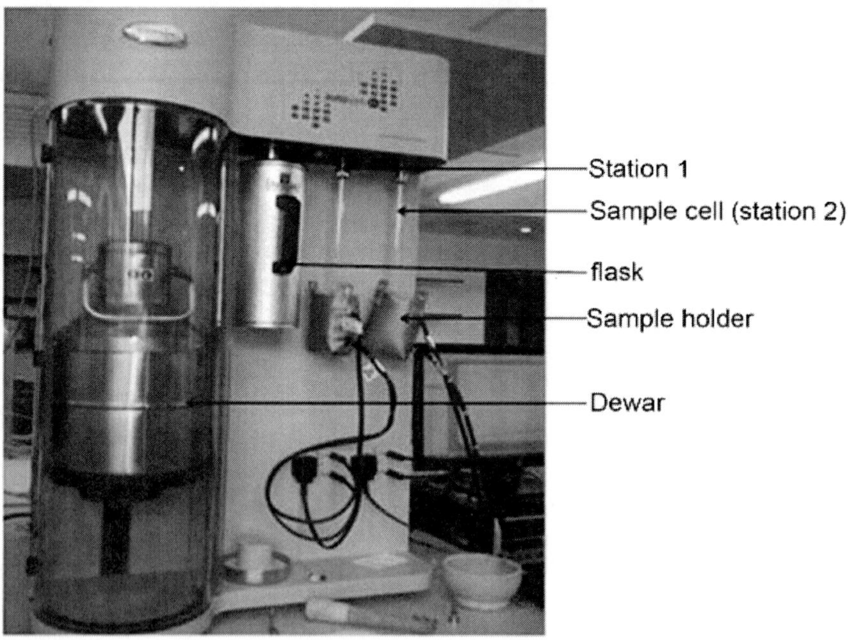

Figure 4. Liquid nitrogen adsorption (Quantachrome 2013 model).

Table 1. Sample and cell weights before and after degassing process.

Sample	Cell weight (g)	Weight of sample + cell before degassing(g)	Actual weight of sample (g)	Weight of sample after degassing (g)
5th dipping	16.8	21.1	4.3	21.0
6th dipping	27.7	31.9	4.2	31.8

2.5. Scanning Electron Microscopy/Energy Dispersive X-Ray Analyser (SEM/EDAX)

The surface examination of the membranes was carried out the Zeiss EVO LS10 SEM/EDAX between the scale range of 10 - 100 μm. The images were analysed at the magnification of 100× at the working distance of 8.5 mm while the chamber pressure was set between the range of 100 - 101 Pa. Figure 5 shows the SEM/EDAX instrument that was used for the membrane morphological characterisation.

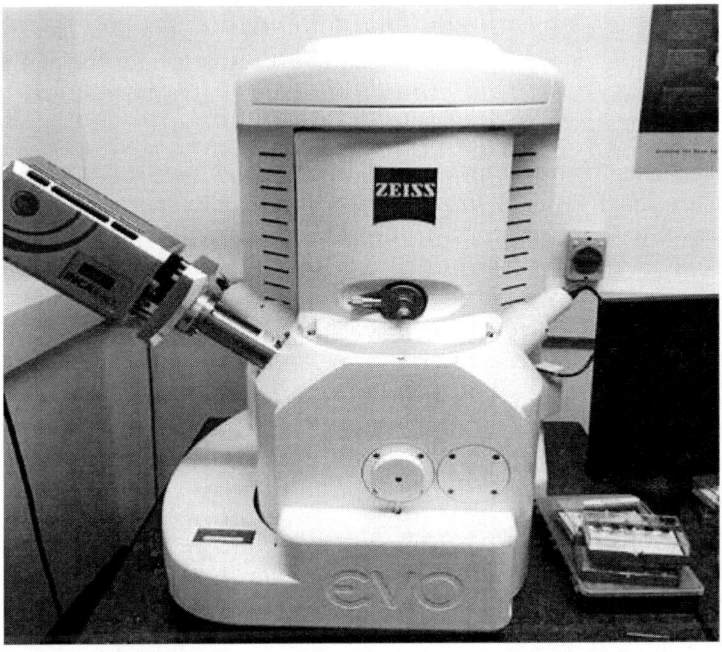

Figure 5. Scanning electron microscope couple with energy X-ray dispersive analyser (SEM/EDAX) (the Zeiss EVO LS10) instrument.

3. RESULT AND DISCUSSION

3.1. Permeation Test

Figure 6 shows the relationship between the flow rate of the different gases and the gauge pressure at 60°C. From Figure 6, it can be seen that the flow rate of the gases increases with an increase in the gauge pressure. However, at a particular pressure the flow difference was explained based on the molecular weight of the different gases. It was found that helium gas with the least molecular weight (2 g/mol) exhibited the highest flow in contrast to other gases. It was assumed that the gas flow was dominated by Knudsen mechanism of gas transport which explains that gases with the less molecular weight permeate through the membrane faster than the gases with higher molecular weight. It was also observed that the flow rate for Ar, N_2 and CO_2 were close to each other over the pressure range investigated. Although they exhibited a similar flow, N_2 with the least a lower molecular weight (28 g/mol) showed a higher flow than Ar(40 g/mol) and CO_2 (44 g/mol) and can be seen that the gas flow rate was based on their respective molecular weight in accordance with the Knudsen mechanism. The order of the gas flow rate with respect to a particular gauge pressure was He (2 g/mol) > N_2 (28 g/mol) > Ar(40 g/mol) > CO_2 (44 g/mol). This also confirms the fact that helium could be a suitable carrier gas for the analysis of the esterification reaction product using gas chromatograph for improved ethyl lactate conversion.

Figure 7 describes the relationship between the gas permeance and the gauge pressure (bar). From Figure 7, it was found that the permeance of the gases decreases with respect to the gauge pressure but the rate of decrease was less pronounced at a gauge pressure of greater than 0.4 bars. A similar result was also observed in our previous study [21] . Also from Figure 7, it was found that the statistical error bars for the gases showed good significant value of the experimental data.

The permeance of Ar, He, CO_2 and N_2 was calculated using Equation (5):

$$Q_i = \frac{J}{\Delta P} \tag{5}$$

where Q_i = gas permeance ($\mathrm{mol \cdot m^{-2} \cdot s^{-1} \cdot Pa^{-1}}$), ΔP= pressure drop across the membrane (bar) and J = Gas flux ($\mathrm{mol \cdot m^{-2} \cdot s^{-1}}$). The

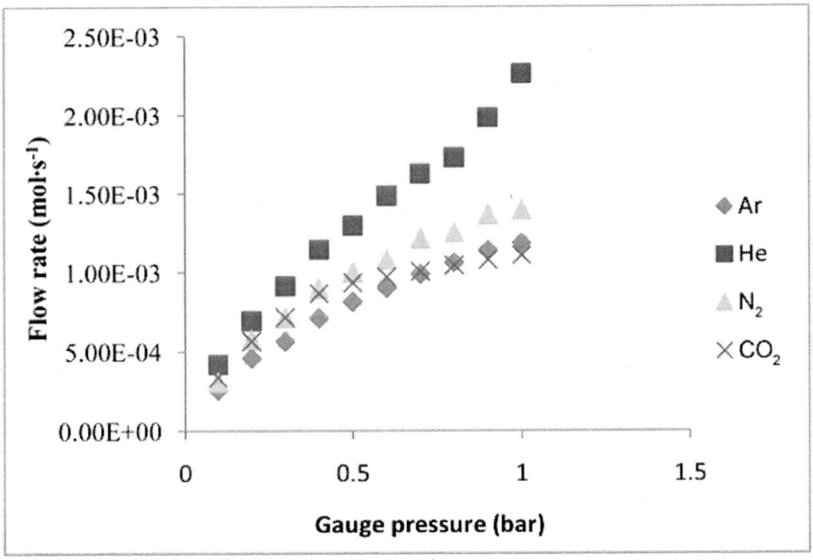

Figure 6. Gas Flow rate ($mol \cdot s^{-1}$) against gauge pressure (bar) at 60°C.

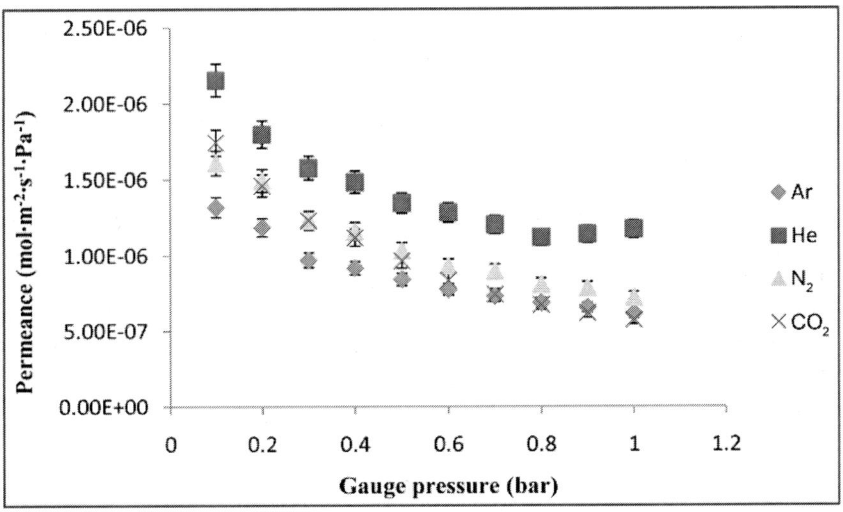

Figure 7. Gas permeance ($mol \cdot m^{-2} \cdot s^{-1} \cdot Pa^{-1}$) against gauge pressure (bar) at 60°C.

mean free paths of the four gases were calculated using Equation (2). The permeability of the membrane was further calculated by multiplying the permeance (Equation (5)) with the thickness of the membrane that was calculated using Equation (4). The formula for the membrane permeability is given by Equation (6):

$$F = \frac{J}{\Delta P} \times L \qquad (6)$$

The pore size of the membrane was calculated using Equation (7):

$$\tau_p = \frac{16 \cdot Ao \cdot \mu}{3 \cdot B} \sqrt{\frac{8RT}{\pi M}} \qquad (7)$$

where τ_p = membrane pore radius (m), Ao = constant representing viscous flow from the permeability graph, B = constant representing Knudsen flow from the permeability graph, μ = gas viscosity ($Pa \cdot s^{-1}$), π = 3.142, M = gas molecular weight (g/mol), T = temperature (K), R = gas molar constant ($K \cdot J \cdot mol$).

The membrane permeability, mean free path, and pore radius for four gases were calculated as shown in Table 2. From the result obtained in Table 2, it was found that the pore radius of the membrane was smaller than the mean free path which confirms the Knudsen flow mechanism of transport. This result was in accordance with a report by Benito et al. [22] which explains that Knudsen diffusion is dominant if the membrane pore radius is smaller than the mean free path of the gas molecules and this is also significant for membrane with small pore radius of <10 nm and defect-free membranes [14] [22] [23] . It can also be seen from Table 2 that He and CO_2 gases recorded a higher permeability values in contrast to N_2 and Ar gases although the permeability value of CO_2 and N_2 gases were quite close.

Table 2. Calculated values of permeability, mean free path and pore radius of the silica membrane with Ar, He, CO_2 and N_2 gases.

Gas molecule	Mean free path (λ) m	Pore radius (m)	Permeability ($mol \cdot m \cdot m^{-2} \cdot s^{-1} \cdot Pa^{-1}$)
Ar	3.15E−04	4.72E−12	3.67E−08
He	3.63E−04	1.09E−11	5.92E−08
N_2	2.96E−04	4.45E−12	4.48 E−08
CO_2	1.11E−04	2.22E−12	4.64E−08

Figure 8 presents the relationship between the gas permeance and the inverse square root of the gas molecular weight at 0.30 bar and 60°C. From the result obtained, it was found that Ar, N_2 and CO_2 gases showed a linear proportionality with the inverse square root of the gas molecular weight which was in accordance with the Knudsen law of diffusion of gases through the pores of the membrane, whereas helium gas exhibited a deviation from the linear line [21] , suggesting that another flow mechanism was in operation for He gas at 0.30 bar and 60°C. The suggested flow mechanism was viscous plus surface diffusion mechanism as these two mechanisms are also said to be controlling rate mechanism of transport in mesoporous membrane. Since the helium gas did not display the evidence of Knudsen flow mechanism, it was suggested that viscous plus surface diffusion mechanism was the additional transport mechanism. The order of the gas permeance with respect to the inverse square root of the gas molecular weight was given as He > N_2 > CO_2 > Ar.

3.2. Liquid Nitrogen Results

A summary of the various parameters obtained from the liquid nitrogen characterisation for the 5thand 6th dip-coating are presented in Table 3. Figure 9 and Figure 10 describe the relationship between the volume and the relative pressure for the modified silica membrane. From the results obtained in Figure 9 and Figure 10 it was found that the BET of the 5th dip-coated membrane exhibited a higher surface area (2.916 m^2/g) in contrast to the 6th (0.088 m^2/g) as shown in Table 3. It was also observed that there was hysteresis on the curves although the hysteresis on the 6thdip-coated membrane was more pronounced in contrast to that of the 5th dip-coated membrane [24] . The BJH of the twice-dipped membranes were also compared. From the BJH results of the twice dipped membranes, it was found that there was reduction on the pore size of the membrane after the modification process [25] . This was attributed to the silica that was used in the membrane modification process. Although the pore of the 6th dip-coated membrane was found to have reduced, the total pore size of the 5th and 6th dip-coated membranes were believed to be within the mesoporous range according to the IUPAC classification.

3.3. SEM/EDAX of the Membrane

Figure 11 shows the SEM/EDAX of the membrane. From Figure 11, it can be seen that the silica was evenly distributed on the surface of the membrane after the modification process. The whitish background on the surface

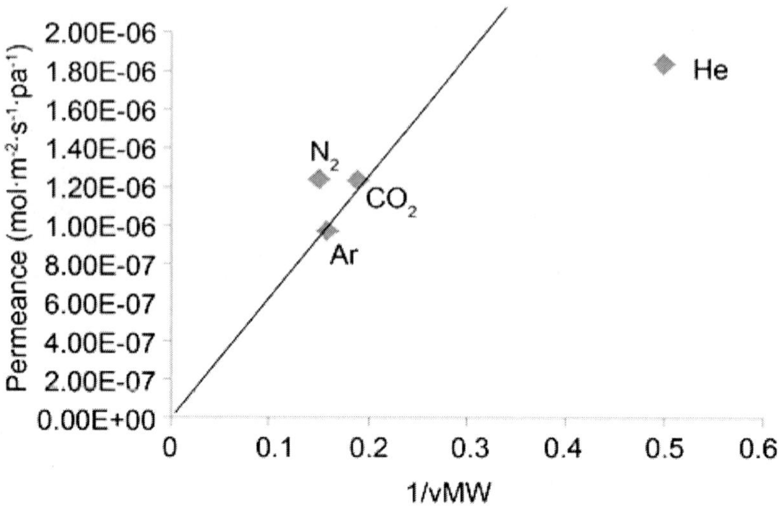

Figure 8. Gas permeance ($\mathbf{mol \cdot m^{-2} \cdot s^{-1} \cdot Pa^{-1}}$) against inverse square root of the gas molecular weight (g/mol) at 0.40 bar and 60°C.

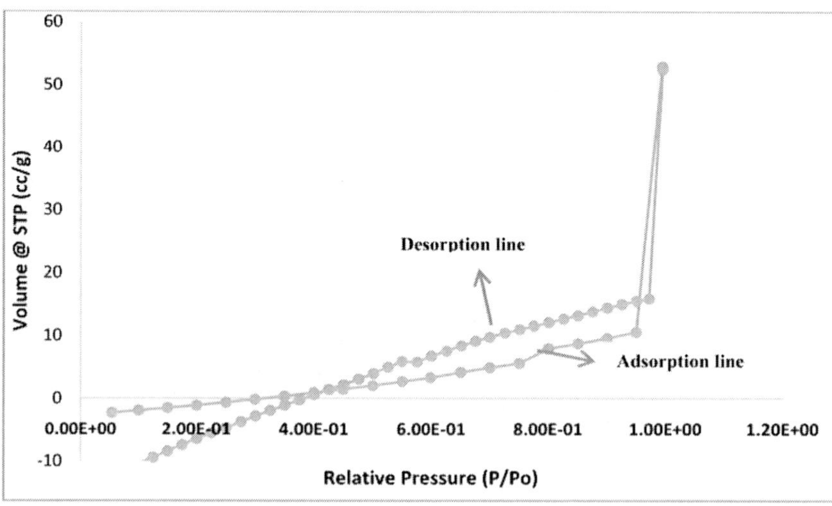

Figure 9. BET surface area of the 5th dip-coated membrane.

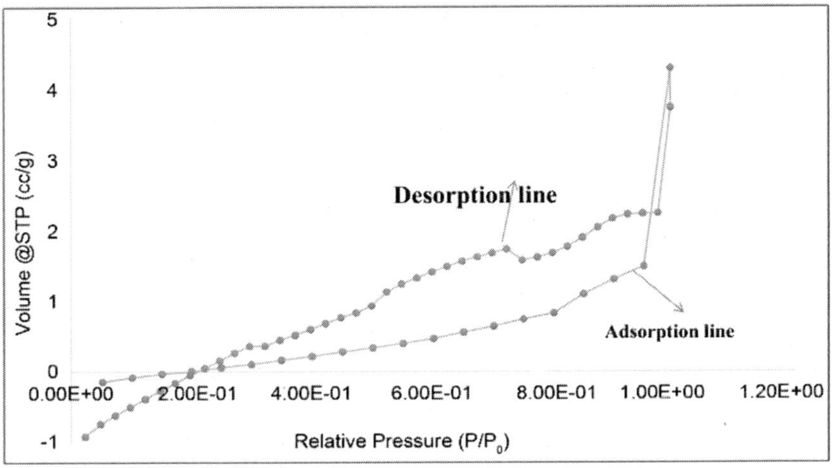

Figure 10. BET surface area of the 6th dip-coated membrane.

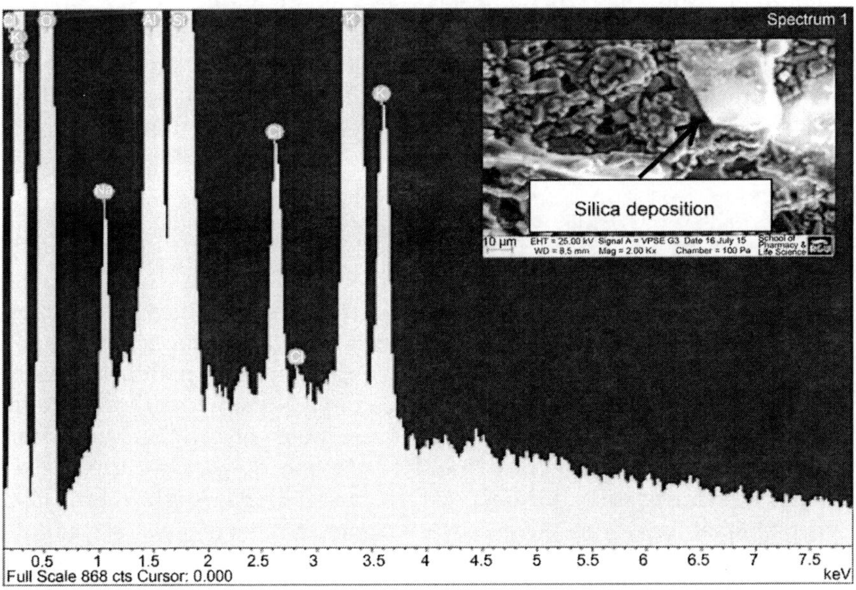

Figure 11. SEM image and EDAX of the silica membrane that was used for the characterisation.

Table 3. BET curve summary for the 5th and 6th dip-coated membranes.

Parameters	5th dip-coating	6th dip-coating
Slope	3005.934	521,655.829
Intercept	−1.812e + 03	−1.245e + 04
Correlation coefficient (R^2)	0.71414	1.00000
C, constant	−0.659	−3.191
Surface Area (m^2/g)	2.916	0.088

image was suggested to be the silica used for the membrane modification. However, the membrane also exhibited a clear surface which also indicated that it was defect-free. The result was further confirmed by the EDAX result for the membrane sample. From the result obtained in Figure 11, it was found that the membrane exhibited different elements on the spectra including oxygen, carbon, potassium, chlorine, aluminium and silicon. The fresh support is made up of Al_2O_3, used for the fabrication of the support. However due to the silica modification, Si, O, C and other elements were identified from the EDAX spectra. So a combination of Si and O was suggested to be SiO_2 originating from the solution that was used for coating the surface of the support. From the EDAX result, it was suggested that the fresh support was initially coated with aluminium oxide and then subsequently coated with silicon oxide after the modification process.

4. CONCLUSION

A repeat dip-coating process has been used to prepare silica membranes on alumina support. The morphology and hydrodynamic characteristics of the silica membrane was carried out using different analytical methods including liquid nitrogen adsorption, scanning electron microscopy/energy dispersive analyser and gas permeation respectively. Helium gas with the highest permeation rate was found as the most suitable carrier gas for analysis esterification product. Ar, N_2 and CO_2 gases showed a linear proportionality with the inverse square root of the molecular weight of the different gases. The permeance of the four gases was found to decrease with increase in gauge pressure confirming Knudsen flow mechanism. The average pore size of the silica membrane was found to reduce after the modification process. The calculated pore radius of the four gases was found to be smaller than the mean free path. The membrane pore size was in accordance with IUPAC classification. The order of the gas flow rate with respect to the gauge pressure was He (2 g/mol) > N_2 (28 g/mol) > Ar (40 g/mol) > CO_2 (44 g/mol). The SEM surface morphology of the membrane showed the distribution of silica on the membrane surface. The EDAX of the silica membrane showed different elements including Al, Si and O. The gas flow rate showed an increase with respect to gauge pressure.

ACKNOWLEDGEMENTS

The author wishes to express sincere thanks to CPIMT for supplying the membrane and other infrastructures used in the study, school of pharmacy and life science, RGU for the SEM/EDAX examination. Additionally, PTDF is great fully acknowledged for their financial support towards the research work.

CITE THIS PAPER

Okon, E., Shehu, H., Orakwe, I. and Edward, G. (2017) Advanced Catalytic Membrane Characterisation and Gas Permeation Properties for Enhanced Ethyl Lactate Conversion. Journal of Materials Science and Chemical Engineering, 5, 1-16.

REFERENCES

1. Calvo, J.I., Bottino, A., Capannelli, G. and Hernández, A. (2008) Pore Size Distribution of Ceramic UF Membranes by Liquid–Liquid Displacement Porosimetry. Journal of Membrane Science, 310, 531-538.
2. Smart, S., Liu, S., Serra, J.M., Diniz da Costa, J.C., Iulianelli, A. and Basile, A. (2013) 8-Porous Ceramic Membranes for Membrane Reactors. In: Basile, A., Ed., Handbook of Membrane Reactors, Woodhead Publishing, Cambridge, 298-336.
3. Bai, C., Jia, M., Falconer, J.L. and Noble, R.D. (1995) Preparation and Separation Properties of Silicalite Composite Membranes. Journal of Membrane Science, 105, 79-87.
4. Jiang, H., Meng, L., Chen, R., Jin, W., Xing, W. and Xu, N. (2013) Progress on Porous Ceramic Membrane Reactors for Heterogeneous Catalysis over Ultrafine and Nano-Sized Catalysts. Chinese Journal of Chemical Engineering, 21, 205-215.
5. Abedini, R. and Nezhadmoghadam, A. (2001) Application of Membrane in Gas Separation Processes: Its Suitability and Mechanisms. Petroleum and Coal, 52, 69-80.
6. Javaid, A. (2005) Membranes for Solubility-Based Gas Separation Applications. Chemical Engineering Journal, 112, 219-226.
7. Lee, D. and Oyama, S.T. (2002) Gas Permeation Characteristics of a Hydrogen Selective Supported Silica Membrane. Journal of Membrane Science, 210, 291-306.
8. Markovic, A., Stoltenberg, D., Enke, D., Schlünder, E. and Seidel-Morgenstern, A. (2009) Gas Permeation through Porous Glass Membranes: Part II: Transition Regime between Knudsen and Configurational Diffusion. Journal of Membrane Science, 336, 32-41.

9. Wang, M., Wang, Z., Li, N., Liao, J., Zhao, S., Wang, J., et al. (2015) Relationship between Polymer–Filler Interfaces in Separation Layers and Gas Transport Properties of Mixed Matrix Composite Membranes. Journal of Membrane Science, 495, 252-268.
10. Phattaranawik, J., Jiraratananon, R. and Fane, A. (2003) Effect of Pore Size Distribution and Air Flux on Mass Transport in Direct Contact Membrane Distillation. Journal of Membrane Science, 215, 75-85.
11. Singh, R.P., Way, J.D. and McCarley, K.C. (2004) Development of a Model Surface Flow Membrane by Modification of Porous Vycor Glass with a Fluorosilane. Industrial & Engineering Chemistry Research, 43, 3033-3040.
12. Li, K. (2007) Ceramic Membranes for Separation and Reaction. John Wiley & Sons, Hoboken.
13. Ismail, A.F. and David, L. (2001) A Review on the Latest Development of Carbon Membranes for Gas Separation. Journal of Membrane Science, 193, 1-18.
14. Pandey, P. and Chauhan, R. (2001) Membranes for Gas Separation. Progress in Polymer Science, 26, 853-893.
15. Basu, A., Akhtar, J., Rahman, M. and Islam, M. (2004) A Review of Separation of Gases Using Membrane Systems. Petroleum Science and Technology, 22, 1343-1368.
16. Weidenthaler, C. (2011) Pitfalls in the Characterization of Nanoporous and Nanosized Materials. Nanoscale, 3, 792-810.
17. Edward, G. (2006) Apparatus and Method for Separating Gases. US Patent No. 7,048,778 B2.
18. Edward, G. (2007) Apparatus and Method for Separating Gases. US Patent No. 7,048,778 B2.
19. Zhu, J., Fan, Y. and Xu, N. (2011) Modified Dip-Coating Method for Preparation of Pinhole-Free Ceramic Membranes. Journal of Membrane Science, 367, 14-20.
20. Vospernik, M., Pintar, A., Bercic, G., Batista, J. and Levec, J. (2004) Potentials of Ceramic Membranes as Catalytic Three-Phase Reactors. Chemical Engineering Research and Design, 82, 659-666.
21. Okon, E., Kajama, M.N. and Gobina, E. (2014) Gas Permeation on Silica Membrane for Lactic Acid Esterification Applications. Journal of Energy, 1, 265-269.
22. Benito, J.M., Conesa, A., Rubio, F. and Rodriguez, M.A. (2005) Preparation and Characterisation of Tubular Ceramic Membranes for Treatment of Oil Emulsions. Journal of the European Ceramic Society, 25, 1853-1903.
23. Okon, E., Shehu, H. and Gobina, E. (2015) Gas Transport through Inorganic Ceramic Membrane and Cation-Exchange Resins Characterization for Ethyl Lactate Separation. Transaction on Engineering Technologies, 6, 403-414.
24. Okon, E., Shehu, H. and Gobina, E. (2015) Characterisation of Inorganic Composite Ceramic Membrane for Lactic Acid Esterification Processes. Advanced Materials Research, 1102, 99-102.
25. Markovic, A., Stoltenberg, D., Enke, D. and Schlünder, E. (2009) Gas Permeation through Porous Glass Membranes: Part I. Mesoporous Glasses—Effect of Pore Diameter and Surface Properties. Journal of Membrane Science, 336, 17-31.

CHAPTER 9

Ground State Thermodynamic and Response Properties of Electron Gas in a Strong Magnetic and Electric Field: Exact Analytical Solutions for a Conventional Semiconductor and for Graphene

Georgios Konstantinou, Konstantinos Moulopoulos

Department of Physics, University of Cyprus, Nicosia, Cyprus

ABSTRACT

Consequences of an exceedingly strong electric field (E field) on the ground state energetics and transport properties of a 2D spinless electron gas in a perpendicular magnetic field (a Quantum Hall Effect (QHE) configuration) are investigated to all orders in the fields. For a conventional semiconductor, we find fractional values of the Hall conductivity and some magnetoelectric coefficients for certain values of E and B fields that do not result from interactions or impurities, but are a pure consequence of a strong enough in-plane E field. We also determine analytically the ground state energy, and response properties such as magnetization and polarization as functions of the electromagnetic field in the strong E field limit. In the case of Graphene, we obtain more complex behaviors leading to the possibility of irrational Hall values. The results are also qualitatively discussed in connection to various mechanisms for the QHE-breakdown.

Keywords: Strong Electric Field, 2D Electron Gas, Quantum Hall Effect, Graphene

1. INTRODUCTION

The Quantum Hall Effect (QHE), discovered by von Klitzing et al. in 1980, is the quantized response of a nearly 2D electronic system subjected to a perpendicular (to the 2D surface) magnetic field B when a small electric field E runs through the system in a longitudinal direction transverse to B. This quantized response, transverse to the longitudinal direction and reflected via the plateau formation of the Hall conductivity (observed at room temperature in Graphene) has led to considerable work with the aim of exploring the underlying physical mechanisms that describe the universality of the quantization. In the usual theoretical formulations of QHE, the electric field is treated as a perturbation to the basic Physics that is mostly governed by the magnetic field. Thermodynamic properties have also been investigated for the 2D electronic gas in a perpendicular magnetic field alone, or in a combination of electric and magnetic fields, limited to the low electric field regime only. This paper follows a thermodynamic methodology that tries to shed some light on both thermodynamic and transport properties of 2D conventional and quasi-relativistic electronic systems when placed inside a strong electric field that is transverse to the magnetic field; the electric field, when sufficiently strong, seems to cause energetic mixtures and seems to have the potential to lead to fractional quantization of the Hall conductivity. The paper is summarized as follows: In Section 2, the thermodynamic and transport properties of a conventional 2D non-interacting electron gas in the low E field regime are described in more quantitative detail than usual, these properties also being derived in the strong E field limit, while in Section 3, a 2D Graphene sheet is investigated with respect both to low and strong E field limit. All thermodynamic properties are given with respect to global internal energy, mag- netization and polarization expressed as functions of the electromagnetic field when a thermodynamic equilibrium has been established. A corresponding calculation of the Hall conductivity in the strong E field regime shows a possibility of fractional, stable values, in a certain range of B values.

2. CONVENTIONAL SYSTEM (A SEMICONDUCTOR)

2.1. Low E-Field Strength

Consider an ideal nonrelativistic two-dimensional spinless electron gas in a perpendicular and homogeneous magnetic field $\boldsymbol{B}$ directed along the positive z-axis. The dimensions of the plane are taken to be macroscopically large $(L_x \times L_y)$. In addition to $\boldsymbol{B}$ there is also an in-plane, homogeneous electric field $\boldsymbol{E}$ pointing in the y-direction that is very weak, i.e. it doesn't cause any overlap of different Landau Levels (L.Ls). In this manner, the Fermi energy can always be situated in the interior of an energy gap (i.e. for certain areal density), causing, as is well-known, the universality of quantization of Hall conductance σ_H. If the electric field is further increased from its first critical value (to be determined below) then the gap closes and the system is highly nonlinear, the general consensus being that this results to destruction of the quantization of σ_H, although from what we will see below several refined behavioral patterns remain (in this strong electric field case), that can even lead to the survival or even a different type of integer quantization in the nonlinear regime. This strong electric field case will be separately studied in the next subsection.

We choose to work in the Landau gauge $A_x = -By$, $A_y = A_z = 0$ along with a scalar potential $V = eEy$ (we take the charge of the electron to be ?e) resulting in the following Hamiltonian:

$$H = \frac{1}{2m}\left(\boldsymbol{p} + \frac{e}{c}\boldsymbol{A}\right)^2 + eEy,$$

with energy spectrum [1]

$$\varepsilon = \hbar\omega(n + 1/2) + \frac{mV_D^2}{2} + eEY_0 \tag{2.1}$$

with $\omega = eB/mc$ the cyclotron frequency, $n = 0,1,2\cdots$ the L.L. index, $V_D = c|\boldsymbol{E}\times\boldsymbol{B}|/B^2 = cE/B$ the modulus of the drift velocity and

$$Y_0 = \frac{cp_x}{eB} - \frac{mc^2 E}{eB^2} \tag{2.2}$$

the guiding center operator eigen value in the y-direction (restricted to the area $-L_y/2 \le Y_0 \le L_y/2$). Now, in specifying the electric field's strength E, if we first want to avoid any overlap among different L.Ls, the single particle energy $\varepsilon(n-1, Y_0 = L_y/2)$ must be lower than (or equal to) $\varepsilon(n, Y_0 = -L_y/2)$, a criterion which leads to the following inequality:

$$E \le \frac{\hbar\omega}{eL_y} \tag{2.3}$$

Therefore, this case of no-overlap involves a limitation of drift velocity's values that depends on both field strengths and is equivalent to either of two criteria: (i) $mV_D L_y \le \hbar$ (the angular momentum of an electron in one edge with respect to a point in the other edge is $\le \hbar$), or (ii) $\frac{mV_D^2}{2} \le \frac{\hbar^2}{2mL_y^2}$ (the drift kinetic energy is $\le a$ a confinement energy along the y-direction (due to the uncertainty principle)).

In this case all L.Ls can be filled independently (such a case is later shown i.e. in Figure 3): according to the least energy principle, at zero temperature $T = 0$, all electrons occupy states labeled by small quantum numbers (n,l), (with n the above mentioned Landau Level index, a non-zero positive integer, and with l another integer specifying the eigenvalue of p_x due to periodic boundary conditions along the x-direction, the eigenvalue being $p_x = hl/L_x$), starting from L.L $n = 0$ and varying all possible values of l, and then successively following $n = 1$, $n = 2$ and so on. Each L.L (indexed by quantum number n) may host up to Φ/Φ_0 spinless electrons (which is the total number of distinct values of l in the thermodynamic limit) with $\Phi = BL_x L_y$ being the magnetic flux across the 2D plane and $\Phi_0 = hc/e$ the flux quantum, in order to be consistent with

Pauli principle. This means that without loss of generality we always have, let us say, $\rho(=1,2,3\cdots)$ L.Ls that are occupied for fixed particle number N and for a certain value of B, and due to the minimal energy criterion, we must have ($\rho-1$) fully occupied L.Ls and the last L.L. either partially or fully occupied. Of course, if N is fixed, the entire occupational procedure is uniquely determined only by the magnetic field strength B and it can be described naturally by the inequalities

$$(\rho-1)\Phi/\Phi_0 \leq N \leq \rho\Phi/\Phi_0 \text{ or } \frac{1}{\rho}n_A\Phi_0 \leq B \leq \frac{1}{\rho-1}n_A\Phi_0 \quad , (2.4)$$

with $n_A = N/L_xL_y$ the 2D areal density of electrons. The last L.L. is occupied by $N-(\rho-1)\Phi/\Phi_0$ electrons allowing us to determine analytically the maximum $Y_{0\text{MAX}}$ of the last electron located at L.L. index value of $n=\rho-1$:

$$Y_{0\text{MAX}} = \frac{chN}{eBL_x} + \left(\frac{1}{2}-\rho\right)L_y \quad (2.5)$$

which gives the expected results that when $N=\rho\frac{\Phi}{\Phi_0}$, $Y_{0\text{MAX}}=L_y/2$ and when $N-(\rho-1)\Phi/\Phi_0$, $Y_{0\text{MAX}}=-L_y/2$.

2.1.1. Thermodynamic Properties

The total internal energy of the system at $T=0$ is a sum over all occupied quantum numbers n and Y_0:

$$E = \sum_n \sum_{Y_0}\left[\hbar\omega(n+1/2)+\frac{mV_D^2}{2}+eEY_0\right] \quad (2.6)$$

In the macroscopic (continuum) limit $L_y \to \infty$ we may approximate the sum with respect to Y_0 with an integral:

$$\sum_{Y_0=-L_y/2}^{L_y/2} f(Y_0) = \sum_{l=l\min}^{l\max} f(Y_0) \to \frac{BL_x}{\Phi_0}\int_{-L_y/2}^{L_y/2}(f(Y_0))\mathrm{d}Y_0 \quad (2.7)$$

with $f(Y_0)$ an arbitrary function of Y_0, given that Y_0 is linearly related to p_x, and p_x is defined through the quantum number l as $p_x = hl/L_x$. The above is valid because in the continuum infinite space limit the momentum in the x direction becomes quasi-continuous (along with Y_0). If we then use (2.2) with the restriction that: $-L_y/2 \le Y_O \le L_y/2$, we can find the available range of values for l:

$$-\frac{eBL_xL_y}{2hc} + \frac{mcL_xE}{Bh} \le l \le \frac{eBL_xL_y}{2hc} + \frac{mcL_xE}{Bh}$$

or $l\min \le l \le l\max$ .We then prove (2.7) as follows:

$$\lim_{L_y \to \infty} \sum_{l=l\min}^{l\max} f(Y_0) = \frac{BL_x}{\Phi_0} \int_{-L_y/2}^{L_y/2} \mathrm{d}Y_0 f(Y_0)$$

$$= \frac{c}{eB}\frac{BL_x}{\Phi_0} \int_{p_{x\min}}^{p_{x\max}} \mathrm{d}p_x f(Y_0) = \int_{l\min}^{l\max} \mathrm{d}l f(Y_0),$$

which implies that in the continuous limit the following holds:

$$\lim_{L_y \to \infty} \sum_{l=l\min}^{l\max} = \int_{l\min}^{l\max} \mathrm{d}l \, ,$$

which completes our proof. We therefore have:

$$E = \frac{BL_x}{\Phi_0} \sum_{n=0}^{\rho-2} \int_{-L_y/2}^{L_y/2} \mathrm{d}Y_0 \left[\hbar\omega(n+1/2) + \frac{mV_D^2}{2} + eEY_0 \right]$$

$$+ \frac{BL_x}{\Phi_0} \int_{-L_y/2}^{Y_{OMAX}} \mathrm{d}Y_0 \left[\hbar\omega(\rho-1/2) + \frac{mV_D^2}{2} + eEY_0 \right]$$

,with $\omega = eB/mc$, and after a number of algebraic manipulations we reach a closed analytical expression for the total minimal energy per electron, namely

$$\frac{E}{N} = -\frac{e^2B^2}{4\pi mc^2 n_A}\rho(\rho-1) + eEL_y\left[\frac{n_A\Phi_0}{2B} - \left(\rho - \frac{1}{2}\right) + \frac{\rho B}{2n_A\Phi_0}(\rho-1)\right]$$
$$+ \frac{mc^2E^2}{2\mathrm{B}^2} + \frac{\hbar eB}{mc}(\rho - 1/2) \qquad (2.8)$$

valid in the magnetic field range $n_A\Phi_0/\rho \le B \le n_A\Phi_0/(\rho-1)$, or, in units of the Fermi energy $\varepsilon_F = \hbar^2 4\pi n_A/2m$ (the one defined for 2D spinless electron gas in the absence of E and B),

$$\frac{E}{N\varepsilon_F} = -\frac{B^2}{2n_A^2\Phi_0^2}\rho(\rho-1) + \frac{B}{n_A\Phi_0}\left(\rho-\frac{1}{2}\right) + \frac{eEL_y n_A\Phi_0}{2B\varepsilon_F}$$
$$-eEL_y\left(\rho-\frac{1}{2}\right) + \frac{e^2E^2n_A\Phi_0^2}{4\pi\varepsilon_F^2B^2} + \frac{eEL_y}{2\varepsilon_F}\frac{B}{n_A\Phi_0}\rho(\rho-1) \tag{2.9}$$

Now, let B be equal to $B = n_A\Phi_0/k$, with $\rho-1 \le k \le \rho$, so that from (2.3) we have $eEL_y/\varepsilon_F \equiv y \le 1/k$. Rewriting then (2.8) in terms of y and $x = B/n_A\Phi_0$ we obtain

$$\frac{E}{N\varepsilon_F} = \frac{\rho(\rho-1)x^2}{2} + x\left(\rho-\frac{1}{2}\right) + \frac{y}{2x}$$
$$+\frac{1}{4\pi n_A L_y^2}\frac{y^2}{x^2} - y\left(\rho-\frac{1}{2}\right) + \frac{yx}{2}\rho(\rho-1)$$

In Figure 1, see the graphs of energy and magnetization per electron, for $n_A L_y^2 = 1$ (a good value so that the internal structure of these quantities are shown in sufficient detail

Using the second thermodynamic law we can also determine analytically the equilibrium magnetization and polarization per electron, which turn out to be

$$\frac{M}{N} = -\frac{\partial\left(\frac{E}{N}\right)}{\partial B} = \frac{e^2B}{2\pi mc^2 n_A}\rho(\rho-1) - eEL_y\left[-\frac{n_A\Phi_0}{2B^2} + \frac{\rho}{2n_A\Phi_0}(\rho-1)\right]$$
$$+\frac{mc^2E^2}{B^3} - \frac{\hbar e}{mc}(\rho-1/2)$$

$$\frac{P}{N} = -\frac{\partial\left(\frac{E}{N}\right)}{\partial E} = -eL_y\left[\frac{n_A\Phi_0}{2B} - \left(\rho-\frac{1}{2}\right) + \frac{\rho B}{2n_A\Phi_0}(\rho-1)\right] - \frac{mc^2E}{B^2} \tag{2.10}$$

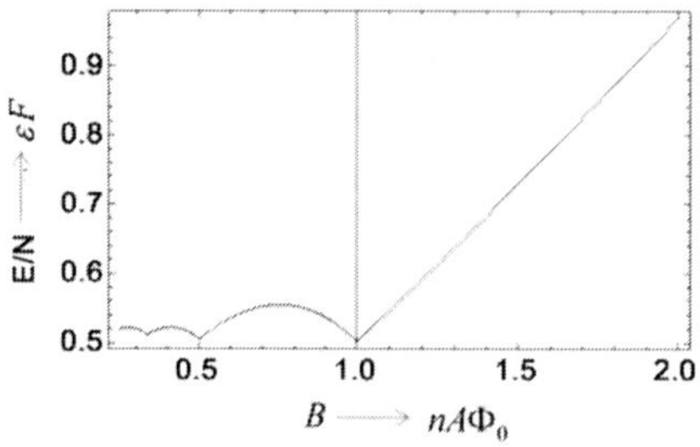

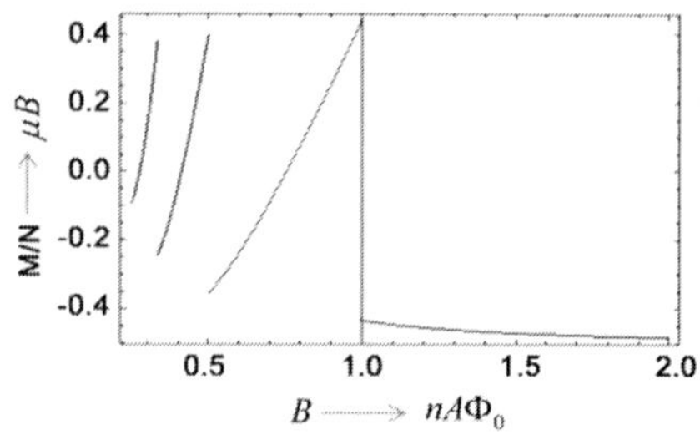

Figure 1. Energy and Magnetization per electron as functions of the magnetic field B.

These thermodynamic expressions demonstrate the effect of the electric field which is non-linear (with respect to the variable E), even for the case of relatively weak electric field. Now we proceed by examining some limits: when all L.Ls are fully occupied, namely when $N = \rho \dfrac{\Phi}{\Phi_0}$, then we obtain:

$$\frac{M}{N} = -\frac{e\hbar}{2mc} + \frac{L_y}{n_A c}\left(\frac{\sigma_H}{2}\right)E + \frac{m}{n_A^3 C}\sigma_H^3 E^2 \tag{2.11}$$

$$\frac{P}{N} = -\frac{m}{n_A^2}\sigma_H^2 E \tag{2.12}$$

with $\sigma_H = \rho e^2 / h$ the Hall conductivity (see below). It is interesting to note a characteristic half of the Hall conductance in the coefficient that connects the so-called magnetoelectric effects with the fields that cause them (that in a problem with chiral properties, usually in Topological Insulator materials, correspond to an extra magnetization caused by a parallel electric field (as in (2.11) above) and an extra polarization caused by an extra magnetic field (see [2] and references therein); here we note such a trend even in a Quantum Hall system (which is not unexpected, and is actually justified based on general Physics arguments [2]). One can actually see such a trend in the polarization as well, for a general B, in (2.10), where in the last linear term we can again see $\sigma_H / 2$ appearing. [It should be added that all these behaviors originate from the final term of (2.9) that describes an EB-coupling, something that we will also see later for the relativistic case.] In addition to all this, note also that for $E = \hbar\omega / eL_y$ the gap closes, and (2.8) becomes:

$$\frac{E}{N} = \frac{1}{2}\varepsilon_F\left(E=0,B=0\right) + \frac{\hbar^2}{2mL_y^2} \tag{2.13}$$

with $\varepsilon_F\left(E=0,B=0\right) = \hbar^2 4\pi n_A/2m$ being the expected Fermi energy (in the absence of fields)—and when all L.Ls are fully occupied, with $B = n_A\Phi_0/\rho$, and with any E (now less than its first critical value) (2.8) becomes:

$$\frac{E}{N} = \frac{1}{2}\varepsilon_F\left(E=0,B=0\right) + \frac{mc^2E^2}{2B^2},$$

i.e. the total energy per electron reduces to 2D energy of free electron gas plus a drift kinetic term. These last simple results are not quite unexpected and may be justified with proper semiclassical considerations.

2.1.2. Hall Conductivity

The Hall conductivity is defined as $\sigma_H = en_Ac/B$. The usual plot representing the QHE [3] shows σ_H as a function of B, and in this we have plateau formation for certain values of B. For B varying in the range defined by the inequaltities (2.4):

$$n_A\Phi_0/\rho \leq B \leq n_A\Phi_0/\left(\rho-1\right) \tag{2.14}$$

then σ_H is quantized in units of $\rho e^2/h$ with ρ being an integer, which counts the number of fully occupied L.Ls. Note that each plateau has maximum width:

$$\frac{\Phi_0}{\left(\rho-1\right)} - \frac{n_A\Phi_0}{\rho} = \frac{1}{\rho\left(\rho-1\right)}n_A\Phi_0 \tag{2.15}$$

In the following we show that, in the very strong E-field case, fractional filling factors may also occur, by variation of electron number N with E and B fixed. [In that case, the analogous width of the corresponding plateau is expected to be reduced, although a serious consideration of plateau-observability should include a study in the presence of disorder, a subject that is beyond the scope of the present article, or in the presence of edge states (see Section 2.2.4 for a qualitative discussion).

2.2. High E-Field Strength

When the electric field exceeds its first critical value, $E > \hbar\omega/eL_y$, inter L.L. overlap occurs. As E gets stronger, more and more L.Ls overlap and degenerate states that belong to different values of the quantum number n appear (in the previous case of a weak electric case, the standard Landau degeneracy had been completely lifted). Energy gap closes, and Fermi energy is always located on a single quantum state, with a significant number of available nearby states. Although Fermi energy is no longer in an energy gap, it will make jumps from one L.L to another by varying the magnetic field (or the particle number N). There are some critical values of B (the transition points) where a jump occurs, at which the Hall conductivity takes fractional values, as we shall see.

We now assume a general strength E field determined by

$$E = z\hbar\omega/eL_y, \tag{2.16}$$

with z a continuous number that describes "overlap percentage": $j \le z < j+1$ with $j = 0,1,2\cdots$. When $j = 0$, or $z \le 1$, overlap vanishes and L.Ls can be filled independently, with the usual energy gap restored. This is just the case discussed in previous section. Equation (2.16) can actually be derived for a certain electric field that obeys the following relation:

$$\begin{aligned} \varepsilon\left(n = j, Y_0 = -L_y/2\right) &\le \varepsilon\left(n = 0, Y_0 = L_y/2\right) \\ &\le \varepsilon\left(n = j+1, Y_0 = -L_y/2\right), \end{aligned}$$

i.e. the electric field has a strength such that the single particle energy of the last electron $\left(Y_0 = L_y/2\right)$ in L.L. $n = 0$ is greater than single particle energy of the first particle $\left(Y_0 = -L_y/2\right)$ in L.L. $n = j$ and lower than the single particle energy of the first particle in L.L. $n = j+1$. Then,

$$\hbar\omega\left(j+1/2\right) - eEL_y/2 \le \hbar\omega/2 + eEL_y/2 \le \hbar\omega\left(j+3/2\right) - eEL_y/2$$

that concludes in

$$j\frac{\hbar\omega}{eL_y} \le E \le \left(j+1\right)\frac{\hbar\omega}{eL_y}, \text{ or } E = z\frac{\hbar\omega}{eL_y}, \text{ with } j \le z < j+1,$$

which is indeed (2.16).

Now we proceed to occupancies: For a constant E and B field (i.e. a constant energetic configuration) varying electron number N results in variation of Fermi energy and a corresponding variation of occupied L.Ls. The goal here is to examine under what N-variations the Fermi level remains at a single (the topmost) L.L level. To achieve this, we fix Fermi level at the highest occupied L.L, $n = \rho - 1,\ \rho \geq 1$, (see Figure 2) with energy

$$\varepsilon_F = \hbar\omega(\rho - 1/2) + mV_D^2/2 + eEY_m, \tag{2.17}$$

with Y_m the guiding center position of the last electron (with highest single particle energy) located at $n = \rho - 1$ (see the isolated dot in Figure 3). In this way, we ensure that variations of N will result in moving Y_m in the interval $Y_L \leq Y_m \leq Y_R$, with $Y_L = -L_y/2$, the left edge of $n = \rho - 1$ L.L. and Y_R is a critical guiding center above which L.L. $n = \rho$ is occupied. This can be determined by equating the Fermi energy to the single particle energy $\varepsilon(n = \rho, Y_0 = -L_y/2)$:

$$\hbar\omega(\rho - 1/2) + mV_D^2/2 + eEY_R = \hbar\omega(\rho + 1/2) + mV_D^2/2 - eEL_y/2,$$

resulting in: $Y_R = -L_y/2 + \hbar\omega/eE$, or, using Equation (2.1): $Y_R = -L_y/2 + L_y/z$. Any number N that takes Y_R above this critical value will result in a nonzero occupation of $n = \rho$ L.L. Therefore, to ensure that exactly ρ L.Ls are occupied, Y_m must vary in the following window:

$$-L_y/2 \leq Y_m \leq -L_y/2 + L_y/z, \tag{2.18}$$

guaranteeing that exactly ρ L.Ls are occupied. Naturally this equation becomes pathological when z is smaller than unity—the weak electric field case—Y_R becomes infinite when $z \to 0$. This of course happens because the above equation is not valid in this case, where no inter-L.L overlaps occur. Limitations must therefore be imposed here. The smallest value that z can take, must be above 1, with $z = 1$ defining the non-overlap limit. If $z < 1$ the above relation just becomes:

$$-L_y/2 \leq Y_m \leq L_y/2 \quad (2.19)$$

as earlier. Having now defined the Fermi level's exact location analytically, we proceed with our method by counting how many L.Ls are fully occupied and how many are partially occupied. An L.L. is completely filled with electrons only when the Fermi energy is greater than the single electron energy located at $Y_0 = L_y/2$ in each L.L. Using this information, we may write down the number of fully occupied L.Ls (i_F) as a sum of theta functions over all L.Ls:

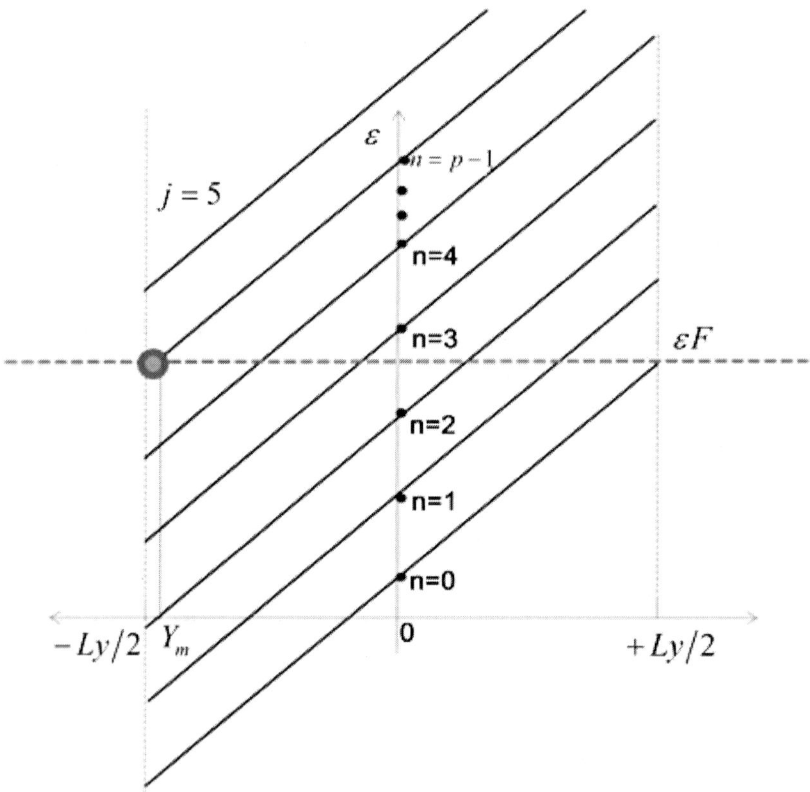

Figure 2. An example of different L.L mixture with $j = 5$ and $z = 5.1$ a possible position of Fermi Level ($\rho = 6$), with only one full L.L. occupied and five L.Ls partially occupied. Ground state demands that the chemical potential of each distinct L.L (highest single-electron energy) must be equal, resulting in a purely horizontal Fermi energy.

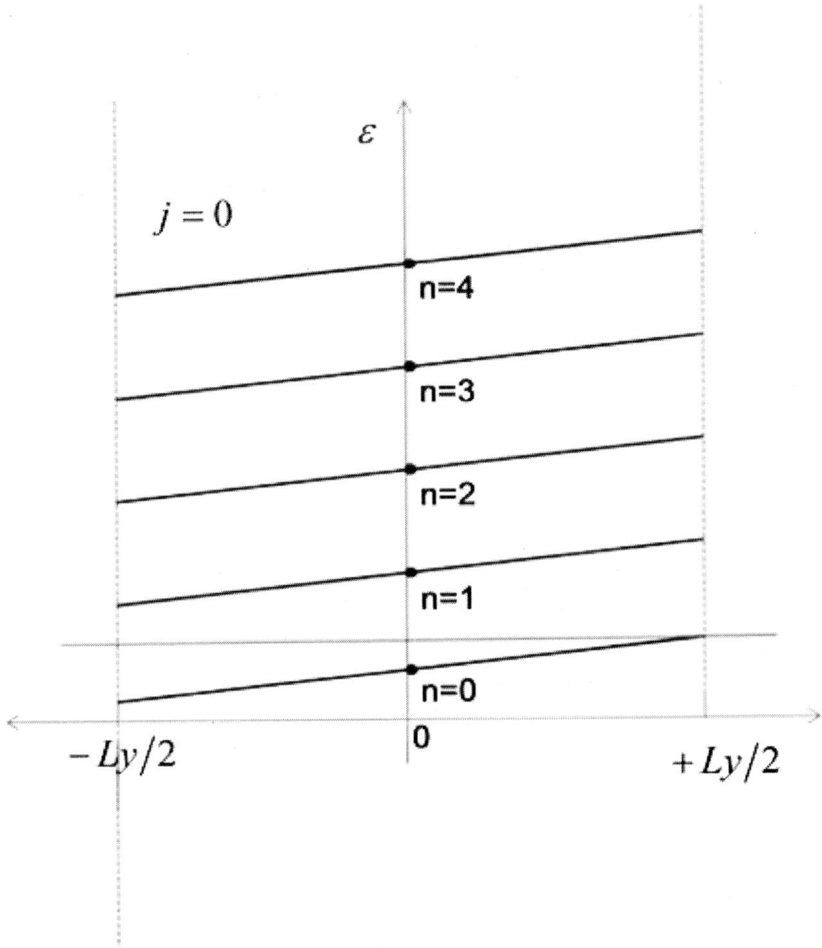

Figure 3. A schematic representation of the energy states (L.Ls) when the electric field is weak: there is no inter-L.L. overlap. All L.Ls can be filled independently, and energy gaps between L.Ls can result in quantization of the Hall conductivity.

$$i_F(Y_m,\rho,z)=\sum_{n=0}^{+\infty}\theta\left[\varepsilon_F-\varepsilon\left(n,\frac{L_y}{2}\right)\right]=\sum_{n=0}^{+\infty}\theta\left[\rho-1-n+z\left(\frac{Y_m}{L_y}-\frac{1}{2}\right)\right]\text{ for } z>1 \tag{2.20}$$

where $\theta[x]$ is a step function, $\theta[x]=1$ if $x \geq 0$ and $\theta[x]=0$ otherwise. Variations of Y_m (i.e. for fixed E and B) according to Equation (2.20) result in variations in i_F as follows:

$$\underbrace{i_F^L(\rho,z)}_{Y_m=Y_L} \le i_F(Y_m,\rho,z) \le \underbrace{i_F^R(\rho,z)}_{Y_m=Y_R}, \tag{2.21}$$

with $i_F^L(\rho,z)=\sum_{n=0}^{+\infty}\theta[\rho-1-z-n]$ and $i_F^R(\rho,z)=\sum_{n=0}^{+\infty}\theta[\rho-z-n]$ are the numbers of completely filled L.Ls calculated at the edges of Equation (2.21). Let's see now some examples:

In Figure 4, z is equal to unity, $z=1$ and $\rho=2$, then

$i_F^L=\sum_{n=0}^{+\infty}\theta[-n]=\theta[0]=1$, and $i_F^R=\sum_{n=0}^{+\infty}\theta[1-n]=\theta[1]+\theta[0]=2$. In Figure 5,

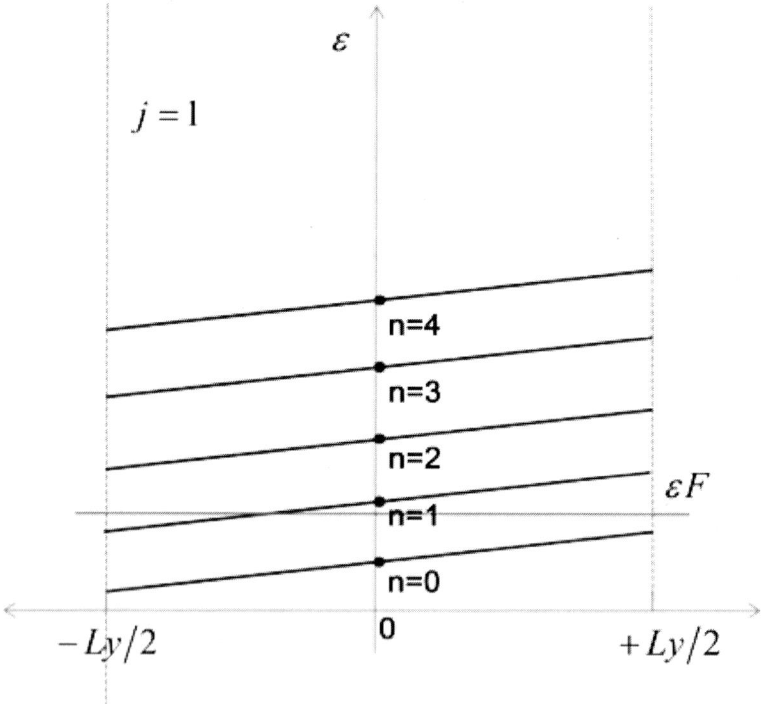

Figure 4. Example of 2 fully occupied Landau Levels with their slope having the highest value corresponding to no overlap (z = 1).

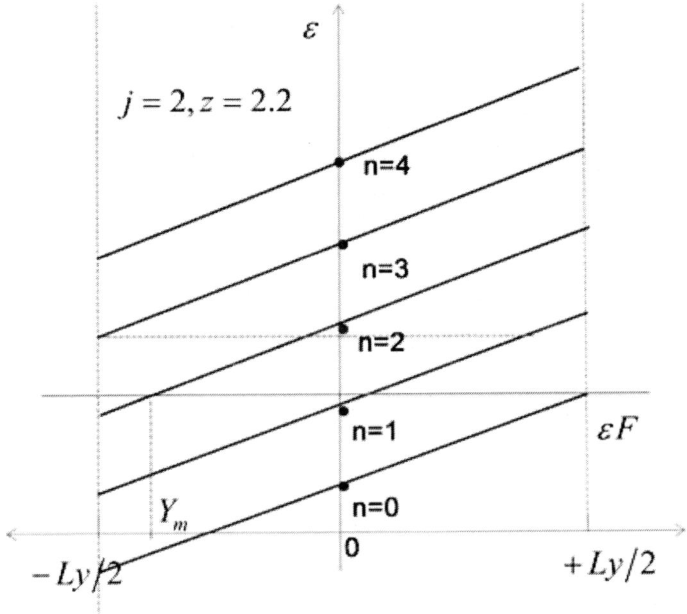

Figure 5. Example of 3 fully occupied Landau Levels with "overlap ratio" z = 2.2.

$z = 2.2$ and $\rho = 3$, $i_F^L = 0$ and $i_F^R = 1$. It is interesting to note that for any value of the variable z, i_F^L and i_F^R always differ by one, which means that varying N or equivalently, Y_m, in the interval defined by (2.19), one full L.L. is added to the system. This resembles the $E = 0$ case, where variations of Y_m in the same interval (but for $z = 1$) the Fermi energy passes through all available states in the last L.L. until it reaches the next L.L., where an extra full L.L. increases the number i_F by 1 (see also Figure 6). [This is actually consistent with the well-known corresponding result in the case we fold the system (in the x- direction) into a cylinder, that is a key result in the Laughlin argument [4] that gives the integral quantization of the Hall conductivity from general gauge arguments—or equivalently to the charge-pumping picture of Thouless [5] or a more general property of the so-called spectral flow [6] in topologically nontrivial systems, such as topological insulators [7] . It is interesting to see that when z reaches its lowest possible value, $z = 1$, $i_F^L = \rho - 1$, meaning that the

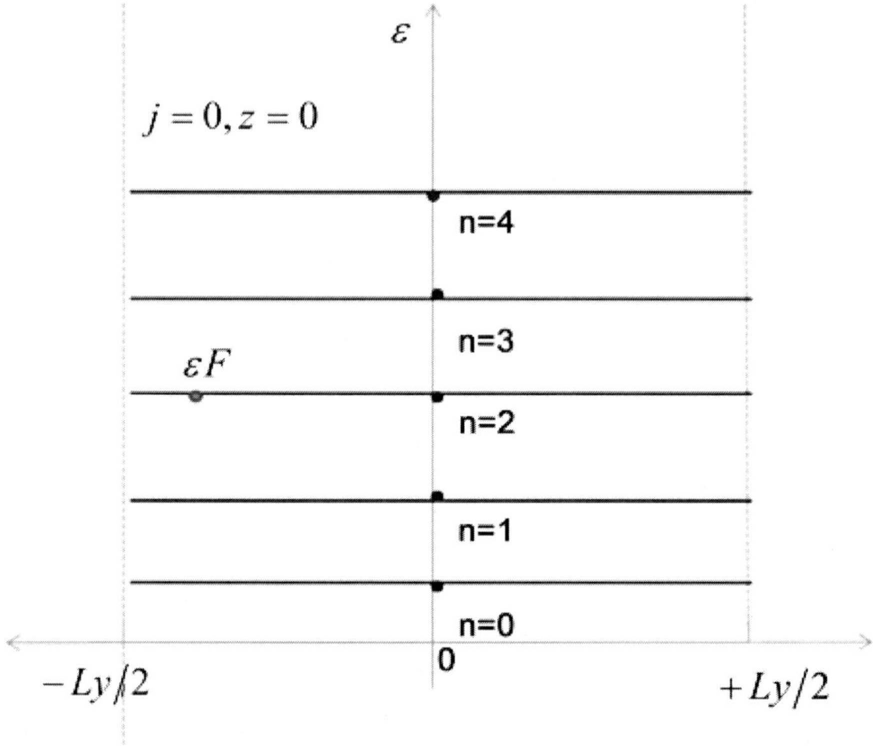

Figure 6. For z = 0 case (identical to $z \leq 1$ case), with no E-field, $i_F^L = \rho - 1$ and $i_F^R = \rho$. Left figure shows $\rho = 3$ occupied L.Ls, with two of them fully occupied and the last one partially occupied. An addition of electrons will stimulate Fermi energy to pass through all available states in its right side until right end of L.L. $n = 2$, where a full L.L. will be added to i_F.

maximum number of partially occupied L.Ls is always 1. Needless to say that z remains a constant only when E and B fields are constant too, (or in a special case that both fields vary in the exactly same rate) and the only variable is the electron number. This happens because of Equation (2.16), $E = z\hbar\omega/eL_y$, which relates B to E through the variable z. For a constant electric field, varying B will result in a variation of z such that E remains a constant. The same principle holds for keeping B constant and varying E.

Generally, from definition of i_F^R and i_F^L we have that: $i_F^R = \rho - \text{Int}[z] + \delta$ and $i_F^L = \rho - \text{Int}[z] - 1 + \delta$, where $\delta = 1$ if z is an integer and $\delta = 0$ otherwise. These equalities hold for $\rho > \text{Int}[z] - \delta$. If $\rho = \text{Int}[z] - \delta$ or lower, then $i_F^R = 0$ and $i_F^L = 0$.

Now, about partially filled L.Ls, these must intersect at certain Y_0 the Fermi energy. Optimal energy requires that states (per L.L.) starting with $Y_0 = -L_y/2$ until the intersection with the Fermi energy must be filled, while states with higher Y_0s must be empty. These intersections can be easily found at the following points:

$$Y_0(l) = \frac{1}{eE}\left[\varepsilon_F - \hbar\omega(i_F + l + 1/2) - \frac{1}{2}mV_D^2\right], \quad \text{with } l = 0,1,2\cdots\rho-1-i_F,$$

(2.22)

where $\rho - i_F$ is the number of partially occupied L.Ls.

2.2.1. Number of States under Fermi Energy

Each L.L contains Φ/Φ_0 available quantum states. Each state corresponds to a single spinless electron, in accordance to Pauli Exclusion Principle. The number of states under Fermi energy is a sum of i_F full states and a sum term that counts all states in the partially occupied L.Ls from $-L_y/2$ until the intersection with Fermi energy:

$$\#\text{states} = i_F\frac{\Phi}{\Phi_0} + \sum_{l=0}^{\rho-1-i_F}\left[\frac{L_x B\varepsilon_F}{Ehc} - \frac{\omega B L_x}{2\pi cE}\left(i_F + l + \frac{1}{2}\right) - \frac{mV_D^2 L_x B}{2Ehc} + \frac{\Phi}{2\Phi_0}\right]$$

(2.23)

This number, in a canonical ensemble, is exactly equal to the constant particle number N:

$$N=i_F\frac{\Phi}{\Phi_0}+(\rho-i_F)\left[\frac{\varepsilon_F}{E}\frac{BL_x}{ch}-mc\frac{L_x}{h}\frac{E}{2B}+\frac{eBL_x}{ch}\frac{L_y}{2}-\frac{eB^2}{4\pi mE}\frac{L_x}{c^2}[\rho+i_F]\right]$$

or

$$N=i_F\frac{\Phi}{\Phi_0}+(\rho-i_F)\left[\frac{1}{4\pi E}\frac{eB^2L_x}{mc^2}(\rho-1-i_F)+\frac{BL_x}{ch}eY_m+\frac{eBL_x}{ch}\frac{L_y}{2}\right], \tag{2.24}$$

if one replaces the Fermi energy in (2.24) with the corresponding expression from (2.17), with $i_F=i_F(\varepsilon_F,E,B)$ also a function of ε_F. This equation determines particle number N when ε_F is also known. Unfortunately, it is rather difficult to solve directly with respect to ε_F, but it can be easily determined numerically. Further simplification of (2.24) can be made using the window of values of ε_F, given by Equations ((2.17) and (2.18)):

$$\hbar\omega\left(\rho-\frac{1}{2}\right)+\frac{mV_D^2}{2}-eE\frac{L_y}{2}\le\varepsilon_F\le\hbar\omega\left(\rho-\frac{1}{2}\right)+\frac{mV_D^2}{2}+eE\left(-\frac{L_y}{2}+\frac{L_y}{z}\right), \tag{2.25}$$

which results in the following window of values for N:

$$\frac{B^2L_x}{Emc^2 4\pi}e(\rho-i_F^L)[\rho-1-i_F^L]+\frac{\Phi}{\Phi_0}i_F^L$$
$$\le N\le\frac{B^2eL_x}{4\pi Emc^2}(\rho-i_F^R)[\rho-1-i_F^R]+\frac{\Phi}{\Phi_0}\left[i_F^R+\frac{(\rho-i_F^R)}{z}\right] \tag{2.26}$$

with $i_F^L=i_F^L(\rho,E,B)$ and $i_F^R=i_F^R(\rho,E,B)$. The above relation defines windows of values for B (if it is considered as a variable) for constant E field, and constant electron number N. Equivalently, one may prepare an experiment, where electron number is not conserved (i.e. an electric circuit) and keep E and B fixed. Note that the above relation shrinks to Equation (2.14) when $z=1$:

$$\rho\frac{\Phi}{\Phi_0}\ge N\ge(\rho-1)\frac{\Phi}{\Phi_0} \tag{2.27}$$

2.2.2. Thermodynamics

We now proceed to ground state energy calculation, which is a sum over all single electron states, until reaching the Fermi energy:

$$E=\frac{BL_x}{\Phi_0}\sum_{n=0}^{i_F-1}\int_{-L_y/2}^{L_y/2}\mathrm{d}Y_0\left[\hbar\omega\left(n+1/2\right)+\frac{1}{2}mV_D^2+eEY_0\right]$$
$$+\frac{BL_x}{\Phi_0}\sum_{l=0}^{\rho-1-i_F}\int_{-L_y/2}^{Y_0(l)}\mathrm{d}Y_0\left[\hbar\omega\left(i_F+l+1/2\right)+\frac{1}{2}mV_D^2+eEY_0\right]$$

The left term describes the energy due to fully occupied L.Ls, while the right term describes the energy of partially occupied L.Ls. After a number of algebraic manipulations, and using Equaitonk (2.22) we conclude to the following result for the total energy per electron:

$$\frac{E}{N}=-\frac{e^2B^2}{4\pi mc^2n_A}i_F\rho+\frac{e\hbar B}{2cm}\left(\rho+i_F\right)+\frac{L_y}{2\left(\rho-i_F\right)}\frac{En_Ach}{B}$$
$$-eEL_y\left(\frac{\rho+i_F}{2\left(\rho-i_F\right)}\right)+\frac{mc^2E^2}{2B^2}+\left[\frac{\rho i_F}{2\left(\rho-i_F\right)}\right]\frac{e^2}{hcn_A}L_yEB$$
$$-\frac{e^2\hbar B^3}{48\pi n_Am^2c^3EL_y}\left(\rho-i_F-1\right)\left(\rho-i_F\right)\left(\rho-i_F+1\right) \tag{2.28}$$

that consists of a sum of several terms which are not all symmetric (with respect to E and B). The last term vanishes in weak E field limit ($i_F=\rho-1$). Note the interesting fact that the presence of the coupling term EB requires fully occupied L.Ls, i.e. $i_F\neq 0$. This is a probably expected pattern resulting from axionic considerations in an electromagnetic field where the coefficient of coupling term involves fully occupied L.Ls (for the weak E field case, the corresponding coefficient is related to the Hall conductivity, and it was briefly mentioned in the last section, originating from the EB-coupling term).

When $i_F=0$, i.e. when all L.Ls are partially occupied ($\rho\leq \mathrm{Int}\left[z\right]-\delta$), we have (from (2.28)):

$$\frac{E}{N}=\frac{e\hbar B}{2cm}\rho+\frac{L_y}{2\rho}\frac{En_Ach}{B}-\frac{eEL_y}{2}+\frac{mc^2E^2}{2B^2}-\frac{e^2\hbar B^3}{48\pi n_Am^2c^3EL_y}\rho\left(\rho^2-1\right)$$

and from (2.26):

$$\frac{B^2eL_x}{Emc^2 4\pi}\rho(\rho-1)\leq N\leq\frac{B^2eL_x}{4\pi Emc^2}\rho(\rho-1)+\frac{\Phi}{\Phi_0}\frac{\rho}{z}$$

Or, substituting $E = z\hbar\omega/eL_y$ we get

$$\frac{\rho}{2z}\frac{\Phi}{\Phi_0}(\rho-1)\leq N\leq\frac{\rho}{2z}\frac{\Phi}{\Phi_0}(\rho+1) \tag{2.29}$$

i.e. there are no quadratic terms with respect to B appearing in the total energy, and there are no coupling terms either. Also, when the Fermi energy travels from one edge of n = ρ − 1 to the other, it passes exactly through $\rho\Phi z/\Phi_0 < \Phi/\Phi_0$ states. On the other hand, when $\rho > \mathrm{Int}[z]-\delta$, i.e. $i_F \neq 0$, the Fermi energy passes through exactly Φ/Φ_0 states.

Now, we want to plot Equation (2.28) by keeping E and B fixed and vary N in the interval given by (2.26). Using E as in (2.16) we rewrite (2.28) and (2.26) as

$$\frac{E}{\frac{e^2B^2L_xL_y}{4\pi mc^2}} = -i_F\rho+\frac{N\Phi_0}{\Phi}(\rho+i_F)+\frac{zN^2}{(\rho-i_F)}\frac{\Phi_0^2}{\Phi^2}-z\frac{N\Phi_0}{\Phi}\left(\frac{\rho+i_F}{\rho-i_F}\right)+Nz^2\frac{\Phi_0^2}{\Phi^2}$$
$$+\frac{z\rho i_F}{\rho-i_F}-\frac{1}{12z}(\rho-i_F-1)(\rho-i_F)(\rho-i_F+1)$$

and

$$\frac{1}{2z}(\rho-i_F^L)[\rho-1-i_F^L]+i_F^L\leq N\frac{\Phi_0}{\Phi}\leq\frac{1}{2z}(\rho-i_F^R)[\rho+1-i_F^R]+i_F^R$$

and the corresponding Graphs of the total energy as function of $y = N\Phi_0/\Phi$ for different z-values are shown in Figure 7 and Figure 8.

A few remarks are in order about these figures: It is clear that when the E field is exceedingly strong, in contrast to the case of a weak E field, there appears a global minimum with respect to particle number variations. This minimum is a consequence of the E field effect on the thermodynamic properties; as E gets stronger, more and more states will gain negative energy, (see for example Figure 5). As these states get occupied by electrons, the total energy will become negative at first, then will rise up to positive values, because states with positive energy will begin to fill up. The result of this competition is this minimum, which can be fully controlled by examining the corresponding z-value. The greater z gets, the more negative-energy states will be occupied, and the minimum will move to greater particle numbers (or larger particle densities). In comparison with the low E field case, where the positive states are more, in this case the total energy may be even lower.

We now plot the total energy (see Figure 9) as a function of variable z, for a constant E field with $y = eEL_y/\varepsilon_F$ (a suitable measure for the E field in units of 2D Fermi energy in the $E = 0,\ B = 0$ case) and $B = mcL_yE/\hbar z$ from Equation (2.16):

$$\frac{E}{N\varepsilon_F} = -\frac{y^2}{2z^2}i_F\rho + \frac{y}{2z}(\rho + i_F) + \frac{1}{(\rho - i_F)}\frac{z}{2} - \frac{y}{2}\left(\frac{\rho + i_F}{\rho - i_F}\right) + \frac{z^2}{L_y^2 4\pi n_A} + \left[\frac{\rho i_F}{2(\rho - i_F)}\right]\frac{y^2}{z} - \frac{1}{24}\frac{y^2}{z^3}(\rho - i_F - 1)(\rho - i_F)(\rho - i_F + 1)$$

of E and B) as function of z (analogous to inverse B) for $y = 3$ (Left), $y = 6.3$ (Right, low E field limit).

where z lies in the following window:

$$\frac{1}{2z^2}(\rho - i_F^L)[\rho - 1 - i_F^L] + \frac{1}{z}i_F^L \le \frac{1}{y} \le \frac{1}{2z^2}(\rho - i_F^R)[\rho + 1 - i_F^R] + \frac{1}{z}i_F^R$$

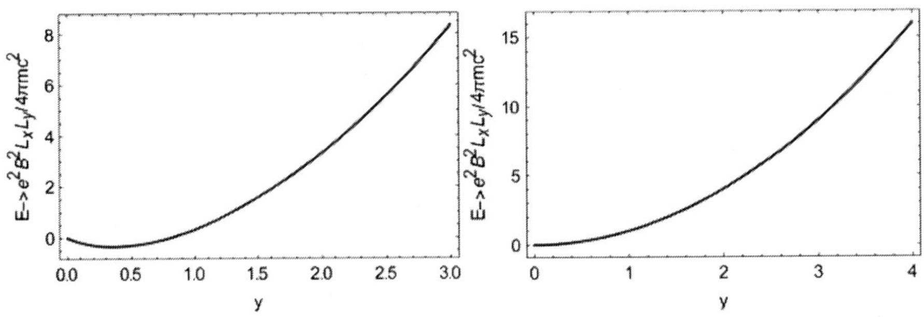

Figure 7. Total energy (in units of $e^2B^2L_xL_y/4\pi mc^2$) as function of particle number (in units of Φ/Φ_0, $y = N\Phi_0/\Phi$) for $z = 3$ (Left), $z = 1$ (Right, low E field limit).

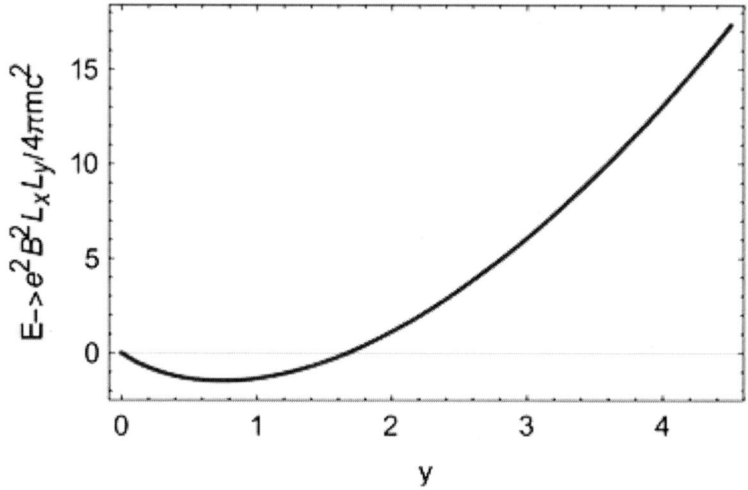

Figure 8. Total energy (in units of $e^2B^2L_xL_y/4\pi mc^2$) as function of particle number (in units of Φ/Φ_0, $y = N\Phi_0/\Phi$) for $z = 6.1$.

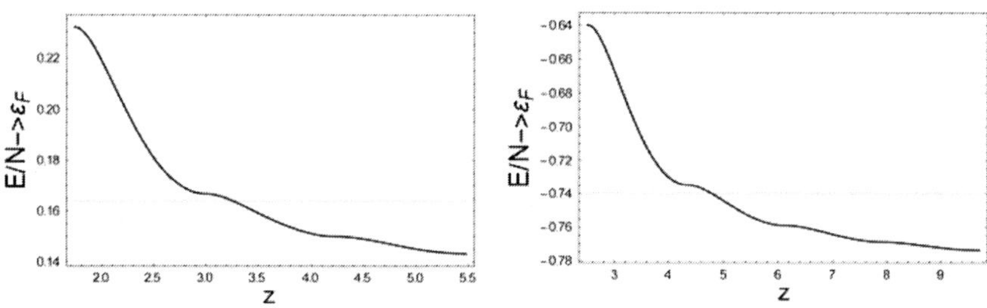

Figure 9. Total energy per particle (in units of Fermi energy in the absence

We should point out here that although the energy is periodic function with respect to inverse B (this is the de Haas-van Alphen effect [8] —see how this can be derived at T = 0 in various cases from rather elementary considerations in ref. [9]), the windows of values of z (or $1/B$) seem to be influenced by the presence of the electric field, namely $\delta(1/B) = \hbar/mcL_yE$ or, by using $y = eEL_y/\varepsilon_F$ we find that

$$\delta(1/B) = \hbar e/ymc\varepsilon_F = 1/yn_A\Phi_0 \tag{2.30}$$

which deviates from the standard semiclassical periodicities

$\delta(1/B) = 1/n_A\Phi_0$ for a 2D system [9] .

In addition, it should be pointed out that the correctness of our results is witnessed by the fact that the total energy turns out to be a smooth (continuous and differentiable) function of z for every z (i.e. the positions of the windows match with the corresponding expressions)—this smoothness of the figures (in their joining through different window-values) strictly testifying for the analytical correctness of the overall expressions that we derived for the total energy.

2.2.3. Hall Conductivity

The Hall conductivity is defined as:

$$\sigma_H = en_A c/B \tag{2.31}$$

and is usually plotted as a function of magnetic field B. For convenience we will make here use of Equation (2.26) and examine σ_H as a function of electron number N instead of B. Starting therefore from

$$\frac{B^2 L_x}{Emc^2 4\pi} e\left(\rho - i_F^L\right)\left[\rho - 1 - i_F^L\right] + \frac{\Phi}{\Phi_0} i_F^L$$

$$\leq N \leq \frac{B^2 e L_x}{4\pi E m c^2}\left(\rho - i_F^R\right)\left[\rho - 1 - i_F^R\right] + \frac{\Phi}{\Phi_0}\left[i_F^R + \frac{\left(\rho - i_F^R\right)}{z}\right] \tag{2.32}$$

we will show that by varying N, fractional filling factors appear, with no interactions and no impurities taken into account. E and B fields are considered as constants (and therefore so is z) throughout all N (and ρ) variations. Substituting Equation (2.16) into Equation (2.32) we may eliminate E:

$$\frac{\Phi}{\Phi_0}\left[\frac{1}{2z}\left(\rho - i_F^L\right)\left(\rho - i_F^L - 1\right) + i_F^L\right] \leq N \leq \frac{\Phi}{\Phi_0}\left[\frac{1}{2z}\left(\rho - i_F^R\right)\left(\rho - i_F^R + 1\right) + i_F^R\right]$$

Because for all numbers ρ and i_F^R the two edges of (2.17) differ by $\frac{\Phi}{\Phi_0}$, we choose to calculate σ_H only at the special values of electron number N given by the maximum edge of (2.32):

$$N = \frac{\Phi}{\Phi_0}\left[\frac{1}{2z}\left(\rho - i_F^R\right)\left(\rho - i_F^R + 1\right) + i_F^R\right],$$

and σ_H is then:

$$\sigma_H = \frac{en_A c}{B} = \frac{e^2}{h}\left[\frac{1}{2z}\left(\rho - i_F^R\right)\left(\rho - i_F^R + 1\right) + i_F^R\right]$$

with $i_F^R(\rho, z) = \sum_{n=0}^{+\infty} \theta[\rho - z - n]$

Let's examine now some cases regarding z-values:

i) z is an integer:

1) $z = 1 \Rightarrow i_F^R = \rho : \sigma_H = \frac{\rho e^2}{h}$

2) $z > \rho \Rightarrow i_F^R = 0 : \sigma_H = \frac{e^2}{h}\left[\frac{1}{2z}\rho(\rho + 1)\right] = \frac{e^2}{h}\frac{1}{z}, \frac{e^2}{h}\frac{3}{z}, \frac{e^2}{h}\frac{6}{z}, \frac{e^2}{h}\frac{10}{z}\cdots$

3) $z = \rho \Rightarrow i_F^R = 1 : \sigma_H = \frac{e^2}{h}\left[\frac{1}{2}(z-1) + 1\right].$

4) $\sigma_H = \frac{e^2}{h}\left[\rho - \frac{1}{2}(z-1)\right] = \frac{e^2}{2h}[z+3], \frac{e^2}{2h}[z+5], \frac{e^2}{2h}[z+7]\cdots$

$z < \rho \Rightarrow i_F^R = \rho - z + 1 : \sigma_H = \frac{e^2}{h}\left[\frac{1}{2}(z-1) + 1\right].$

More compactly, for a fixed integer z, varying ρ will result in the following values for σ_H:

$$\sigma_H = \underbrace{\frac{e^2}{h}\frac{1}{z}, \frac{e^2}{h}\frac{3}{z}, \frac{e^2}{h}\frac{6}{z}, \frac{e^2}{h}\frac{10}{z}\cdots}_{z-1\text{Terms}} \frac{e^2}{h}\left[\frac{1}{2}(z-1) + 1\right], \frac{e^2}{2h}[z+3], \frac{e^2}{2h}[z+5], \frac{e^2}{2h}[z+7]\cdots \tag{2.33}$$

where, for $z = 1$ we get the known results: $\sigma_H = \frac{e^2}{h}, \frac{2e^2}{h}, \frac{3e^2}{h}, \frac{4e^2}{h}\cdots$

ii) z is not an integer:

For a non-integer z we have:

$$\sigma_H = \underbrace{\frac{e^2}{zh},\frac{3e^2}{zh},\frac{6e^2}{zh},\frac{10e^2}{zh}}_{j\,\text{Terms}},\cdots\frac{e^2}{h}\left[1+\frac{j}{2z}(j+1)\right],\frac{e^2}{h}\left[2+\frac{j}{2z}(j+1)\right]\cdots \quad ,(2.34)$$

where $j = \text{Int}(z)$.

We re-emphasize here that the above fractional values in Hall conductivity do not result from any interactions (i.e. they are not related to the Physics of the Fractional Quantum Hall Effect) but are a genuine consequence of a high E-field strength that causes the L.L overlaps.

2.2.4. Plateaux Formation

When the external E field is relatively small, i.e. given by Equation (2.3), the Hall conductance has a plateau-like structure. To see this, imagine that we vary in a continuous matter the particle number, keeping B fixed. As a result, the Fermi energy is continuously moved from the beginning of a Landau Level to the right end, after passing through all states in that L.L. And when Fermi energy is in the interior of a L.L., Hall conductivity remains a constant and it only changes at the critical N values ($N = \rho\Phi/\Phi_0$), namely when Fermi energy has a transition between adjacent L.Ls.

In the strong E field case however, this is not the case. When Fermi energy lies in the interior of the ρth L.L, plateaux are destroyed, because the energy spectrum is no longer discretized. It is as if one has a single L.L. band (all L.Ls form a single band), with an infinite number of neighboring states to accommodate all electrons (a metallic character) and Hall conductivity is more likely to behave as in the classical case, with an eternal increasing (or decreasing) character by varying either N or B.

Inclusion of impurity potentials [10] however may change this expectation in the sense that, there might be a case where Fermi energy will lie in a mobility gap (which is the true criterion for the existence of QHE), and some of the extended states per L.L might be occupied, resulting in a blurred quantization of Hall conductance. This is a matter that needs to be investigated more carefully. In this case, the previous σ_H values will correspond to certain plateaux by varying B or N.

2.2.5. Inclusion of Edge States

In real-life samples, however, electronic systems are confined, and the confinement and impurity potential must be seriously taken into account. QHE manifestation is accomplished by considering both effects on transport properties (the confinement and disorder). In this case, a thermodynamic approach may be of low importance and insufficient to explain the robust quantization of Hall conductivity. Here, we will try to visualize a special case, where the electric field is strong enough, so that it causes an inter-L.L overlap and the system is at the same time confined in y-direction. The key to solve this problem lies in the range of influence of the confinement potential. As one starts from $y=0$, and moving towards lets say, the right end ($L_y/2$) he will experience the confinement potential influence only on a certain distance from the end (a few multiples of magnetic length l_B). From that point on, L.Ls will start to rise abruptly.

Now, if the electric field is low enough (Figure 10(a)) there is no inter-L.L overlap and the Fermi energy will always intersect the same number of edge states at the left end and right end respectively. But in the case of a strong E field (Figure 10(b)), things may change dramatically; if (for a certain E field) the overlap starts at a point before the confinement potential influence, there might be a case where the Fermi energy intersects two points (two L.Ls) at the left end and only one point (L.L) at the right end (where the edge states are located). This is possible only if there is an additional intersection in the bulk of a L.L. (see

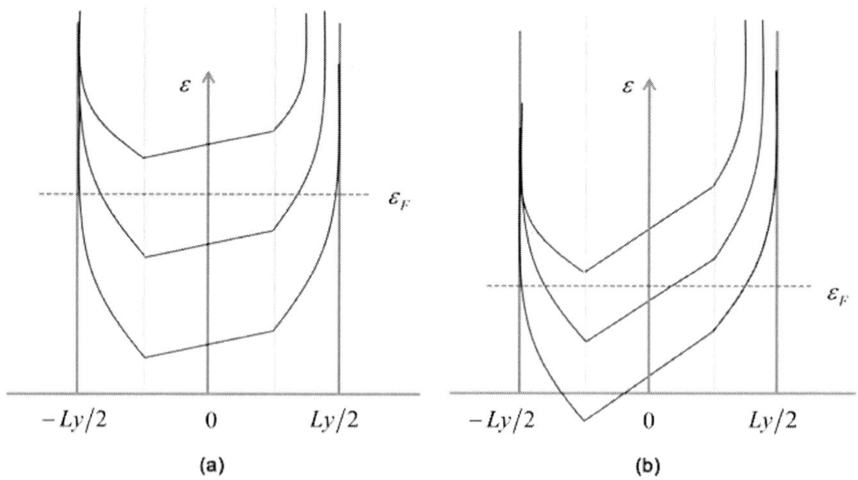

Figure 10. Impact of the edges on Landau Levels. (a) low-E field, (b) strong-E field case.

the middle L.L at Figure 10(b)). Varying then the Fermi energy will not change the number of intersections, until the Fermi energy reaches the bottom of the top L.L. and so on. In other words, the occupations of edge states will remain the same, until another L.L will be occupied. Therefore, there might be some robustness of the Hall values (found above) before this crossover occurs.

3. GENERALIZATION TO RELATIVISTIC SYSTEMS

3.1. Graphene—Low E-Field Strength

Our results can also be applied to Relativistic solid state systems that obey a Dirac-type of equation, such as Graphene [11] and topological insulators [7] . In this Section we give a first analytical study of strong electric field effects in the ground state energetics and also in transport properties of Graphene, i.e. Hall conductivity. We find again that, when the E-field is strong enough, L.Ls with different quantum numbers overlap, and the Hall conductivity becomes electric and magnetic field-dependent, indicating the non-linearity of Hall conductivity in relativistic systems, this having more refined consequences compared to the previous nonrelativistic case; we obtain, for example, the possibility of irrational values (the observability of which is of course up to the uncertainties related to the lack of disorder, already mentioned in the last section).

We start with Graphene's energy spectrum when placed in homogeneous crossed electric $\boldsymbol{E} = E\hat{x}$ and magnetic $\boldsymbol{B} = B\hat{z}$ fields, considering only the po- sitive branch [12] (we continue in this section with SI units):

$$\varepsilon_{n,k_y} = \sqrt{2n\hbar eB}u_F\left(1-\beta^2\right)^{3/4} + \hbar u_F \beta k_y \quad , (3.1)$$

where $\beta = E/u_F B$ is a dimensionless parameter smaller than unity and

$$X_0 = l_B^2 k_y - \operatorname{sgn}(n)\frac{\sqrt{2|n|}l_B\beta}{\left(1-\beta^2\right)^{1/4}} \quad (3.2)$$

is the guiding center eigenvalue, and $l_B = \sqrt{\hbar/eB}$ is the magnetic length. Each L.L (indexed by n) has Φ/Φ_0 available states (independent of E strength) and may host up to $4\Phi/\Phi_0$ electrons, due to spin (a factor of two) and valley degeneracy (another factor of two) except for n = 0 L.L., which has capacity $2\Phi/\Phi_0$, because it is equally shared with holes.

Using (3.2) we may rewrite (3.1) as:

$$\varepsilon_{n,k_y} = \frac{\sqrt{2n\hbar eBu_F}}{\left(1-\beta^2\right)^{1/4}} + eEX_0,\ n = 0,1,2\cdots \tag{3.3}$$

with the restriction $-L_x/2 \le X_0 \le L_x/2$. Using (3.3), we determine the inter-L.L. energy gap as:

$$\delta\varepsilon_{n,k_y} = \frac{\sqrt{2\hbar eBu_F}}{\left(1-\beta^2\right)^{1/4}}\left[\sqrt{n+1}-\sqrt{n}\right] \tag{3.4}$$

This energy gap depends on E-field, in contrast to conventional semiconductor systems (of last section); in addition, the greater L.L. index, the smaller the gap. This means that in some L.L. range there will be unavoidable overlap accompanied with a zero energy gap (and the system is then metallic). This is exactly the case we will examine later. Now, for adjacent L.Ls, no overlap is observed only when the following condition holds:

$$\varepsilon\left(n, L_x/2\right) \le \varepsilon\left(n+1, -L_x/2\right) \tag{3.5}$$

which leads to:

$$\frac{\sqrt{2e\hbar Bu_F}}{\left(1-\beta^2\right)^{1/4}}\left[\sqrt{n+1}-\sqrt{n}\right] \ge eEL_x \tag{3.6}$$

For small enough indexes n, and small enough E-fields, inequality (3.6) will be satisfied. But, for a given E-field, there will be a critical L.L. index i_F above which (i.e. $n = i_F + 1$) the inequality (3.6) changes sign:

$$\frac{\sqrt{2e\hbar Bu_F}}{\left(1-\beta^2\right)^{1/4}}\left[\sqrt{i_F+2}-\sqrt{i_F+1}\right] \le eEL_x \tag{3.7}$$

This i_F describes the topmost L.L. that does not overlap with other L.Ls, while all remaining L.Ls indexed by $n = i_F + 1,\ i_F + 2\cdots$ become continuous, without any energy gap and with always available neighboring states for an electron to be scattered in. It is useful to define a new continuous number z, which obeys:

$$\sqrt{i_F+1}-\sqrt{i_F} \geq z \geq \sqrt{i_F+2}-\sqrt{i_F+1} \tag{3.8}$$

This number, when put in (3.7), results in the following equality:

$$\frac{\sqrt{2e\hbar B}}{\left(1-\beta^2\right)^{1/4}} u_F z = eEL_x \tag{3.9}$$

which will be used later below. We now proceed to our calculation of thermodynamic properties by considering that Fermi energy is always located at an L.L. whose index ($n=\rho-1$, with $\rho=1,2,3\cdots$) always obeys Equation (3.6), i.e. $\rho-1\leq i_F$. In this manner, all L.Ls are occupied independently, and when we have full occupations, Fermi energy will surely lie in an energy gap.

We make the assumption that electron number is such that ρ L.Ls are occupied, with the last L.L. ($n=\rho-1$) partially occupied. Extra care must be taken in the counting of states, recalling the special case of the lowest L.L. $n=0$, which has capacity $2\Phi/\Phi_0$ electrons, while all other L.Ls may accommodate up to $4\Phi/\Phi_0$ electrons. Considering this, we find that the electron number lies in the following window of values (for a constant electromagnetic field):

$$(2\rho-3)\frac{2\Phi}{\Phi_0} \leq N \leq (2\rho-1)2\frac{\Phi}{\Phi_0} \tag{3.10}$$

(Note that when $\rho=1$, Equation (3.10) becomes: $0\leq N\leq 2\Phi/\Phi_0$). For a constant electron number N, we can solve (3.10) with respect to B:

$$\frac{1}{2(2\rho-3)}n_A\Phi_0 \geq B \geq \frac{1}{2(2\rho-1)}n_A\Phi_0 \tag{3.11}$$

What this inequality means is that when B lies in the above range, the first $\rho-1$ ($n=0,1,2\cdots\rho-2$) L.Ls are fully occupied and the final L.L. ($n=\rho-1$) is partially occupied (it is filled with $N-(2\rho-3)2\Phi/\Phi_0$

electrons). Equation (3.11) also describes the well-known unconventional Hall effect in graphene:

$$\sigma_H = \frac{n_A e}{B} = (2\rho-1)2\frac{\Phi}{L_x L_y \Phi_0 B}e = (2\rho-1)2\frac{e^2}{h} = \left(\rho-\frac{1}{2}\right)\frac{4e^2}{h} \tag{3.12}$$

where the Hall conductivity is quantized in half integer multiples of the quantity $4e^2/h$.

We now calculate the total internal energy of the system, which has to be minimalized when the temperature is zero ($T=0$):

$$E = \sum_{n,X_0} \varepsilon_{n,k_y} = \sum_{n,X_0} \frac{\sqrt{2n\hbar eB}u_F}{\left(1-\beta^2\right)^{1/4}} + eEX_0 \tag{3.13}$$

Considering the following transformations:

$$\sum_{X_0} \to \frac{4BL_y}{\Phi_0}\int_{-L_x/2}^{L_x/2} dX_0 \text{ for } n>0 \text{ and } \sum_{X_0} \to \frac{2BL_y}{\Phi_0}\int_{-L_x/2}^{L_x/2} dX_0 \text{ for } n=0 \quad \text{L.L. (3.14)}$$

we can write:

$$E = E_{\text{FULL}} + E_{\text{PART}}, \tag{3.15}$$

where E_{FULL} is the energy of the fully occupied L.Ls and E_{PART} is the energy of the last partially occupied L.L.

$$E_{\text{FULL}} = \frac{4BL_y}{\Phi_0}\sum_{n=1}^{\rho-2}\int_{-L_x/2}^{L_x/2} dX_0 \frac{\sqrt{2n\hbar eB}u_F}{\left(1-\beta^2\right)^{1/4}} + \frac{4BL_y eE}{\Phi_0}\sum_{n=1}^{\rho-2}\int_{-L_x/2}^{L_x/2} dX_0 X_0$$
$$= \frac{4\Phi}{\Phi_0}\frac{\sqrt{2\hbar eB}u_F}{\left(1-\beta^2\right)^{1/4}}\sum_{n=0}^{\rho-2}\sqrt{n} \tag{3.16}$$

To calculate E_{PART}, it is necessary to determine the last electron's guiding center position in L.L $n=\rho-1$.

$$X_0 = l_B^2\frac{2\pi l}{L_y} - \frac{\sqrt{2(\rho-1)}l_B\beta}{\left(1-\beta^2\right)^{1/4}} \tag{3.17}$$

The index l, appearing in k_y, has a starting value l_0 that needs to be determined by the condition $X_0 = -L_x/2$, (left edge):

$$-\frac{\Phi}{2\Phi_0} + \frac{L_y}{2\pi l_B}\frac{\sqrt{2(\rho-1)}\beta}{\left(1-\beta^2\right)^{1/4}} = l_0 \tag{3.18}$$

But the last L.L hosts $\left[N-(2\rho-3)\frac{2\Phi}{\Phi_0}\right]/4$ states for the remaining electrons to be placed inside, so we conclude that the last electron in $n=\rho-1$ L.L. has a guiding center position:

$$X_{0\text{MAX}} = l_B^2\frac{2\pi}{L_y}\left[\frac{N}{4} - \rho\frac{\Phi}{\Phi_0} + \frac{\Phi}{\Phi_0}\right] \tag{3.19}$$

For consistency reasons we check Equation (3.19) for certain values of particle number, i.e. when $N=(2\rho-3)2\Phi/\Phi_0$ then $X_0 = -L_x/2$ and when $N=(2\rho-1)2\Phi/\Phi_0$ we have $X_0 = L_x/2$. So far we are correct.

The energy of the electrons located at the Fermi L.L. E_{PART} is:

$$\begin{aligned}E_{\text{PART}} &= \frac{4BL_y}{\Phi_0}\int_{-L_x/2}^{X_{0\text{MAX}}} \mathrm{d}X_0 \frac{\sqrt{2(\rho-1)\hbar eBu_F}}{\left(1-\beta^2\right)^{1/4}} + eE\frac{4BL_y}{\Phi_0}\int_{-L_x/2}^{X_{0\text{MAX}}} \mathrm{d}X_0 X_0\\ &= \frac{4BL_y}{\Phi_0}\frac{\sqrt{2(\rho-1)\hbar eBu_F}}{\left(1-\beta^2\right)^{1/4}}\left[X_{0\text{MAX}} + L_x/2\right] + eE\frac{4BL_y}{\Phi_0}\left[\frac{X_{0\text{MAX}}^2}{2} - \frac{L_x^2}{4}\right]\end{aligned} \tag{3.20}$$

Substituting Equation (3.19) in Equation (3.20) we get:

$$\begin{aligned}E_{\text{PART}} &= \frac{\sqrt{2(\rho-1)\hbar eBu_F}}{\left(1-\beta^2\right)^{1/4}}\left[N - 2(2\rho-3)\frac{\Phi}{\Phi_0}\right]\\ &\quad + \frac{hN^2E}{8BL_y} - eEN(\rho-1)L_x + e^2E\frac{2BL_y}{h}L_x^2\left[\rho^2 - 2\rho + \frac{1}{2}\right]\end{aligned} \tag{3.21}$$

From Equation (3.16) and (3.21) we obtain the final result for the total energy per electron of the system:

$$\frac{E_{TOT}}{N} = \frac{\varepsilon_F}{\left(1-\beta^2\right)^{1/4}}\left[4\left(\frac{B}{n_A\Phi_0}\right)^{3/2}\left(2\sum_{n=1}^{\rho-2}\sqrt{n}-(2\rho-3)\sqrt{(\rho-1)}\right)+2\left(\frac{B}{n_A\Phi_0}\right)^{1/2}\sqrt{(\rho-1)}\right]$$
$$+\frac{n_A\Phi_0 eEL_x}{8B}-eE(\rho-1)L_x+eE\frac{2BL_x}{n_A\Phi_0}\left[\rho^2-2\rho+\frac{1}{2}\right] \quad (3.22)$$

with $\sum_{n=1}^{\rho-2}\sqrt{n}=-\zeta\left(-\frac{1}{2},\rho-1\right)-\frac{\zeta\left(\frac{3}{2}\right)}{4\pi}$, where $\zeta(-1/2,\rho-1)$ is the Hurwitz zeta function and $\zeta(3/2)$ is the Riemann function. Note that the term proportional to the E field appearing in (3.22) when expressed per electron,

$$e^2E\frac{2B}{n_Ah}L_x\left[\rho^2-2\rho+\frac{1}{2}\right]+\frac{hNE}{8BL_y}, \quad (3.23)$$

gives the following magnetization (which is also proportional to E):

$$-e^2E\frac{2}{n_Ah}L_x\left[\rho^2-2\rho+\frac{1}{2}\right]+\frac{hNE}{8B^2L_y} \quad (3.24)$$

When all ρ L.Ls are fully occupied, then $B=n_A\Phi_0/2(2\rho-1)$, and this term becomes:

$$\frac{L_x}{2n_A}\left([4\rho-1]\frac{e^2}{h}\right)E \,\square\, \frac{L_x}{2n_A}\frac{\sigma_H}{2}E \quad (3.25)$$

which is in accordance to the non-relativistic case (see corresponding term of (1.11)). Although this $\sigma_H=[4\rho-1]e^2/h$ differs from Equation (3.12) where $\sigma_H=(2\rho-1)2e^2/h$ this point needing further clarification. Let us then check some limits: When the electric field is absent, E = 0, the total energy per electron becomes (in units of Fermi energy in the absence of B, $\varepsilon_F=\hbar u_F\sqrt{\pi n_A}$):

$$\frac{E_{\text{TOT}}}{N}=\varepsilon_F\left[4\left(\frac{B}{n_A\Phi_0}\right)^{3/2}\left[2\sum_{n=1}^{\rho-2}\sqrt{n}-(2\rho-3)\sqrt{(\rho-1)}\right]+2\left(\frac{B}{n_A\Phi_0}\right)^{1/2}\sqrt{(\rho-1)}\right] \tag{3.26}$$

This is the result we would have gotten had we solved the problem from the beginning without the E-field.

When B is at a critical value, $B=n_A\Phi_0/2(2\rho-1)$, (all ρ L.Ls fully occupied) then:

$$\begin{aligned}E_{\text{TOT}}&=2N\hbar u_F\frac{1}{(2\rho-1)}\sqrt{\frac{n_A2\pi}{(2\rho-1)}}\left[\sum_{n=1}^{\rho-2}\sqrt{n}+\sqrt{(\rho-1)}\right]\\&=2N\hbar u_F\sqrt{\frac{n_A2\pi}{(2\rho-1)^3}}\sum_{n=1}^{\rho-2}\sqrt{n}+2N\hbar u_F\sqrt{\frac{n_A2\pi(\rho-1)}{(2\rho-1)^3}}\end{aligned} \tag{3.27}$$

If $\rho=1$ then $E_{\text{TOT}}=0$, as one would expected, because all electrons are placed at the lowest L.L. having zero energy.

If $\rho=2$ $E_{\text{TOT}}=\sqrt{2\hbar eB}u_F\left[4\frac{\Phi}{\Phi_0}\right]$ and so on. In all cases, we can clearly see the difference between the case of 2DEG in conventional semiconductors (note, for example, that at the critical values of B the total energy is no longer equal to the Fermi energy but keeps rising up—see Figure 11 and its differences with Figure 1).

3.2. Stronger E-Field

The inequality

$$\frac{\sqrt{2e\hbar B}u_F}{\left(1-\beta^2\right)^{1/4}}\left[\sqrt{i_F+1}-\sqrt{i_F}\right]\geq eEL_x$$

,

is the criterion where no overlap occurs, as long as $\rho-1\leq i_F$. Now, if we have 'more' electrons and have to place them in available states, we will finally reach

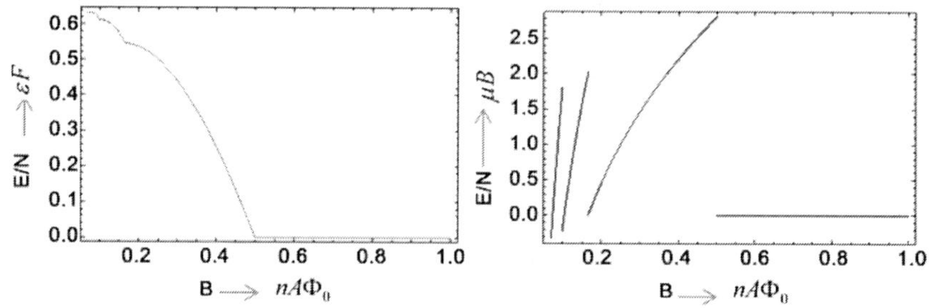

Figure 11. Energy in units of ε_F and Magnetization in units of Bohr magneton μ_B as functions of magnetic field in the absence of an electric field.

the condition $\rho > i_F + 1$. In this case, states indexed by $n = i_F + 2 \cdots \rho - 1$, overlap, and the energy gap closes at Fermi energy. This means that graphene gains a metallic E-field induced character and Hall conductivity is modified (analogous to the non-relativistic case) by a term containing both electric and magnetic field.

Recall from previous work [13] that the following relation also holds:

$$\beta < 1 \Rightarrow \frac{E}{u_F B} < 1 \Rightarrow E < u_F B$$

,

(so that no collapse of L.Ls occurs), which can always be made true as long as we treat particle number as our variable, and keep E and B-fields constant. Or, equivalently, we may keep N and B fixed and treat E as our main variable, starting from zero until reaching maximum value $E_{\text{MAX}} = u_F B$. We remind here the reader of the limitations of this problem regarding also the magnetic field, which has to be such that the magnetic length is much greater than lattice constant (i.e. we will not study any Hofstadter effects in this work). Either way, our results and predictions can be obtained by satisfying all possible limitations. For example, we may prepare an experiment where E is strong enough for inter L.L. overlap to occur, and at the same time B is also strong enough to satisfy the above inequality.

We examine now the case appearing in Figure 12. Landau levels n = 0, 1 and 2 can be occupied independently, while further occupations above $n = 2$ L.L. leads to the unavoidable mixing of states; this makes the problem interesting, in the sense that peculiarities may arise both in thermodynamic and transport properties. We fix Fermi energy ε_F at L.L $n = \rho - 1$:

$$\varepsilon_F = \frac{\sqrt{2(\rho-1)\hbar eBu_F}}{\left(1-\beta^2\right)^{1/4}} + eEX_{0M}$$

(3.28)

with X_{0M} the guiding center position of the last electron in the Fermi level $n=\rho-1$. We consider now the simplest case: From Figure 12, we will examine the case where all states are occupied up to the Fermi energy (Equation 3.28). The Fermi energy is located at n = 4 with $X_{0M}=-L_x/2$:

$$\varepsilon_F = \frac{\sqrt{8\hbar eBu_F}}{\left(1-\beta^2\right)^{1/4}} - \frac{eEL_x}{2}$$

(3.29)

shows ε_{n,k_y} vs X_0. The inter-L.L. energy gap gets smaller by increasing Landau Level index n. States n = 0,1 and 2 do not overlap with each other, but states $n=3,4,5\cdots$ and so on, do. If the Fermi energy is located at n = 0,

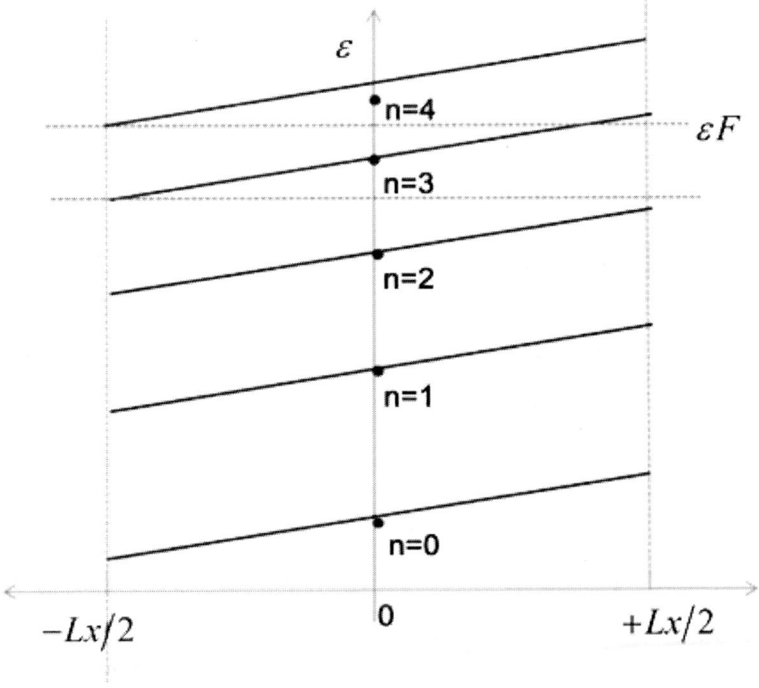

Figure 12. Schematic Representation of energy levels in graphene. Figure

1 or 2, we expect that Hall conductivity will still be quantized in the form $\sigma_H = (2\rho - 1)2e^2/h$, with ρ = 1, 2, and 3.

To determine the number of states in L.L. n = 3, we examine its intersection with Fermi energy (see Figure 12):

$$\frac{\sqrt{6\hbar eBu_F}}{\left(1-\beta^2\right)^{1/4}} + eEX_{0F} = \frac{\sqrt{8\hbar eBu_F}}{\left(1-\beta^2\right)^{1/4}} - \frac{eEL_x}{2}$$

$$\Rightarrow X_{0F} = -\frac{L_x}{2} + \frac{\sqrt{\hbar eBu_F}}{eE\left(1-\beta^2\right)^{1/4}}\left(\sqrt{8}-\sqrt{6}\right) \tag{3.30}$$

Now, we determine the starting (l_0) and final (l_F) point of l for this case:

$$l_0 = -\frac{\Phi}{2\Phi_0} + \frac{L_y}{2\pi l_B}\frac{\sqrt{6}\beta}{\left(1-\beta^2\right)^{1/4}} \tag{3.31}$$

$$X_0 = l_B^2\frac{2\pi l_F}{L_y} - \frac{\sqrt{6}l_B\beta}{\left(1-\beta^2\right)^{1/4}}$$

$$\Rightarrow l_F = -\frac{\Phi}{2\Phi_0} + \frac{L_y}{h}\frac{B\sqrt{\hbar eBu_F}}{E\left(1-\beta^2\right)^{1/4}}\left(\sqrt{8}-\sqrt{6}\right) + \frac{L_yE}{hu_F\sqrt{eB}}\frac{e\sqrt{6\hbar}}{\left(1-\beta^2\right)^{1/4}} \tag{3.32}$$

The number of states in L.L. n = 3 is given by:

$$l_F - l_0 = \frac{L_y}{h}\frac{B\sqrt{\hbar eBu_F}}{E\left(1-\beta^2\right)^{1/4}}\left(\sqrt{8}-\sqrt{6}\right), \tag{3.33}$$

and the total number of states under Fermi energy is:

$$g = \underbrace{\frac{\Phi}{\Phi_0}}_{n=0} + 2\frac{\Phi}{\Phi_0} + \frac{L_y}{h}\frac{B\sqrt{\hbar eBu_F}}{E\left(1-\beta^2\right)^{1/4}}\left(\sqrt{8}-\sqrt{6}\right) \tag{3.34}$$

If all states are filled with electrons, then the electron number is (considering that we have four electrons in each case, except for n = 0, in which we place two electrons in each state):

$$N = 10\frac{\Phi}{\Phi_0} + 4\frac{L_y}{h}\frac{B\sqrt{\hbar eBu_F}}{E\left(1-\beta^2\right)^{1/4}}\left(\sqrt{8}-\sqrt{6}\right) \tag{3.35}$$

This relation modifies Hall conductivity as:

$$\sigma_H = \frac{n_A e}{B} = N\frac{e}{SB} = 10\frac{e^2}{h} + 4\frac{e}{hL_x}\frac{\sqrt{\hbar eBu_F}}{E\left(1-\beta^2\right)^{1/4}}\left(\sqrt{8}-\sqrt{6}\right) \tag{3.36}$$

Now, by directly using Equation (3.9) and substituting in (3.36):

$$\frac{\sqrt{2e\hbar Bu_F}}{\left(1-\beta^2\right)^{1/4}}z = eEL_x \Rightarrow E = \frac{\sqrt{2e\hbar Bu_F}}{eL_x\left(1-\beta^2\right)^{1/4}}z \tag{3.37}$$

we conclude to:

$$\sigma_H = \frac{n_A e}{B} = N\frac{e}{SB} = 10\frac{e^2}{h} + 4\frac{e^2}{zh}\left(\sqrt{4}-\sqrt{3}\right) \tag{3.38}$$

demonstrating the possibility of irrational quantization of the Hall conductance (always if the E-field is sufficiently strong).

4. CONSIDERING THE QHE-BREAKDOWN IN HIGH INJECTED CURRENTS

At this point one may wonder about the relevance of the above results, as these do not seem to take into account mechanisms associated with the breakdown of QHE under non-equilibrium conditions, i.e. due to high injected current densities [14] [15] [16] . Indeed, in the case of hundreds of μA (0.6 - 0.9 mA, taken from ref. [16] , the exact value depending on the filling factor) flowing through the sample in the y-direction, the longitudinal resistivity is different from zero by orders of magnitude, resulting in an inevitable dissipation of energy, and the Hall conductivity deviates from its exact quantized values.

One of the reasons behind this deviation is the thermal instability that happens due to large current densities flowing continuously for a period of time through the material. The amount of heat gained by the electrons per unit time per unit area due to the electric field is given by

$$Q_{in} = \sigma_{yy} E^2$$

This amount of heat is then passed to the lattice structure, ending up in the Helium coolant. This means that there is a gradient of temperature between the electrons and the atoms forming the crystal that relates the heat transferred to the atoms through the following equation:

$$Q_{out} = k(T_e - T_h)$$

When the electric field exceeds a critical value, the amount of heat gained by the electrons (Q_{in}) is always greater than the amount of heat transferred to the atoms (Q_{out}) and the excess in thermal energy results in an energy excitation in the order of the cyclotronic energy, destroying the stability of the quantum Hall conductivity plateaus.

It should be noted that the present work takes into account the applied E-field directly within the energy spectrum, and there is no need to investigate the heating of the electron gas separately. It is contained a priori in the thermodynamic formulation that we have used to derive our results. Furthermore, the results shown in this paper can be relevant at even relatively low E-fields, or low injected current densities as well. To see this, we may assign some numbers to our mathematical relations, starting from Equation (2.1) that gives the limit of strong E-field: $E > \hbar\omega / eL_y$ for $z > 1$. For a strong magnetic field of 40 Tesla, and a macroscopic length L_y of 1 cm, we have:

$$E = \hbar\omega / eL_y = \hbar B / mL_y = 1.054 \times 40/91 = 0.46 \text{ V/m}.$$

Above this critical E-field (and therefore for low Hall E-field) all the inter-L.L. overlaps that we considered can occur and the need for introducing breakdown effects is therefore eliminated. In the non-equilibrium breakdown regime of the QHE state in Graphene, it has been verified that not all L.Ls experience the breakdown [17] . For example, the $n = 0$ L.L. remains robust under the influence of a relatively strong E-field. Therefore, the breakdown in Graphene is a filling factor-dependent phenomenon, with different critical electric fields for each filling factor in a way that we may always impose a low enough E-field to achieve the results derived in this paper. The stronger the B-field is, the weaker the E-field gets, much lower than the breakdown limit.

Apart from the above mentioned electron-heating mechanism we point out, for completeness, that percolation of incompressible regions [18] has also been invoked for the QHE-breakdown, as well as the possible existence of compressible regions in the bulk [19] , which however do not seem to offer insights relevant to the present calculation.

Finally, there exists a small number of publications in the literature that seem to be close to our exact solution (although they deviate from it); they also find fractional quantization of the Hall conductivity by seriously taking into account a conceptual difference between the externally applied E-field and the Hall-electric field (the one that is formed in the (classical) steady state, as opposed to the field being applied). They take into account the non-negligible effect of the Hall E-field on the electronic density of states, which is further broadened by the presence of this Hall E-field, see i.e. ref. [20] . Just like interactions lead to a splitting of L.Ls (as is well-known in the Fractional QHE), the Hall electric field succeeds in a similar manner to divide each Landau Level into many sub-levels without any introduction of impurities or interactions. Fractional quantization is then observed in higher L.Ls (due to the density of states-broadening), but not in the lowest L.L. (where interactions are dominant, and a Composite Fermion approach—hence a passage to the so-called Λ-Levels— must be used instead (see [9] for a practical use of Λ-Levels)). However, strong anisotropies have been observed at these fractions [17] which suggests that they need to be considered more carefully as functions of the Hall E-field.

By way of comparison with the above, we must state that in the present work we have avoided the notion of a Hall electric field, as this is a built-in consequence of the presence of the B-field in the (classical) steady state (or in a stationary state in the corresponding quantum problem), or, alternatively, as this can be viewed as a response of the system to an external voltage (or the externally applied E-field) which is already fully taken into account in our physical picture (Hamiltonian) and in our associated calculations. We use, instead, the "physical broadening" given by the presence of the external E-field in order to describe the fractional quantization. Furthermore, with respect to the possible robustness of our results (such as Equation (2.31)) for a conventional system, that gives the Hall conductivity value for a certain magnetic field and fixed particle number), we recall that topological protection and exact quantization of σ_H in a high E-field is not guaranteed for our system in the thermodynamic limit. This is better clarified at least in a visual manner in Figure 10(a) and Figure 10(b) (where the number of edge states plays a role), but it further needs inclusion of other factors not taken into account here, like the impurity potential; this, however, is actually expected to have a positive effect—it is expected to eventually broaden the electronic density of states, and, at least, assist in establishing the topological stability of the system (in a similar manner as this occurs in the ordinary Integer-QHE).

5. CONCLUSION

We have shown that by finding the optimal energy at zero temperature in the case of conventional semiconductors, fractional Hall values appear at the points of jumps of the Fermi energy. This does not necessarily mean that plateaux appear. Inclusion of impurity potential may lead to quantized plateau structure even in the case of strong E field, due to further broadening of L.Ls, isolating extended from localized states. We should point out that it is not the gap closing that causes plateau disappearance, but rather the fact that we have ignored the impurity potential in our calculations. We have also calculated analytically, using ground state energy considerations the total internal energy, magnetization and polarization as functions of the electromagnetic field. The associated de Haas- van Alphen oscillation periods are also influenced by the presence of the electric field in a specific quantitative manner. A corresponding exact calculation in a pseudo-relativistic system, such as Graphene, is more involved but it has also been carried out here in detail. An immediate result of our toy model is the possibility of irrational Hall values, although further investigation is required (i.e. inclusion of disorder) in order to see if these effects survive under realistic conditions.

CITE THIS PAPER

Konstantinou, G. and Moulopoulos, K. (2017) Ground State Thermodynamic and Response Properties of Electron Gas in a Strong Magnetic and Electric Field: Exact Analytical Solutions for a Conventional Semiconductor and for Graphene. Journal of Applied Mathematics and Physics, 5, 639-666.

REFERENCES

1. Yoshioka, D. (2002) The Quantum Hall Effect. Springer, Berlin, 28-29.
2. Moulopoulos, K. (2015) Gauge Nonlocality in Planar Quantum-Coherent Systems. Quantum Matter, 4, 577-588.
3. Prange, R.E. and Girvin, S.M. (1990) The Quantum Hall Effect. Springer, Berlin.

4. Laughlin, R.B. (1981) Quantized Hall Conductivity on Two Dimensions. Physical Review B, 23, 5632.
5. Thouless, D.J. (1983) Quantization of Particle Transport. Physical Review B, 27, 6083.
6. De Nittis, G. and Schulz-Baldes, H. (2016) Spectral Flows Associated to Flux Tubes. Annales Henri Poincaré, 17, 1-35.
7. Hasan, M.Z. and Kane, C.L. (2011) Colloquium: Topological Insulators. Reviews of Modern Physics, 82, 3045. Qi, X.-L. and Zhang, S.-C. (2011) Topological Insulators and Superconductors. Reviews of Modern Physics, 83, 1057.
8. Shoenberg, D. (1984) Magnetic Oscillations in Metals. Cambridge University Press, Cambridge.
9. Konstantinou, G. and Moulopoulos, K. (2013) Thickness-Induced Violation of de Haasvan Alphen Effect through Exact Analytical Solutions at a One-Electron and a One-Composite Fermion Level. European Physical Journal B, 86, 326.
10. Halperin, B.I. (1982) Quantized Hall Conductance, Current-Carrying Edge States, and the Existence of Extended States in a Two-Dimensional Disordered Potential. Physical Review B, 25, 2185.
11. Katsnelson, M.I. (2012) Graphene: Carbon in Two Dimensions. Cambridge University Press, Cambridge.
12. Peres, N.M.R. and Castro, E.V. (2007) Algebraic Solution of a Graphene Layer in a Transverse Electric and Perpendicular Magnetic Fields. Journal of Physics: Condensed Matter, 19, Article ID: 406231.
13. Lukose, V., Shankar, R. and Baskaran, G. (2007) Novel Electric Field Effects on Landau Levels in Graphene. Physical Review Letters, 98, Article ID: 116802.
14. Kawaji, S., Hirakawa, K. and Nagata, M. (1993) Device-Width Dependence of Plateau Width in Quantum Hall States. Physica B, 184, 17-20.
15. Nachtwei, G. (1999) Breakdown of the Quantum Hall Effect. Physica E, 4, 79.
16. Ebert, G., et al. (1983) Two-Dimensional Magneto-Quantum Transport in GaAs/ AlGaAs Heterostructures under Non-Ohmic Conditions. Journal of Physics C, 16, 5441.
17. Lilly, M.P., Cooper, K.B., Eisenstein, J.P., Pfeiffer, L.N. and West, K.W. (1999) Evidence of an Anisotropic State of Two-Dimensional Electrons in High Landau Levels. Physical Review Letters, 82, 394.
18. Kazarinov, R.F. and Luryi, S. (1982) Quantum Percolation and Quantization of Hall Resistance in Two-Dimensional Electron Gas. Physical Review B, 25, 7626.
19. Tsemekhman, V., et al. (1997) Theory of the Breakdown of the Quantum Hall Effect. Physical Review B, 55, R10201.
20. Kramer, T. (2006) Landau Level Broadening without Disorder, Non-Integer Plateaus without Interactions—An Alternative Model of the Quantum Hall Effect. Revista Mexicana de Fisica, 52, 49.

CHAPTER 10

Natural Gas at Thermodynamic Equilibrium Implications for The Origin of Natural Gas

Frank D Mango*[1], Daniel Jarvie[2] and Eleanor Herriman[1]

[1]Petroleum Habitats, 806 Soboda Ct., Houston, Texas 77079, USA and

[2]Worldwide Geochemistry, 218 Higgins Street, Humble, Texas 77338, USA

ABSTRACT

It is broadly accepted that so-called 'thermal' gas is the product of thermal cracking, 'primary' thermal gas from kerogen cracking, and 'secondary' thermal gas from oil cracking. Since thermal cracking of hydrocarbons does not generate products at equilibrium and thermal stress should not bring them to equilibrium over geologic time, we would not expect methane, ethane, and propane to be at equilibrium in subsurface deposits. Here we report compelling evidence of natural gas at thermodynamic equilibrium. Molecular compositions are constrained to equilibrium,

$$K = [(CH_4) * (C_3H_8)]/[(C_2H_6)^2]$$

and isotopic compositions are also under equilibrium constraints:

$$K_{1,2} = [(^{13}C_2) * (^{12}C_1)]/[(^{13}C_1) * (^{12}C_2)]$$

$$K_{1,3} = [(^{13}C_3) * (^{12}C_1)]/[(^{13}C_1) * (^{12}C_3)]$$

The functions $[(CH_4)*(C_3H_8)]$ and $[(C_2H_6)^2]$ exhibit a strong nonlinear correlation ($R^2 = 0.84$) in which the quotient Q progresses to K as wet gas progresses to dry gas. There are striking similarities between natural gas and catalytic gas generated from marine shales. A Devonian/Mississippian New Albany shale generates gas with Q converging on K over time as wet gas progresses to dry gas at 200°C.

The position that thermal cracking is the primary source of natural gas is no longer tenable. It is challenged by its inability to explain the composition of natural gas, natural gases at thermodynamic equilibrium, and by the existence of a catalytic path to gas that better explains gas compositions.

BACKGROUND

The hydrocarbons in natural gas are believed to come from two sources, one biological ('biogenic gas'), and the other from thermal cracking, 'primary thermal gas' from kerogen cracking and 'secondary thermal gas' from oil cracking [1, 2]. Although there is general agreement on the source of biogenic gas, disagreement persists over the origin of thermal gas. One point of controversy is that thermal cracking does not produce a gas resembling natural gas. Oil and kerogen pyrolysis typically give between 10 and 60% wt methane (C_1–C_4) [3, 4, 5, 6, 7, 8, 9] while natural gas contains between 60 and 95+% methane [10]. None of the explanations that have been offered to explain this [11, 12, 13, 14, 15] are satisfactory [16]. Catalysis by reduced transition metals can, in theory, explain high-methane in natural gas [17], and this hypothesis is supported by experimental results. Crude oils and *n*-alkanes decomposed over reduced nickel and cobalt oxides produce gas resembling natural gas in molecular and isotopic compositions [18]. There is, however, no evidence of metal activity in sedimentary rocks and therefore no compelling reason to question thermal cracking theory. Moreover, recent hydrous pyrolysis experiments with metal-rich Permian Kupferschiefer shale showed little evidence of catalytic activity, seemingly dismissing the possibility of a catalytic path to natural gas [19].

This changed with the recent disclosure of natural catalytic activity in marine shales at temperatures 300° below thermal cracking temperatures [20]. Shales generated gas under anoxic gas flow at 50°C, nearly five times more gas than the same shale would generate at 350°C through thermal cracking. Although there was only indirect evidence for transition metals as the active catalysts, it nevertheless established natural catalytic activity in source rocks believed to be major sources of natural gas. There are, therefore, two possible paths to natural gas, a thermogenic path operating almost exclusively at high temperatures, and a catalytic path operating at much lower temperatures. The latter redefines the time-temperature dimensions of gas habitats opening the possibility of gas generation at subsurface temperatures previously thought impossible.

Thermodynamic equilibrium could shed light on which of these paths might dominate in nature. Thermal reactions are generally under kinetic control and their products removed from thermodynamic equilibrium while catalytic reactions are often under equilibrium control and their products near equilibrium. Hydrocarbons can achieve equilibrium through metathesis where homologues interconvert (Reaction 1).

$$2\ C_nH_m \rightleftharpoons C_{n-1}H_{m-2} + C_{n+1}H_{m+2} \quad (1)$$

Olefin metathesis [21, 22] is a well-known catalytic reaction shown in Reaction 2 for propylene. It is an extraordinary catalytic process because it breaks and makes carbon-carbon double bonds and reshuffles olefinic carbons distributing them randomly to new partners. Catalyzed by a variety of transition metals, it proceeds to equilibrium at low temperatures with conservation of π and σ bonds [23]. Metathesis of methane, ethane, and propane is illustrated in Reaction 3, referred to here as 'gas metathesis'. Although hypothetical, it bears a strong resemblance to olefin metathesis in stoichiometry (1) and to low-temperature gas generation in marine shales [20]. Hydrocarbons decompose over reduced nickel and cobalt oxides to C_1–C_3 compositions near equilibrium, possibly through catalytic gas metathesis [18]. Equilibrium between hydrocarbons in natural environments is not limited to metathesis, however. Metastable equilibria have been reported from the interaction of hydrocarbons, water, and authigenic mineral assemblages [24, 25, 26].

$$2\ C_3H_6 \rightleftharpoons C_2H_4 + C_4H_8 \quad (2)$$

$$2\ C_2H_6 \rightleftharpoons CH_4 + C_3H_8 \quad (3)$$

Gas metathesis without the aid of a catalytic agent is highly unlikely. Hydrocarbon cracking generates methane, ethane, and propane removed from thermodynamic equilibrium [3, 27], and their extraordinary thermal stabilities [28, 29] preclude equilibrium over geologic time. Thus natural gas at metathetic equilibrium (3) would implicate catalytic assistance.

Here we address metathetic equilibrium in shale gas generation and in natural gas deposits, and discuss the genetic implications.

RESULTS AND DISCUSSION

Methane or propane tends to dominate the hydrocarbons emerging from marine shales under anoxic isothermal gas flow [20]. Fig. 1 illustrates two examples, a methane-rich gas from a high-maturity Mississippian Barnett shale (a) and a propane-rich gas from a lower-maturity Devonian/ Mississippian New Albany shale (b).

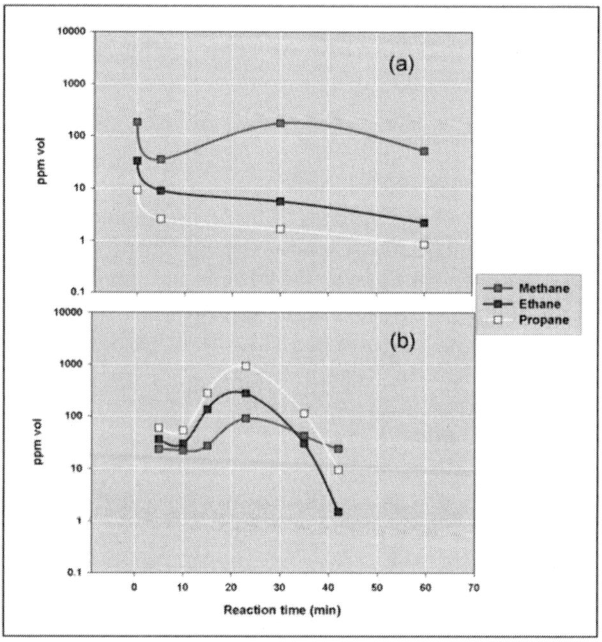

Figure 1. SHALE DECOMPOSITION PRODUCTS UNDER ANOXIC CONDITIONS, 200°C ISOTHERMAL HELIUM FLOW. The figure shows hydrocarbon concentrations (ppm vol) in the effluent gas coming off the indicated shales over time. The experimental procedure and product analysis are described elsewhere [20]. **(A)** Barnett Shale is Mississippian from the Delaware Basin, TX (Reeves County, well cuttings, 3500 m). Yield = 0.04 mg gas/g rock (C_1–C_5). Rock-Eval: TOC = 8.1% wt; S1 = 0.95 mg/g; S2 = 1.1 mg/g; S3 = 0.25 mg/g; Tmax = 548. (b) Upper Dev/L Miss New Albany Shale from the Illinois Basin, KY (side wall core, 1025 m). Rock-Eval: TOC = 6.2% wt; S1 = 2.2 mg/g; S2 = 17 mg/g; S3 = 0.3 mg/g; Tmax = 448. Yield = 1.2 mg gas/g rock (C_1–C_5).

The possibility that these are preexisting hydrocarbons desorbed under isothermal gas flow is inconsistent with the order in which hydrocarbons are released over time. Desorption under isothermal gas flow is a first order process where light hydrocarbons (C_x) will desorb before heavy hydrocarbons (C_y) with concentrations [C_x] and [C_y] in the effluent gas stream falling exponentially over time [30]. In our analysis of exponential desorption, ratios ([C_x]/[C_y]) will also fall exponentially over time irrespective of the concentrations adsorbed on surfaces or in solution. Thus, the relative amounts of heavy and light hydrocarbons adsorbed will effect the ratio ([C_x]/[C_y]), but not its exponential fall over time with isothermal gas flow. It would not be possible for ([C_x]/[C_y]) to remain constant or increase over time, for example.

There is no evidence of desorption in Fig. 1. [C_1]/[C_3] in (a) rises from 20 to over 100 in the first 30 minutes of gas flow, then falls to 60 at the end of data collection. The ratio in (b) is nearly constant over 35 minutes of gas flow (~0.3), then rises sharply to 2.5 at the end of data collection. The remarkable proportionality between ethane and propane sustained throughout both reactions and its independence of methane concentrations are also inconsistent with desorption. Since desorption under isothermal gas flow should not produce [C_2]/[C_3] ratios remaining constant over time and [C_1]/[C_3] ratios increasing over time, it must be dismissed as a possible source of the gases produced in these experiments.

The gases are distinct in one other respect. Barnett gas is near thermodynamic equilibrium while New Albany gas is far removed from equilibrium. Equation 4 shows the equilibrium constant K for methane, ethane, and propane (3) (where $C_1 = CH_4$; $C_2 = C_2H_6$; $C_3 = C_3H_8$).

$$K = (C_1 * C_3)/(C_2)^2 \tag{4}$$

Log K for equilibrium at a reaction temperature of 200°C is 0.90 at one atmosphere [31]. The average composition for the gases in Fig. 1 place the quotient Q, $Q = (C_1 * C_3)/(C_2)^2$, at log Q = 0.40 for Barnett gas and log Q = -1.30 for New Albany gas, where C_1, C_2, & C_3 are % vol.

A catalytic reaction can be metathetic, under equilibrium control, and still yield these hydrocarbons removed from equilibrium. All catalytic reactions tend to equilibrium over time in hydrocarbons that interconvert. Product compositions will therefore change over time, removed from equilibrium initially (short residence times), but approaching equilibrium with time. Thus, in flow reactors where residence times are short, gas metathesis could generate these hydrocarbons removed from equilibrium.

The New Albany reaction was repeated under closed conditions for evidence of gas metathesis over longer residence times. Gas flow was continued for 20 minutes to insure active gas generation, then the reactor was closed and its contents allowed to stand for 200 hours at 200°C. Fig. 2 shows the changes in gas composition over time and Fig. 3 shows the approach to equilibrium that attends these changes.

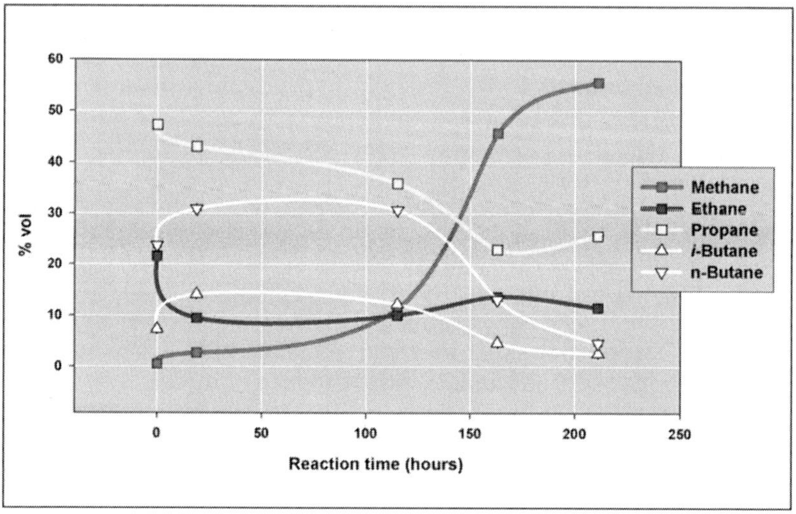

Figure 2. GAS COMPOSITIONS OVER TIME, CLOSED REACTOR, ANOXIC PROCEDURE, NEW ALBANY SHALE (FIG. 1), 200°C, HELIUM. After anoxic helium flow for 20 min., the reactor was closed, and the gas was analyzed (GC) at the indicated times by opening the reactor briefly to allow gas from the reactor to pass into a six-way valve for GC analysis [20].

The New Albany shale generates gas removed from equilibrium under flow conditions (Fig. 1b) and approaching equilibrium under closed conditions (Fig. 3). Equilibrium over time is a signature of catalysis. In this instance, it progresses to equilibrium and gas dryness in concert (Fig. 2). Natural gases might also progress to equilibrium and dryness in concert if the natural process is similarly catalytic.

Two methods were used to approximate the equilibrium constant K (eq. 4) in the subsurface. One approximates Gibbs free energies as a function of temperature and pressure [32] and the other log K values at various temperatures for ideal gases at one atmosphere [31]. For temperatures between 325 K and 575 K, and pressures between 3 MPa and 150 MPa, log K falls between 0.9 and 1.3 [32]. The second approximation [31] also places equilibrium limits within the same region: log K = 0.73 (575 K) and 1.4 (325

K). If natural gas compositions are constrained by equilibrium forces, they should have log Q values near these limits relative to the log Q limits for unaltered thermogenic gases.

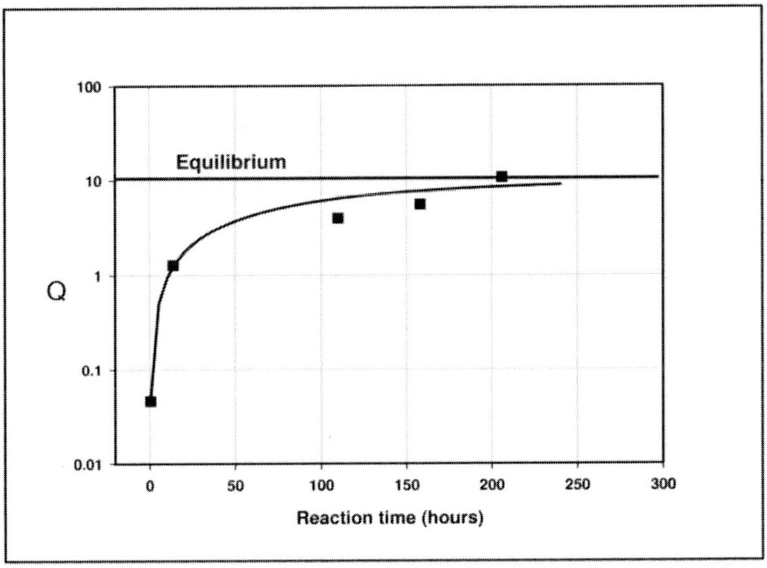

Figure 3. GAS COMPOSITION OVER TIME AND THERMODYNAMIC EQUILIBRIUM. Gas compositions are shown in Fig. 2. Q is the quotient $[(C_1)(C_3)]/[(C_2)^2]$, where concentrations are % vol. The horizontal line is the equilibrium constant K (K = 10) for the reaction conditions [31]. The solid line through the data is from the equation $Q(t) = 10*(1-e^{-\alpha t})$, where Q(t) is the quotient at time t (hours), 10 is Q at infinite time, and the constant α was set to 0.0094 to fit the data.

Figure 4 is a histogram of log Q for offshore Gulf of Mexico gases [33]. These gases were chosen because they are mostly free gases not associated with crude oils or other materials that might compromise their thermodynamic properties. The gases are divided into non-microbial and microbial according to dryness (% wt C_1 in C_1–C_4). The non-microbial gases are largely within the approximated equilibrium limits, while the microbial gases are clearly removed from those limits.

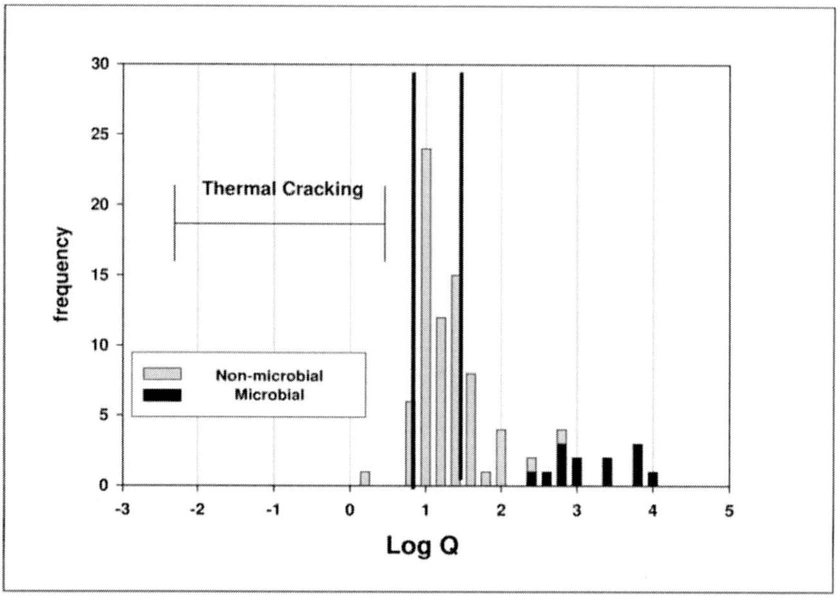

Figure 4. HISTOGRAM LOG Q, 87 OFFSHORE GULF OF MEXICO GASES [29]. % vol was used to calculate Q, the quotient $[(C_1)(C_3)]/[(C_2)^2]$. *Microbial* gases have % wt C_1 (C_1–C_4) > 99% and average log Q = 3.1 ± 0.53; $\delta^{13}C_1$ average -61.7 ± 7.1, a signature considered biogenic. Only one had $\delta^{13}C_1$ below – 50. *Non-microbial* gases have % wt C_1 (C_1–C_4) < 99% and average log Q = 1.2 ± 0.38. The dark vertical lines indicate approximate thermodynamic equilibrium limits for subsurface conditions [31, 31]. The log Q region marked 'Thermal Cracking' represents the products of thermal cracking based on laboratory experiments [3, 27, 34, 35, 36].

Gas products from thermal cracking experiments fall within the region marked 'Thermal Cracking' in Fig. 4[3, 27, 34, 35, 36]. We would expect 'primary thermal gas' from kerogen cracking and 'secondary thermal gas' from oil cracking to fall within this region as well. The displacement of natural gases to the right of this region by two log unit is therefore significant.

Figure 5 shows a similar displacement in 1600 gases from various basins in North America.

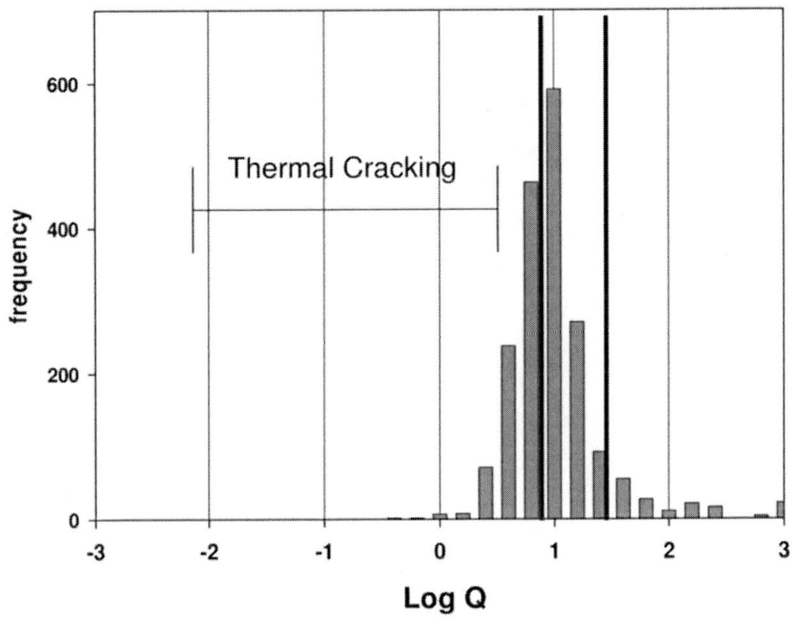

Figure 5. NATURAL GAS COMPOSITIONS AND THEIR RELATIONSHIP TO THERMODYNAMIC EQUILIBRIUM. Histogram of log Q (Q = $[(C_1)*(C_3)/(C_2)^2]$) for 1600 gas compositions obtained from the U.S. Department of Interior; mean log Q = 0.90 ± 0.43 [10]. Hydrocarbon concentrations used to calculate log Q were % vol C_1–C_5. These gases do not include compositions with C_2or C_3 < 0.5% vol. Since concentrations were rounded off to the nearest tenth in the DOI database, values in that range introduced substantial errors in calculating Q. The vertical dark lines and the horizontal bar are defined in Fig. 4.

Figs. 4 &5 challenge the notion that thermal cracking is the source of natural gas irrespective of thermodynamic equilibrium. How do we explain log Q values displaced from where they should be by two log units? The fact that they are displaced towards thermodynamic equilibrium, in this case metathetic equilibrium, raises the possibility that they were generated catalytically under equilibrium constraints, not thermally under kinetic constraint. It is possible, in other words, that there was no displacement; they were generated where they are.

Figure 6 supports this supposition. It shows a strong correlation between $[(CH_4)*(C_3H_8)]$ and $[(C_2H_6)^2]$ consistent with gas compositions constrained to equilibrium. It follows a power function (the solid line) rather than a linear function (lines parallel to the dashed line). The ratio Q

([[(CH_4)*(C_3H_8)]/[$(C_2H_6)^2$]]) thus varies systematically with concentrations, displaced from equilibrium at high concentrations of C_2 and C_3 (wet gas) and at equilibrium at low concentrations C_2 and C_3 (dry gas). Fig. 7 shows the approach to equilibrium with gas dryness. The line through the data is an equilibrium curve for a reversible reaction approaching equilibrium over time (t): $Q = K_{equi}(1\text{-}e^{-\alpha t})$. Time (t) has been replaced with $C_1/(C_2+C_3)$ in Fig. 7, consistent with the generally accepted notion that wet gas converts to dry gas over geologic time [1, 2].

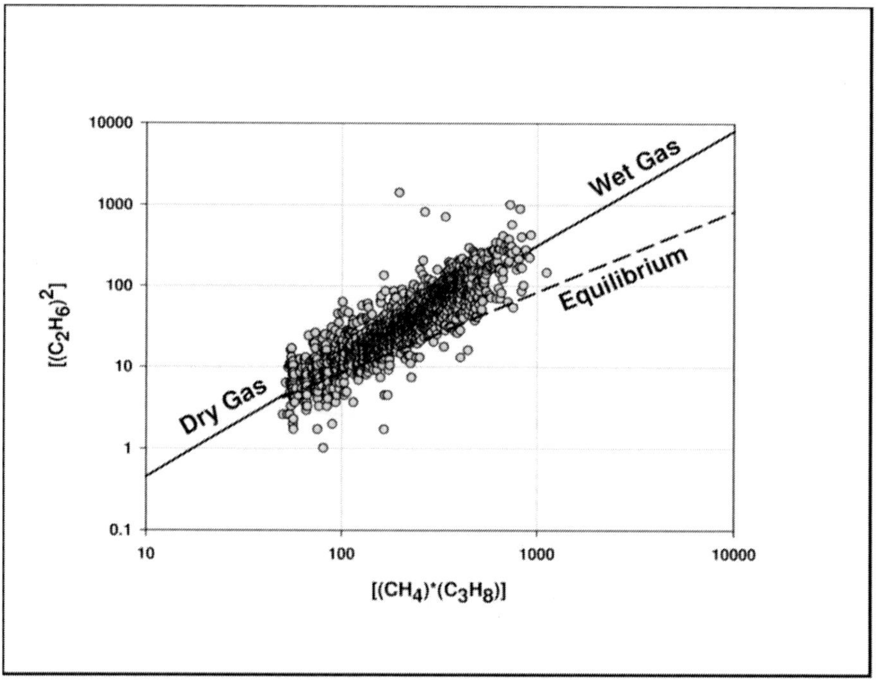

Figure 6. EQUILIBRIUM PLOT OF C_1–C_3 (EQ. 1) IN 1600 NATURAL GASES **(FIG. 5)**. Hydrocarbon concentrations are % vol in C_1 – C_5. These gases do not include compositions with C_2 or C_3 < 0.5% vol since the data, rounded off to the nearest tenth %, injects unacceptable error into the x and y functions. The dark line through the data is the regression line for the power equation $y = 0.0282\ x^{1.308}$, where $y = (C_2H_6)^2$, $x = (CH_4)*(C_3H_8)$, and $R^2 = 0.840$. The dashed line is for x/y = 12.0, thermodynamic equilibrium at 400 K [31]. Gas compositions were obtained from the U.S. Department of Interior [10]. The mean log Q (Q = (x/y) for the data = 0.90 ± 0.43.

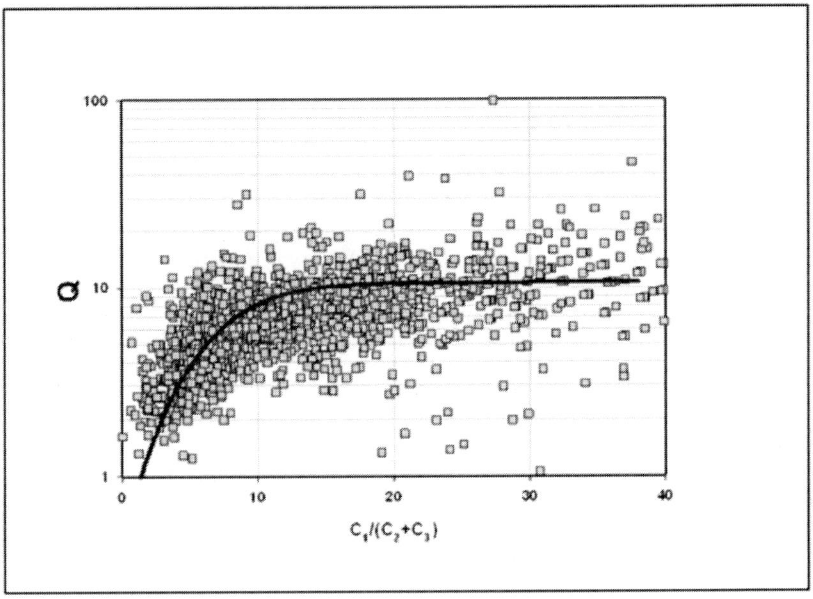

Figure 7. THERMODYNAMIC EQUILIBRIUM AND GAS DRYNESS. $Q = (C_1)*(C_3)/(C_2)^2$. The data is taken from Fig. 6. The black line passing through the data is the equilibrium curve, where Q approaches the equilibrium constant K (10.4) with gas dryness: $Q = 10.4(1-e^{-\alpha(C1/C2+C3)})$, α was set to 0.1 to fit the data.

Isotope ratios ($^{13}C/^{12}C$) in petroleum hydrocarbons are believed to be functions of primary biological inputs and isotope effects in gas generation and decomposition [2]. Isotopic equilibrium is another factor that can alter primary biological isotope ratios. Replacing ^{12}C with ^{13}C lowers the free energy of hydrocarbons because the carbon and hydrogen bonds to ^{13}C are stronger than the same bonds to ^{12}C. Bond energy enhancement increases with carbon number. Replacing ^{12}C with ^{13}C in ethane yields more additional bond energy than it does in methane, for example. Thus, at isotopic equilibrium, ^{13}C will be distributed preferentially in the higher hydrocarbons according to their respective lower free energies [37]. Original ^{13}C input will control the amount of ^{13}C shared between hydrocarbons at equilibrium, but their respective free energies will control how ^{13}C is distributed between them. The distribution of ^{13}C at equilibrium will therefore be independent of original input, rates of gas generation and rates of gas decomposition.

The isotopic equilibrium reaction for methane and ethane is shown in Reaction 5 and for methane and propane in Reaction 6 (the position of ^{13}C in C_3H_8 is unspecified).

$$^{13}CH_4 + ^{12}C_2H_6 \rightleftharpoons ^{12}CH_4 + ^{13}CH_6 \quad (5)$$

$$^{13}CH_4 + ^{12}C_3H_8 \rightleftharpoons ^{12}CH_4 + ^{13}C^{12}C_2H_8 \quad (6)$$

Eqs. 7 and 8 are the isotopic equilibrium equations with $K_{1,2}$ (7) the carbon isotopic equilibrium constant for methane and ethane, and $K_{1,3}$ (8) the carbon isotopic equilibrium constant for methane and propane: $^{12}C_n$ and $^{13}C_n$are fugacities; $^{13}C_2 = ^{13}C^{12}CH_6$, and $^{13}C_3 = ^{13}C^{12}C_2H_8$.

$$(^{13}C_2) * (^{12}C_2) = K_{1,2}(^{13}C_1) * (^{12}C_2) \quad (7)$$

$$(^{13}C_3) * (^{12}C_1) = K_{1,3}(^{13}C_1) * (^{12}C_3) \quad (8)$$

Table 1 shows the molar isotope ratios and the quotients Q for methane, ethane, and propane in 285 natural gases and it includes data for catalytic gases [18] for comparison, to be discussed below. The quotients $Q_{1,2}$ and $Q_{1,3}$ are very close to theoretical equilibrium values at 400 K: $K_{1,2} = 2.039$; $K_{1,3} = 3.101$ [37]. All ratios show substantial invariance. The variance in $Q_{1,2}$ and $Q_{1,3}$ is one half that in the molar isotope ratios within them reflecting the correlations between molar isotope ratios shown in Fig. 8. Table 1 also displays the extraordinary match between catalytic gases and natural gases.

Table 1. Statistical properties of molar isotope ratios and isotopic equilibrium constants in 285 natural gases and 5 catalytic gases.

	Natural Gases			Catalytic Gases		
	mean	sd × 10^4	V × 10^6	Mean	sd × 10^4	v × 10^7
$^{13}C_1/^{12}C_1$	0.00991	0.55	5.8	0.00991	0.40	0.30
$^{13}C_2/^{12}C_2$	0.02031	1.07	5.2	0.02030	0.63	1.78
$^{13}C_3/^{12}C_3$	0.03090	1.50	4.4	0.03090	0.97	1.9
$Q_{1,2}$	2.0486	70.6	2.2	2.0481	35.7	5.7
$Q_{1,3}$	3.1175	130	3.3	3.1172	30.6	1.8

Natural gas data was taken from USGS Energy Resource Organic Geochemistry Data Base, http://energy.cr.usgs.gov/prov/og/. Catalytic gases are from octadecene decomposition over reduced nickel oxide (180 – 210°C) [18]. C_1–C_3 compositions were normalized to % wt carbon. $\delta^{13}C$ values were converted to mass ratios which were used to calculate wt ^{13}C at each carbon number: x_1 at C_1, x_2 at C_2, and x_3 at C_3. Wt $^{13}C_1 = x_1$; Wt $^{13}C_2$ is wt C_2 with composition $^{13}C^{12}C = x_2+((12/13.00335)x_2)$; wt $^{13}C_3$ is wt C_3 with composition $^{13}C^{12}C_2 = x_3+((24/13.00335)x_3)$. Wt ^{12}C at C_2 was calculated as the total wt ^{12}C at C_2 minus the wt ^{12}C in $^{13}C_2$. Wt ^{12}C at C_3 was also the total wt ^{12}C at C_3 minus the wt ^{12}C in $^{13}C_3$. Weights (per 100 g) were converted to moles/(100 g C_1–C_3) which were used throughout our analysis. The quotient $Q_{1,2} = (^{13}C_2)*(^{12}C_1)/(^{13}C_1)*(^{12}C_2)$ and $Q_{1,3} = (^{13}C_3)*(^{12}C_1)/(^{13}C_1)*(^{12}C_3)$, where concentrations are moles/100 g. Variance (v) is $(sd)^2$ for log (ratio).

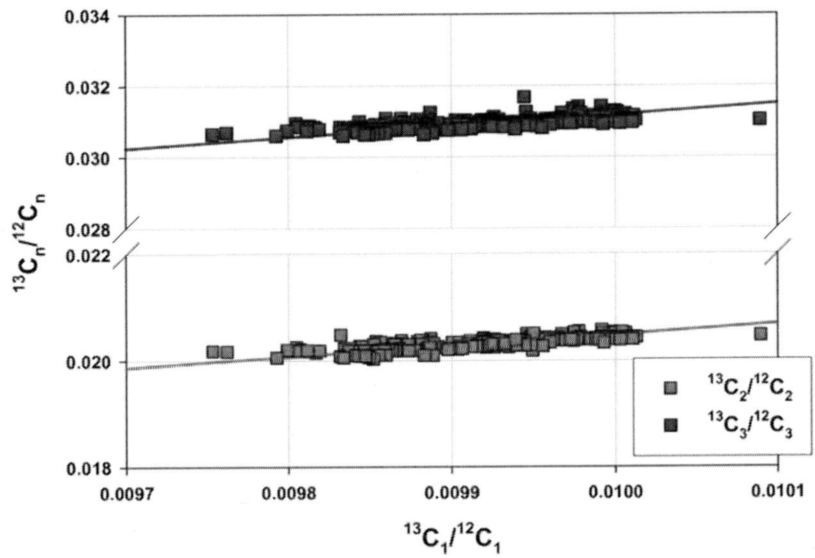

Figure 8. THE CORRELATIONS BETWEEN MOLAR ISOTOPE RATIOS IN METHANE, ETHANE, AND PROPANE IN 285 GASES (USGS, TABLE 1). Ratios are molar, calculated as described in Table 1. The lines are linear regression lines with a coefficients of correlation $R^2 = 0.638$ for $^{13}C_2/^{12}C_2$ with slope (zero intercept) = 2.028, and $R^2 = 0.47$ for $^{13}C_3/^{12}C_3$ with slope (zero intercept) = 3.055.

Fig. 9 shows the proximities of natural gases and catalytic gases to isotopic thermodynamic equilibrium on a log K scale.

The approach to equilibrium with dryness (Figs. 6 & 7) mirrors the experimental results shown in Figs. 2 &3. The isotopic results (Figs. 8 &9) reinforce the position that hydrocarbons in natural gas are generated under equilibrium constraints. It is a metathetic equilibrium and therefore a catalytic equilibrium.

Other explanations are less satisfactory. It is difficult to explain dry gas generation through thermal cracking [28, 29] and harder to explain gas metathesis through thermal stress. Equilibrium requires the facile exchange of carbon atoms between methane, ethane, and propane. Carbon-carbon and carbon-hydrogen σ bonds are broken and reformed with overall bond conservation. This is unprecedented in thermal hydrocarbon reactions and inconceivable without catalytic assistance.

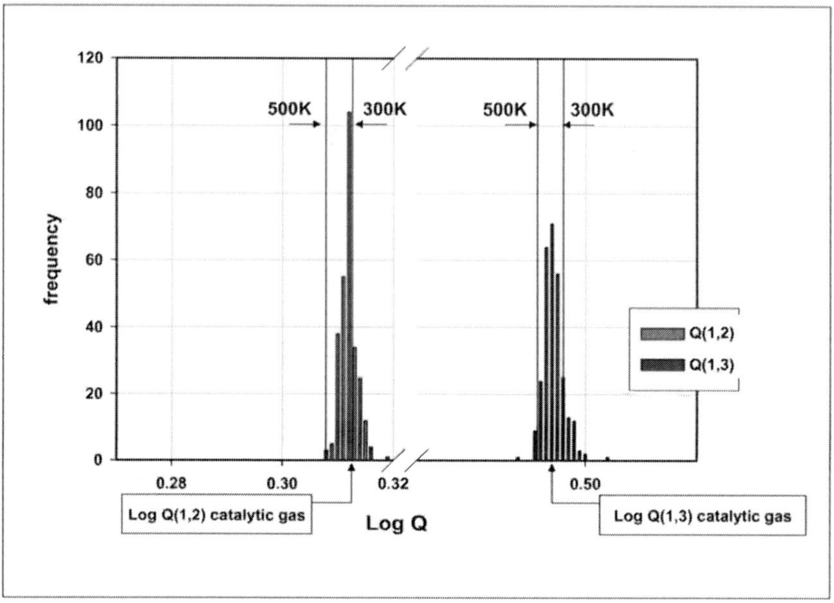

Figure 9. HISTOGRAM OF ISOTOPIC QUOTIENTS (LOG $Q_{1,2}$ AND LOG $Q_{1,3}$) FOR 285 NATURAL GASES (TABLE 1). All values of $Q_{1,2}$ and $Q_{1,3}$were calculated as described in Table 1. The arrows beneath the chart (catalytic gas) mark the positions of log $Q_{1,2}$ and log $Q_{1,3}$ for Catalytic Gases in Table 1. The vertical lines mark isotopic equilibrium constants at 300 K, log $K_{1,2}$ = 0.31259; log $K_{1,3}$ = 0.49602, and at 500 K, log $K_{1,2}$ = 0.30786; log $K_{1,3}$ = 0.49142 [37].

CONCLUSION

The following results support our position that natural gases are at or near thermodynamic equilibrium:

1) Gas compositions are significantly displaced from thermogenic compositions to equilibrium compositions (Figs. 5&6).

2) There is a strong nonlinear correlation between $[(C_1)*C_3)]$ and $[(C_2)^2]$ in which the quotient Q converges on equilibrium as wet gas progresses to dry gas (Figs. 6) consistent with an approach to equilibrium over geologic time (Fig. 7).

3) The isotopic compositions of methane, ethane, and propane are also constrained to equilibrium compositions (Figs 8 &9).

We propose catalytic gas metathesis as the source of equilibrium in natural gas. The natural catalytic activity in marine shales [20], or some similar activity in other sedimentary rocks, is probably the source of equilibrium in natural gas deposits. This view is supported by the New Albany shale experiment in which Q progressed to metathetic equilibrium over time as wet gas progressed to dry gas (Figs. 2 &3). A mechanistic connection between degradation to methane [20] and metathesis is suggested.

Catalysis by low valent transition metals [16, 17, 18] may be the source of gas metathesis and degradation in the origin of natural gas. The match in isotope ratios between catalytic gases and natural gases (Table 1) supports this position and the sensitivity of marine shales to oxygen poisoning [20] supports it as well.

The position that thermal cracking adequately explains the origin of natural gas [1, 2] is no longer tenable. It cannot explain the high concentrations of methane in natural gas [16], the distribution of light hydrocarbons [38, 39], and the associations with thermodynamic equilibrium reported here. Of the two possible pathways to natural gas, the catalytic path [20] appears the more attractive. It is simple, economic, and does not suffer the now mounting deficiencies challenging thermal cracking theory.

ACKNOWLEDGEMENTS

We thank Stephen Garcia for experimental assistance.

COMPETING INTERESTS

The authors declare that they have no competing interests.

AUTHORS' CONTRIBUTIONS

FDM formulated theory, experimental design and supervised the experimental work. DJ contributed to the experimental work and helped in shaping the paper. EH was instrumental in early strategy and in writing the ms.

REFERENCES

1. Tissot BP, Welte DH: Petroleum Formation and Occurrence. 1984, Springer, New York
2. Hunt JM: Petroleum Geochemistry and Geology. 1995, Freeman, New York
3. Voge HH, Good GM: Thermal cracking of higher paraffins. Journal American Chemical Society. 1949, 71: 593-597. 10.1021/ja01170a059.
4. McNab JG, Smith PV, Betts RI: The evolution of petroleum. Petroleum and Engineering Chemistry. 1952, 44: 2556-2563.
5. Evans RJ, Felbeck GT: High temperature simulation of petroleum formation I. The pyrolysis of Green River Shale. Organic Geochemistry. 1983, 4: 135-144.
6. Saxby JD, Riley KW: Petroleum generation by laboratory-scale pyrolysis over six years simulating conditions in a subsiding basin. Nature. 1984, 308: 175-177.
7. Espitalie J, Ungerer P, Irwin I, Marquis F: Primary cracking of kerogens. Experimenting and modeling C1, C2-C5, C6-C15, and C15+ classes of hydrocarbons formed. Organic Geochemistry. 1987, 13: 893-899.
8. Hikita T, Takahashi I, Yoshimichi T: Hydropyrolysis of heavy oils. Fuel. 1989, 68: 1140-1144.
9. Horsfield B, Schenk HJ, Mills N, Welte DH: An investigation of the in-reservoir conversion of oil to gas: compositional and kinetic findings from closed-system programmed-temperature pyrolysis. Organic Geochemistry. 1992, 19: 191-204.
10. Mango FD: The light hydrocarbons in petroleum: a critical review. Organic Geochemistry. 1997, 26: 417-440.
11. Price LC, Schoell M: Constraints on the origins of hydrocarbon gas from compositions of gases at their site of origin. Nature. 1995, 378: 368-371.
12. McNeil RI, BeMent WO: Thermal stability of hydrocarbons: Laboratory criteria and field examples. Energy and Fuels. 1996, 10: 60-67. 10.1021/ef9501399.
13. van Aarssen DGK, Bastow TP, Alexander R, Kagi RI: Distribution of methylated naphthalenes in crude oils: indicators of maturity, biodegradation and mixing. Organic Geochemistry. 1999, 30: 1213-1227.
14. Inan S: Gaseous hydrocarbons generated during pyrolysis of petroleum source rocks using unconventional grain size. Organic Geochemistry. 2000, 31: 1409-1418.
15. Snowdon LR: Natural gas composition in the geological environment and the implications for the processes of generation and preservation. Organic Geochemistry. 2001, 32: 913-931.
16. Mango FD: Methane concentrations in natural gas: the genetic implications. Organic Geochemistry. 2001, 32: 1283-1287.
17. Mango FD: Transition metal catalysis in the generation of petroleum and natural gas. Geochimica et Cosmochimica Acta. 1992, 56: 553-555.
18. Mango FD, Elrod LW: The carbon isotopic composition of catalytic gas: A comparative analysis with natural gas. Geochimica et Cosmochimica Acta. 1999, 63: 1097-1106.
19. Lewan MD, Kotarba MJ, Wieclaw D, Piestrzynski A: Evaluating transition-metal catalysis in gas generation from the Permian Kupferschiefer by hydrous pyrolysis. Geochimica et Cosmochimica Acta. 2008, 72: 4069-4093.
20. Mango FD, Jarvie DM: Low temperature gas from marine shales. Geochemical Transactions. 2009, 10: 3
21. Cotton FA, Wilkinson G, Murillo CA, Bochmann M: Advanced Inorganic Chemistry. 1999, John Wiley & Sons, NY, 1282-1285. 6
22. Rouhi AM: Olefin metathesis: Big-deal reaction. Chemical and Engineering News. 2002, 80 (51): 29-83.
23. Mango FD: Olefin Metathesis. Coordination Chemistry Reviews. 1975, 15: 152-163.

24. Fabuss BM, Smith JO, Lait RI, Borsanyi AS, Satterfield CN: Rapid thermal cracking of n-hexadecane at elevated temperatures. Industrial and Engineering Chemistry, Product Develoment. 1962, 1: 293-299.
25. Helgeson HC, Knox AM, Owens CE, Shock EI: Petroleum oil field waters, and authigenic mineral assemblages: Are they in metastable equilibrium in hydrocarbon reservoirs. Geochimica Cosmochimica Acta. 1993, 57: 3295-3339.
26. Seewald JS: Evidence for metastable equilibrium between hydrocarbons under hydrothermal conditions. Nature. 1994, 370: 285-287. 10.1038/370285a0.
27. Seewald JS: Organic-Inorganic interactions in petroleum-producing sedimentary basins. Nature. 2003, 426: 323-333. 10.1038/nature02132.
28. Laidler KJ, Wojciechowske BW: Kinetics and mechanisms of the thermal decomposition of ethane. I. The uninhibited reaction. Proceedings Royal Society. 1961, A260: 91-102. 10.1098/rspa.1961.0015.
29. Laidler KJ, Sagert NH, Wojciechowske BW: Kinetics and mechanisms of the thermal decomposition of propane. Proceedings Royal Society. 1962, A270: 242-253.
30. Tokoro Y, Misono M, Uchijima T, Yoneda Y: Analysis of thermal desorption curves for heterogeneous surfaces. I. A linear variation of the activation energy of desorption. Bulletin of the Chemical Society of Japan. 1978, 51: 85-89.
31. Stull DR, Westrum EF, Sinke GC: The chemical thermodynamics of organic compounds. 1969, Johy Wiley & Sons, New York
32. Lemmon EW, Huber ML, McLinden MO: NIST Standard Reference Database 23: Reference Fluid Thermodynamic and Transport Properties – REFPROP 8.0. 2007, National Institute of Standards and Technology, Standard Reference Data Program, Gaithersburg, Maryland 20899, USA
33. Rice DD, Threlkeld CN: Natural gas analyses from Offshore Gulf of Mexico. From Gulf Coast Oils and Gases, Their Characteristics, Origin, Distribution, and Exploration and Production Significance. Edited by: Schumacher D, Perkins BF. 1990, Proceeding of the Ninth Annual Research Conference, Gulf Coast Section Society of Economic Paleontologists and Mineralologists Foundations
34. Behar F, Ungerer P, Kressmann S, Rudkiewiez JL: Thermal evolution of crude oils in sedimentary basins: experimental simulation in a confined system and kinetic modeling. Revue de l'Institut Francais du Pétrole. 1991, 46: 151-181.
35. Horsfield B, Schenk HJ, Mills N, Welte DH: An investigation of the in-reservoir conversion of oil to gas: compositional and kinetic findings from closed-system programmed-temperature pyrolysis. Organic Geochemistry. 1992, 19: 191-204.
36. Burkle-Vitzum V, et al: Kinetics of alkyl aromatics on the thermal stability of hydrocarbons under geologic conditions. Organic Geochemistry. 2004, 35: 3-33.
37. Galimov EM, Ivlev AA: Thermodynamic isotope effects in organic compounds. I. Carbon isotope effects in straight-chain alkanes. Russian Journal of Physical Chemistry. 1973, 47: 1564-66.
38. Mango FD: An invariance in the isoheptanes of petroleum. Science. 1987, 237: 514-517.
39. Mango FD: The stability of hydrocarbons under the time-temperature conditions of petroleum genesis. Nature. 1991, 352: 146-

CHAPTER 11

Gas Flow Behavior of Nanoscale Pores in Shale Gas Reservoirs

Weijun Shen [1,2,*], Xizhe Li [3], Yanmei Xu [3], Yuping Sun [3] and Weigang Huang [3]

[1]*Institute of Mechanics, Chinese Academy of Sciences, Beijing 100190, China*

[2]*School of Engineering Science, University of Chinese Academy of Sciences, Beijing 100049, China*

[3]*PetroChina Research Institute of Petroleum Exploration and Development, Beijing 100083, China*

ABSTRACT

The gas transport in shale nanopores is always one of the major concerns in terms of the development of shale gas reservoirs. In this study, the gas flow regimes in shale nanopores were classified and analyzed according to Knudsen number. Then the gas flow model considering Darcy flow, slip flow, transition flow, molecular free flow and adsorption effect was proposed to evaluate the gas flow behavior in shale nanopores. The result shows that the contributions of Darcy flow, slip flow and transition flow in shale nanopores are reciprocal, and are mainly dominated by pore radius and pressure. The adsorption effect greatly influences the total mass flux. The total mass flux will increase as Langmuir pressure and temperature increase while it will decrease with reservoir pressure and the adsorption thickness. These results can provide insights for a better understanding of gas flow in the shale nanopores so as to optimize the production performance of shale gas reservoirs.

Keywords: shale gas reservoirs; nanoscale pores; Knudsen number; mathematical model; flow behavior

1. INTRODUCTION

In the last few years, shale gas is playing an important role among energy sources and has attracted wide attention [1,2]. According to a survey from U.S. Energy Information Administration, 34% of gas production in 2011 in the U.S.A is from shale and the proportion will reach 45% by 2035. Unlike conventional gas reservoirs, shale gas reservoirs are characterized by extremely low porosity, ultra-low permeability and high clay content [3,4]. Moreover, the pore structures in shale gas reservoirs are varied and heterogeneous, including organic matter, nonorganic matrix, natural fractures and pore space induced by hydraulic fractures [5,6,7]. The giant variation of pores scales makes gas flow in shale gas reservoirs very complex. Thus an understanding of gas flow and transport in shale pores is great significance for gas productivity and for optimizing the hydraulic-fracturing design in shale gas reservoirs.

The gas in shale gas reservoirs includes gas adsorbed on organic matter and inorganic mineral surfaces and free gas in fractures and pores [8,9]. Using Scanning Electron Microscopy (SEM) methods Javadpour et al. [10], Locks et al. [11] found that the main pore size of shale matrix is in the range from 1 to 200 nm. Due to the existence of nanopores, the widely used Darcy's law cannot accurately describe gas flow in shales. Based on the work from Beskok and Karniadakis [12], Swami et al. [13] divided gas flow in shales into four regimes, including viscous flow ($K_n \leq 0.001$), slip flow ($0.001 \leq K_n \leq 0.1$), transition flow ($0.1 \leq K_n \leq 10$) and Knudsen flow ($K_n \geq 10$). The gas transport in shale nanopores is a complex combination of all these flow mechanisms, which cannot be described by Darcy's law [14]. In addition, the adsorption and desorption process also exist in the pores with adsorbed gas on organic matter and inorganic mineral surfaces.

Some transport models have been developed to understand and quantify gas flow in shale gas reservoirs over the past years. Javadpour [15] and Ozkan et al. [16] proposed a gas transport model in shales including viscous flow and Knudsen flow. Moridis et al. [17] and Freeman et al. [18] presented a very comprehensive review of flow models in shale gas reservoirs considering the four regimes. In order to analyze production dynamics in shale gas reservoirs, an extended dynamic-slippage model was put forward by Clarkson et al. [19,20]. Civan et al. [21], Sakhaee and

Bryant [22] extended Knudsen diffusion into slippage factor to describe gas transport in shales. Anderson et al. [23] proposed a model to describe the transport mechanisms, which is related to the experiment empirical coefficients. Although there are lots of models proposed to describe gas transport in shale nanopores, a detailed comparison and analysis of different flow patterns is lacking. Moreover, the effects of these parameters on gas flow patterns, such as pressure, temperatures, and adsorption parameters, were also not studied systematically. Therefore, it is extremely necessary to understand gas flow behavior in shale nanopores and the effects on different flow patterns so as to predict gas production and optimize the fracturing treatment for shale gas reservoirs.

In this work, the gas flow regimes were described according to Knudsen number in shale gas reservoirs. The gas flow mathematical models under different flow patterns were analyzed and proposed to understand the flow behavior. Then the effects of pressure and temperature on flow pattern versus pore diameters were analyzed. Besides, a series of sensitivity studies to quantify the effects of these parameters such as adsorption, adsorption thickness, Langmuir pressure, temperature, pressure on the gas flow behavior were performed. This work should provide insights for a better understanding of gas flow behavior in shale nanopores.

2. GAS FLOW REGIMES

2.1. Knudsen Number

Compared with conventional gas reservoirs, shale gas reservoirs are characterized by low porosity and ultra-low permeability. With decreasing shale pore radius, the gas transport in shales will change from Darcy flow to other flow patterns. Thus the Darcy flow equation does not correctly describe gas transport in shale gas reservoirs. When the shale pore radius is less than the gas molecular mean free path, the collisions between gas molecules and hole-walls are more frequent than those between gas molecules, and the Knudsen diffusion will occur. In order to divide gas flow patterns, Hadjiconstanttinou [24] used Knudsen number (K_n) to distinguish gas flow regions in the process of micro-nano pipe flow. It can be expressed as follows:

$$K_n = \frac{\lambda}{r} \tag{1}$$

where K_n is Knudsen number; λ is mean free path of gas molecular; and r is pore radius.

Knudsen number is a characteristic parameter of gas flow in different scale channels [25]. It is very important to correctly determine the mean free path of gas molecules. The mean free path of gas molecules can be written as:

$$\lambda = \frac{k_B T}{\sqrt{2}\pi\delta^2 P} \tag{2}$$

where k_B is the Boltzmann constant, $k_B = 1.3805 \times 10^{-23}$ J/K; T is temperature; P is pressure; and δ is collision diameter.

2.2. Gas Flow Regimes

According to Knudsen number, the gas flow in shales can be divided four regimes, including continuous flow (known as Darcy flow), slip flow, transition flow and molecular free flow, as shown in Figure 1. At a low Knudsen number ($K_n < 0.001$), the gas transport in shales is Darcy flow. Compared with the pore throat size, the mean free path can be ignored. The molecules' collisions with pore-walls are negligible but the intermolecular collisions are important, so the gas flow is controlled by gas viscosity while Knudsen diffusion can be ignored. The widely used Darcy flow equation and Navier-Stokes equation without slippage both describe gas flow in porous media. With increasing Knudsen number ($0.001 < K_n < 0.1$), the regime is slip flow. For this Knudsen number range, the flow velocity at the pore boundary cannot be ignored and the slippage effect need to be considered in viscous flow. For the mean free path of gas molecules becomes important, the molecules' collisions with pore-walls need to be considered. The Navier-Stokes equation considering slippage effect can describe gas transport in porous media.

When K_n ranges from 0.1 to 10, this is the transition flow. The mean free path and pore throat size are the same in scale. The molecules' collisions with pore-walls and the intermolecular collisions are both important. In this condition of flow regime, the continuity assumption and Navier-Stokes equation cannot describe gas flow, which is the most difficult regime. Slip flow and Knudsen diffusion must be considered. When K_n is over 10, this is the molecular free flow. For this range, the mean free path of gas molecules becomes predominant, the molecules' collisions with pore-walls are important and the intermolecular collisions can be

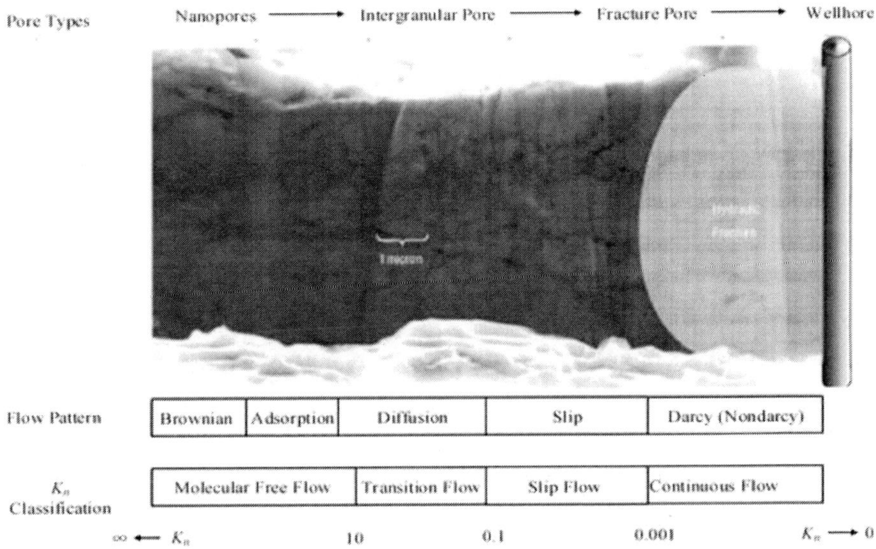

Figure 1. Gas flow regimes in shale reservoirs according to Knudsen number.

ignored. The gas flow is controlled by Knudsen diffusion while gas viscosity are negligible. Thus the Knudsen diffusion equation can be used to describe gas flow in this regime.

3. GAS FLOW MATHEMATICAL MODEL

3.1. Darcy Flow

When K_n is less than 0.001, the gas flow in shale nanopores can be described by the Darcy equation, which is expressed as:

$$\nu_g = -\frac{k_g}{\mu_g}\nabla P \tag{3}$$

where ν_g is gas velocity; k_g is gas permeability; μ_g is gas viscosity; and ∇P is pressure gradient.

According to the Hagen-Poiseuille equation, the Darcy permeability can be expressed as:

$$k_g = \frac{r^2}{8} \tag{4}$$

Substituting Equation (4) into Equation (3), the mass flux in Darcy flow can be expressed as:

$$J_V = \frac{\rho_g r^2}{8\mu_g} \nabla P \tag{5}$$

where ρ_g is gas density; and r is radius.

3.2. Slip Flow

When K_n ranges from 0.001 to 0.1, the molecules' collisions with pore-walls are not negligible and there exists slip flow. The gas velocity along pore-walls is not zero, so the slippage effect needs to be considered. An equation proposed by Klinkenberg [26] can be expressed as:

$$k_g = k_l(1 + \frac{4c\lambda}{r}) = k_l(1 + \frac{b_k}{P}) \tag{6}$$

where k_l is equivalent liquid permeability; c is constant; P is pore pressure; and b_k is Klinkenberg coefficient.

For the slip flow regime, Roy et al. [25] presented a similar equation as follows:

$$k_g = k_l[1 + 4(\frac{2 - \sigma_v}{\sigma_v})\frac{\lambda}{D}] \tag{7}$$

where σ_v is tangential momentum coefficient, $\sigma_v \approx 0.9$; and D is pore diameter.

Then the permeability in slip flow can be expressed as:

$$k_g = k_l(1 + 5K_n) \tag{8}$$

Thus the mass flux in the slip flow regime can be written as:

$$J_s = F\frac{\rho_g r^2}{8\mu_g}\nabla P \tag{9}$$

where F is slip flow coefficient, $F = 1 + 5K_n$.

3.3. Transition Flow

When K_n is in the range between 0.1 and 10, the mean free path of gas molecules is the same order as the pore size of porous media. The regime is the transition flow. It not only considers the slippage effect, but also includes the Knudsen diffusion, which is a complex flow process. For the transition flow regime, Ho and Webb [27] used the dusty gas model to describe gas flow, which is expressed as:

$$k_g = k_l(1 + \frac{b_k}{P}) = k_l(1 + \frac{\mu D_k}{k_l}\frac{1}{P}) \tag{10}$$

$$D_k = \frac{\varphi}{\tau_h}\frac{2r}{3}(\frac{8RT}{\pi M})^{0.5} \tag{11}$$

where D_k is the Knudsen diffusion coefficient.

Based on the model from Beskok and Karniadakis [28], Civan [29] revised the rarefaction coefficient to describe the gas flow in this regime as follows:

$$k_t = \left[(1 + \alpha K_n)(1 + \frac{4K_n}{1 - bK_n})\right]k_l \tag{12}$$

$$\alpha = \frac{1.358}{1 + \frac{0.17}{K_n^{0.4348}}} \tag{13}$$

where α is the rarefaction coefficient; and b is the slip coefficient.

Shi et al. [30] used a ratio coefficient to describe the slippage and diffusion effect in the transient gas flow, and the mass flux can be expressed as:

$$J_t = fJ_s + (1 - f)J_{K_n} \tag{14}$$

where f is the ratio coefficient; J_s and J_{Kn} are the mass flux in the slip and Knudsen diffusion, separately.

3.4. Molecular Free Flow

When K_n exceeds 10, the gas transport is the molecular free flow. This regime is controlled by Knudsen diffusion. For this regime, the mass flux can be expressed as:

$$J_M = -D_{K_n} \nabla n_i \tag{15}$$

$$\nabla n_i = \frac{-M}{10^3 RT} \nabla P \tag{16}$$

where D_{Kn} is the Knudsen diffusivity; and ∇n_i is the gas density gradient of gas component i.

Based on a circular tube, Roy et al. [25] proposed the Knudsen diffusivity which can be written as follows:

$$D_{K_n} = \frac{d}{3}\left(\frac{8RT}{\pi M}\right)^{0.5} \tag{17}$$

Thus the mass flux in the molecular free flow can be expressed as:

$$J_M = \frac{2rM}{3 \times 10^3 RT}\left(\frac{8RT}{\pi M}\right)^{0.5} \nabla P \tag{18}$$

4. MATHEMATICAL MODEL WITH MULTI-FLOW REGIMES

4.1. Gas Mathematical Model

According to the above analytical result, the gas flow in shale gas reservoirs is a complex flow process, including the Darcy flow, slip flow, transition flow and molecular free flow. For the gas transport in shale nanopores, these flow patterns may exist at the same time, and it is not comprehensive to consider one or two flow process. Thus the total mass flux with multi-flow regimes may be expressed as:

$$J = J_V + J_S + J_T + J_M \tag{19}$$

Substituting Equations (5), (9), (14), (18) and (19), we obtain:

$$J = \left[\left(1 + \frac{2 + \left(\frac{K_n}{4.5}\right)^4}{1 + \left(\frac{K_n}{4.5}\right)^4} \cdot (1 + 5K_n) \right) \frac{\rho_g r^2}{8\mu_g} + \frac{1 + 2\left(\frac{K_n}{4.5}\right)^4}{1 + \left(\frac{K_n}{4.5}\right)^4} \frac{2rM}{3 \times 10^3 RT} \left(\frac{8RT}{\pi M}\right)^{0.5} \right] \nabla p \tag{20}$$

When considering the effect of gas adsorption, the adsorption thickness based on the Langmuir equation can be expressed as:

$$d = d_0 \frac{P}{P + P_L} \tag{21}$$

where d is the adsorption thickness; d_0 is the max adsorption thickness; and P_L is Langmuir pressure.

And the effective radius in the shale nanopores and Knudsen number considering the adsorption effect can be written as:

$$r_e = r - d = r - d_0 \frac{P}{P + P_L} \tag{22}$$

$$K_{ne} = \frac{\lambda}{r_e} \tag{23}$$

Thus the total mass flux with multi-flow regimes and adsorption effect can be written as:

$$J = \left[\left(1 + \frac{2 + \left(\frac{K_{ne}}{4.5}\right)^4}{1 + \left(\frac{K_{ne}}{4.5}\right)^4} \cdot (1 + 5K_{ne}) \right) \frac{\rho_g r_e^2}{8\mu_g} + \frac{1 + 2\left(\frac{K_{ne}}{4.5}\right)^4}{1 + \left(\frac{K_{ne}}{4.5}\right)^4} \frac{2r_e M}{3 \times 10^3 RT} \left(\frac{8RT}{\pi M}\right)^{0.5} \right] \nabla P \tag{24}$$

4.2. Case Study

According to the Equations (20) and (24), it can be seen that the total mass flux in shale gas flow is influenced by lots of properties and parameters, such as Knudsen number, gas density, pore size, gas viscosity temperature, gas molar mass and pressure gradient. In order to understand the gas flow behavior in shale nanopores, here we will analyze different flow patterns under different conditions. Table 1 presents the mole fraction, mole mass and collision diameter for each component in a typical shale gas reservoir. Table 2 shows the properties of gas flow in shale nanopores. According to the calculations, the ratio mass flux in different flow patterns versus pore size is illustrated in Figure 2. From the result of Figure 2, with increasing pore size, the mass flux ratio of Darcy flow rises while that of slip flow slows down. The transition flow increases firstly, and then decreases after the pore size reaches 10 nm. When the pore size is less than 10 nm, the slip flow is predominant. In the range from 10 nm and 1000 nm, the slip and transicnt flow dominate the gas transport, and the Darcy flow plays a role in the process. It is noteworthy although the molecular free flow has an effect on gas flow, its mass flux is very small.

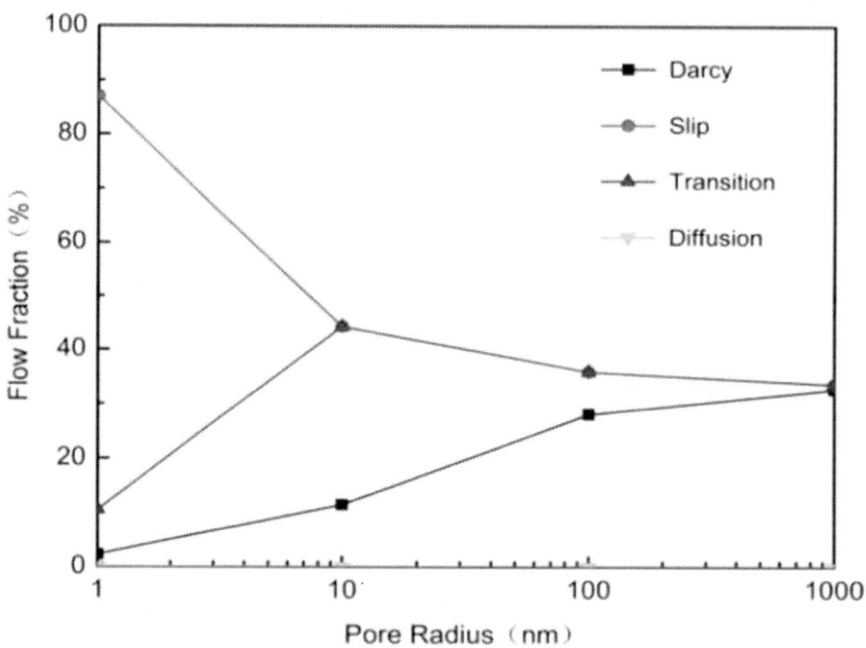

Figure 2. Gas flow fraction versus pore radius.

Table 1. Mole fraction, mole mass and collision diameter for each component in typical shale gas reservoir.

Gas Components	Mole Fraction (%)	Mole Mass (kg/kmol)	Collision Diameter (δ, $\times 10^{-9}$ m)
CH_4	87.4	16	0.40
C_2H_6	0.12	30	0.52
CO_2	12.48	44	0.45
Average		19.50	0.41

Table 2. Properties of gas flow in shale nanopores.

Property	Value	Unit
Gas density (ρ_g)	8.03	kg/m^3
Gas viscosity (μ_g)	0.12	$\times 10^{-4}$ Pa·s
Average mean-free-path (λ)	5.55	nm
Maximum adsorption thickness (d_0)	0.50	nm
Knudsen number (K_n)	5.55	1
Temperature (T)	300	K
Pressure (P)	10	MPa
Pressure gradient (∇P)	5	MPa
Langmuir pressure (P_L)	10	MPa
Universal gas constant (R)	8.31	$Pa \cdot m^3/(mol \cdot K)$
Gas mole mass (M)	19.50	$\times 10^{-3}$ kg/kmol

5. SENSITIVITY ANALYSIS

5.1. Effect of Pressure on Flow Regimes

Figure 3 presents the effects in different pressures where four cases were calculated by varying pressure from 5 MPa to 40 MPa to study different flow patterns in shale nanopores. From the results of Figure 3, it is known that the flow fraction of Darcy and molecular free flow increases with increasing pressure, but the slip and transition flow decreases. When pressures are constant, with increasing pore radius, the slip and transition flow slows down while the Darcy and molecular free flow gradually rise.

5.2. Effect of Temperature on Flow Regimes

The relationship between different flow patterns and pore radius for different temperatures from 320 K to 400 K is shown in Figure 4 where it is seen that as temperatures increase, the ratio of the Darcy, slip, transition and molecular free flow changes little at given pore radius. When temperatures are given, with the increasing of pore radius, the Darcy and molecular free flow increases while the slip and transition flow decreases.

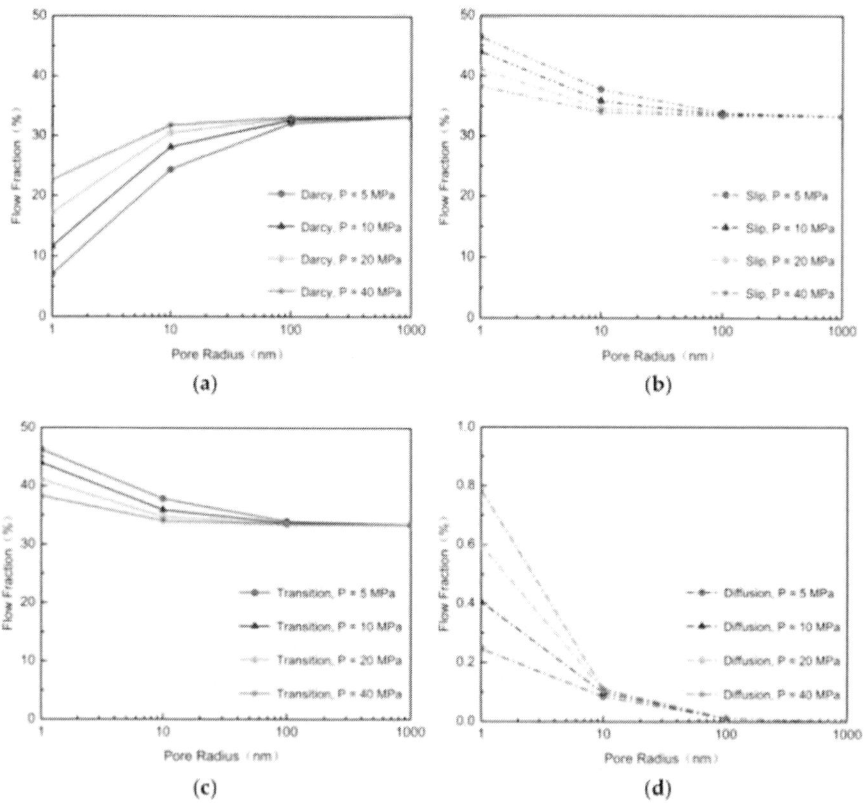

Figure 3. Gas flow fraction versus pore radius for various pressures. (**a**) The fraction of Darcy flow; (**b**) the fraction of Slip flow; (**c**) the fraction of Transition flow; and (**d**) the fraction of Diffusion flow.

5.3. Effect of Langmuir Pressure

Four cases of the Langmuir pressure values from 10 MPa to 25 MPa are selected to study the gas flow in shale nanopores, as shown in Figure 5. The result shows the $J_{ad}/J_{no\text{-}ad}$ (J_{ad} and $J_{non\text{-}ad}$ are the mass flux with adsorption and without adsorption, respectively) ratio will increase with Langmuir pressure when the pore radius is constant. That means that the total mass flux rises with increasing Langmuir pressure. For a given Langmuir pressure, with the pore radius increasing, the $J_{ad}/J_{no\text{-}ad}$ ratio will rise. It says that the total mass flux increases with the pore radius.

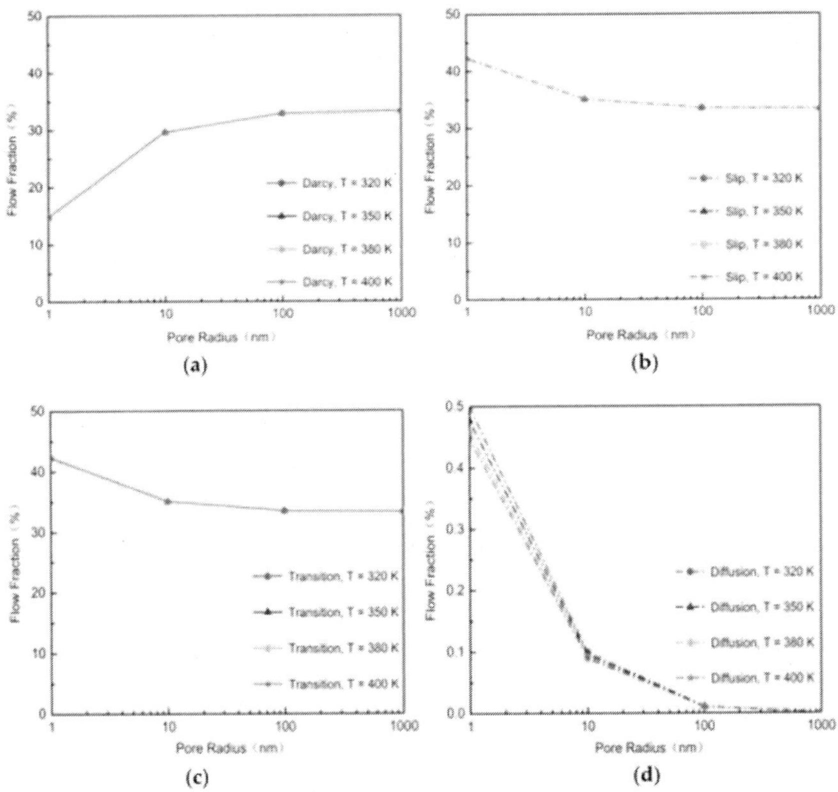

Figure 4. Gas flow fraction versus pore radius for various temperatures. (**a**) The fraction of Darcy flow; (**b**) the fraction of Slip flow; (**c**) the fraction of Transition flow; and (**d**) the fraction of Diffusion flow.

5.4. Effect of Adsorption Thickness

The adsorption thickness values from 0.3 nm to 1 nm are chosen to study the effect of adsorption thickness on gas flow in shale nanopores. The $J_{ad}/J_{no\text{-}ad}$ ratio versus pore radius is illustrated in Figure 6. From the results of Figure 6, it is seen that the $J_{ad}/J_{no\text{-}ad}$ ratio will decrease with adsorption thickness for a given pore size. In other words, with increasing adsorption thickness, the total mass flux will decrease. When pressure is constant, the $J_{ad}/J_{no\text{-}ad}$ ratio will increase with the rising pore radius. It means that the total mass flux will increase as the pore radius increases.

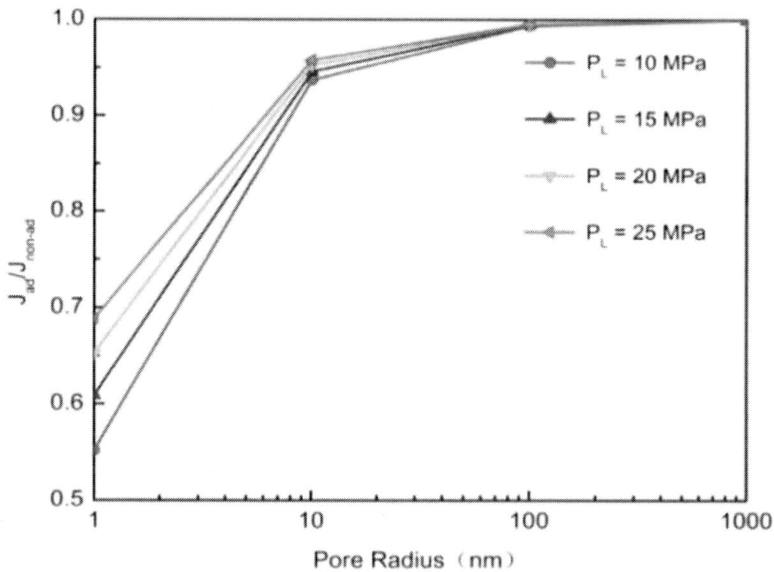

Figure 5. $J_{ad}/J_{non\text{-}ad}$ ratio versus pore radius for various Langmuir pressures.

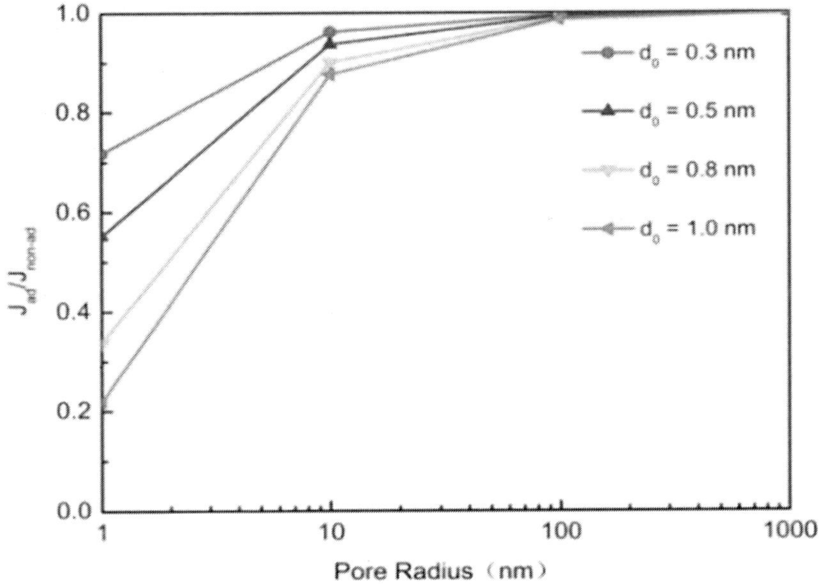

Figure 6. $J_{ad}/J_{non\text{-}ad}$ ratio versus pore radius for various adsorption thickness.

5.5. Effect of Pressure

Figure 7 shows the effects in different pressures where four cases from 5 MPa to 40 MPa are considered to study gas flow in shale nanopores. From the result of Figure 7, with increasing pressure in shale nanopores, the $J_{ad}/J_{no\text{-}ad}$ ratio will decrease when the pore radius is constant. The total mass flux decreases as pressure increases. However, for a given pressure the $J_{ad}/J_{no\text{-}ad}$ ratio will increase with the increasing pore radius. It means that the total mass flux will increase as pore size increases when pressure is given.

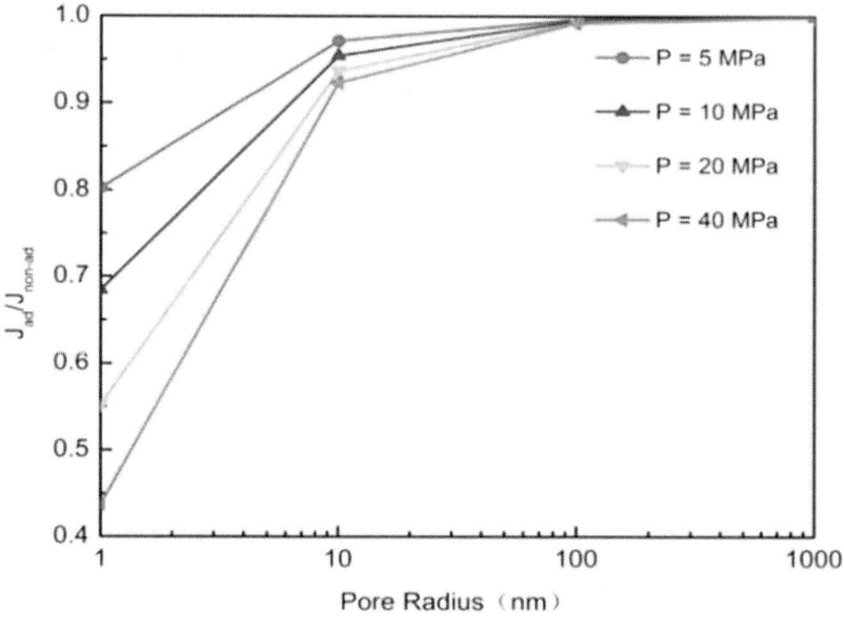

Figure 7. $J_{ad}/J_{non\text{-}ad}$ ratio versus pore radius for various pressures.

5.6. Effect of Temperature

The values of different temperatures between 320 K and 400 K are conducted to investigate the effect of adsorption thickness on the gas flow in shale pores. The variation of the $J_{ad}/J_{no\text{-}ad}$ ratio versus pore radius is shown in Figure 8. As illustrated in Figure 8, it is seen that as temperatures increase, the $J_{ad}/J_{no\text{-}ad}$ ratio increases for a given pore radius. It means that the elevated temperature favors total mass flux. When pressure is constant, the $J_{ad}/J_{no\text{-}ad}$ ratio rises with increasing pore radius. That is, the corresponding total mass flux increases as the pore radius increases.

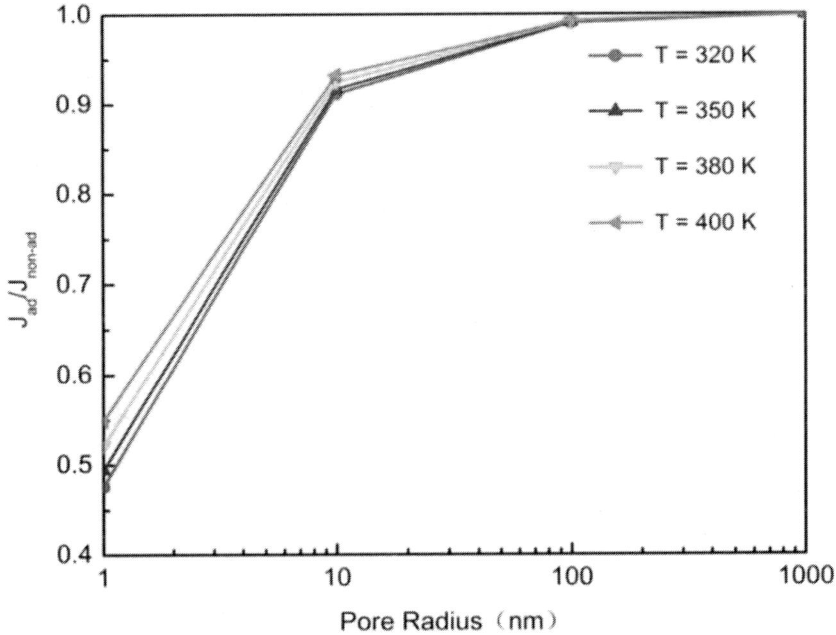

Figure 8. $J_{ad}/J_{non\text{-}ad}$ ratio versus pore radius for various temperatures.

6. CONCLUSIONS

Based on Knudsen number, the gas flow regimes in shale nanopores were classified. Then the gas flow model considering Darcy flow, slip flow, transition flow, molecular free flow and adsorption effect was proposed to describe the gas flow behavior in shale nanopores. The proportion of flow patterns and the effects on flow patterns in shale pores were analyzed. According to the above analytical results, the following conclusions can be drawn: (1) As the pore radius increases, the Darcy flow rises steadily in shale nanopores while the slip flow decreases. The transition flow increases, and then decreases after the pore radius reaches 10 nm. When the pore radius is less than 10 nm, the slip flow is predominant. In the range from 10 nm and 1000 nm, the slip and transient flow dominates the gas transport, and the Darcy flow plays a role in the process. It is noteworthy although the molecular free flow has an effect on gas flow, its mass flux is very small; (2) For a given pore radius, with increasing pressure, the Darcy and mole free flow will increase while the slip and transition flow will decrease. The temperature has little effect on the four flows in shale nanopores; (3) The adsorption process has a strong influence on the total mass flux in shale pores. When the pore radius is constant, the total mass flux will increase with increasing Langmuir pressure and temperature. However, it will decrease as reservoir pressure and the adsorption thickness increase.

ACKNOWLEDGMENTS

This work was supported by the Youth Foundation of Key Laboratory for Mechanics in Fluid Solid Coupling Systems, Chinese Academy of Sciences, and by National Science and Technology Major Project of the Ministry of Science and Technology of China Project (NO. 50150503-12 and NO. 2016ZX05037006), and by the Project of PetroChina Research Institute of Petroleum Exploration & Development (NO. RIPED-LFFY-2017-5S-118). We thank the support from the Foundation of China Scholarship Council (CSC).

AUTHOR CONTRIBUTIONS

Weijun Shen, Xizhe Li and Yanmei Xu analyzed and proposed the mathematical models; Weijun Shen, Yuping Sun and Weigang Huang performed and analyzed the result; Weijun Shen wrote the paper.

CONFLICTS OF INTEREST

The authors declare no conflict of interest.

REFERENCES

1. Shen, W.J.; Zheng, L.G.; Oldenburg, C.M.; Cihan, A.; Wan, J.M.; Tokunaga, T.K. Methane diffusion and adsorption in shale rocks—A numerical study using the dusty gas model in Tough2/EOS7C-ECBM. In Proceedings of the TOUGH Symposium, Lawrence Berkeley National Laboratory, Berkeley, CA, USA, 28–30 September 2015.
2. Sutton, R.P.; Cox, S.A.; Barree, R.D. Shale gas plays: A performance perspective. In Proceedings of the SPE Tight Gas Completions Conference, San Antonio, TX, USA, 2–3 November 2010.
3. Shen, W.J.; Wan, J.M.; Kim, Y.; Li, X.Z. Porosity calculation, pore size distribution and mineral analysis within shale rocks: Application of scanning electron microscopy. Electron. J. Geotech. Eng. 2015, 20, 11477–11490.

4. Shen, W.J.; Xu, Y.M.; Li, X.Z.; Huang, W.G.; Gu, J.R. Numerical simulation of gas and water flow mechanism in hydraulically fractured shale gas reservoirs. J. Nat. Gas Sci. Eng. 2016, 35, 726–735.
5. Davies, D.K.; Bryant, W.R.; Vessel, R.K.; Burkett, P.J. Porosity, Permeabilities and Microfabrics of Devonian Shales. In Microstructure of Fine-Grained Sediments; Bennett, R.H., Bryant, W.R., Hulbert, M.H., Eds.; Springer Verlag: New York, NY, USA, 1991; pp. 109–119.
6. Bustin, R.M.; Bustin, A.M.; Cui, X.; Ross, D.; Pathi, V.M. Impact of Shale Properties on Pore Structure and Storage Characteristics. In Proceedings of the SPE Shale Gas Production Conference, Fort Worth, TX, USA, 16–18 November 2008.
7. Reed, R.M.; Loucks, R.G.; Ruppel, S.C. Comment on "Formation of nanoporous pyrobitumen residues during maturation of the Barnett Shale (Fort Worth Basin)" by Bernard et al. (2012). Int. J. Coal Geol. 2014, 127, 111–131.
8. Hill, C.K.; Nelson, C.R. Gas Productive Fractured Shales: An Overview and Update; Gas Tips of Gas Research Institute: Des Plaines, IL, USA, 2010; Volume 6, pp. 4–13.
9. Strapoc, D.; Mastalerz, M.; Schimmelmann, A.; Drobniak, A.; Hasenmueller, N.R. Geochemical constrains on the origin and volume of gas in the New Albany Shale (Devonian-Mississippian), eastern Illinois Basin. AAPG Bull. 2010, 94, 1713–1740.
10. Javadpour, F.; Fisher, D.; Unsworth, M. Nanoscale gas flow in shale gas sediments. J. Can. Pet. Technol. 2007, 46, 16–21.
11. Loucks, R.G.; Reed, R.M.; Ruppel, S.C.; Jarvie, D.M. Morphology, genesis and distribution of nano-scale pores in siliceous mudstones of the Mississippian Barnett Shale. J. Sed. Res. 2009, 79, 848–861.
12. Beskok, A.; Karniadakis, G.E. Simulation of heat and momentum transfer in complex micro geometries. J. Thermophys. Heat Transf. 1994, 8, 647–653.
13. Swami, V.; Clarkson, C.R.; Settari, A. Non-darcy flow in shale nanopores: Do we have a final answer. In Proceedings of the SPE Canadian Unconventional Resources Conference, Calgary, AB, Canada, 30 October–1 November 2012.
14. Bird, R.B.; Stewart, W.E.; Lightfoot, E.N. Transport Phenomenon, 2nd ed.; John Wiley & Sons, Inc.: New York, NY, USA, 2012.
15. Javadpour, F. Nanopores and apparent permeability of gas flow in mudrocks (shales and siltstone). J. Can. Pet. Technol. 2009, 48, 16–21.
16. Ozkan, E.; Brown, M.; Raghavan, R.; Kazemi, H. Comparison of fractured horizontal-well performance in conventional and unconventional reservoirs. SPE Reserv. Eval. Eng. 2009, 14, 248–259.
17. Moridis, G.J.; Blasingame, T.A.; Freeman, C.M. Analysis of mechanisms of flow in fractured tight gas and shale gas reservoirs. In Proceedings of the SPE Latin American and Caribbean Petroleum Engineering Conference, Lima, Peru, 1–3 December 2010.
18. Freeman, C.M.; Moridis, G.J.; Blasingame, T.A. A Numerical study of microscale flow behavior in tight gas and shale gas reservoir systems. Transp. Porous Media 2011, 90, 253–268.
19. Clarkson, C.R.; Nobakht, M.; Kaviani, D.; Ertekin, T. Production analysis of tight gas and shale gas reservoirs using the dynamic-slippage concept. SPE J. 2012, 17, 230–242.
20. Clarkson, C.R.; Qanbari, F.; Nobakht, M.; Heffner, L.E. Incorporating geomechanical and dynamic hydraulic fracture property changes into rate-transient analysis: Example from the haynesville shale. SPE Reserv. Eval. Eng. 2013, 16, 303–316.
21. Civan, F.; Rai, C.S.; Sondergeld, C.H. Shale-gas permeability and diffusivity inferred by improved formulation of relevant retention and transport mechanisms. Transp. Porous Media 2011, 86, 925–944.
22. Sakhaee, P.A.; Bryant, S.L. Gas permeability of shale. SPE Reserv. Eval. Form. Eval. 2012, 8, 401–409.
23. Anderson, J.M.; Moorman, M.W.; Brown, J.R.; Hochrein, J.M.; Thornberg, S.M. Isothermal mass flow measurements in microfabricated rectangular channels over a very wide knudsen range. J. Micromech. Microeng. 2014, 24, 055013.

24. Hadjiconstantinou, N.G. The limits of navier-stokes theory and kinetic extensions for describing small-scale gaseous hydrodynamics. Phys. Fluids 2006, 18, 111301–111320.
25. Roy, S.; Raju, R.; Chuang, H.F.; Cruden, B.A.; Meyyappan, M. Modeling gas flow through microchannels and nanopores. J. Appl. Phys. 2003, 93, 4870–4879.
26. Klinkenberg, L.J. The Permeability of Porous Media to Liquids and Gases. Drilling and Production Practice; American Petroleum Institute: Washington, DC, USA, 1941; pp. 200–213.
27. Ho, C.K.; Webb, S. Gas Transport in Porous Media; Springer Netherlands: Houten, The Netherlands, 2006.
28. Beskok, A.; Karniadakis, G.E. A model for flows in channels, pipes, and ducts at micro and nano scales. Microscale Thermophys. Eng. 1999, 3, 43–77.
29. Civan, F. Effective correlation of apparent gas permeability in tight porous media. Transp. Porous Media 2010, 82, 375–384.
30. Shi, J.T.; Zhang, L.; Li, Y.S.; Wang, T. Diffusion and flow mechanisms of shale gas through matrix pores and gas production forecasting. In Proceedings of the SPE Unconventional Resources Conference, Calgary, AB, Canada, 5–7 November 2013.

CHAPTER 12

Thermal Electromagnetic Radiation of Rarefied Gas

Andrey N. Volobuev, Eugene S. Petrov, Eugene L. Ovchinnikov

Department of Physics, Samara State University, Samara, Russia

ABSTRACT

The Boltzmann kinetic equation for rarefied radiating gas is found. It is shown, that process of radiation is defined by excitation of atoms at their collision, and also spontaneous radiation of quantums at transition of electrons to the basic power level and the compelled radiation of quantums at collision of the excited atoms. It is shown, that distributions on velocities of the excited and not excited atoms submit to various laws. Distinctions in laws of distribution of the excited and not excited atoms define power parameters of radiating gas, and also a share of radiating molecules in gas.

Keywords: Rarefied Gas; Boltzmann Kinetic Equation; Electromagnetic Radiation; Laws of Atoms Distribution in the Velocities

1. INTRODUCTION

The theory of the kinetic equations is rather widely applied to the analysis of various physical processes [1]. The interaction of molecules of rarefied gas in gas laser [2], particles in plasma, phonons in crystals, etc. have been successfully analyzed with the help of this theory. Most strict deduction of

Maxwell distribution of ideal gas molecules by the speeds (or their kinetic energy) is carried out with the help of Boltzmann kinetic equation [3] on consideration of system of interacting atoms in a stationary condition. Boltzmann kinetic equation, first of all, reflects non-stationary process of interaction of particles among themselves, influence of this interaction on the establishment of corresponding distribution of molecules and influence on examined system of external factors.

The method of the kinetic equations with use of a principle of detailed equilibrium is applied also at research of transition of the atoms interacting with electromagnetic radiation, from one condition in another [3].

The theory of thermal radiation represents one of the most advanced physical theories. Actually this theory has begun development of quantum mechanics. The matter is that even before creation of quantum mechanics the formula for spectral density of energy of equilibrium thermal radiation, so-called Rayleigh-Jeans law has been found. The deduction of this law was based on the wave nature of light. Concentration of electromagnetic standing waves in a cavity has been found and multiplied on the average energy of a standing wave kT, where kBoltzmann constant, T-absolute temperature [3].

In spite of the fact that the deduction of this law is completely transparent, it was in the absolute contradiction with experiment, showing, that the density of energy of electromagnetic radiation monotonously grows with increase in frequency. It has induced M. Planck to researches which have led him to Planck's to well-known formula for radiation of absolutely black body and further to development of quantum statistics.

All events of this history were repeatedly discussed as in scientific, and scientific-popular literature.

We shall consider some questions of thermal radiation, on which researchers usually take insufficient attention.

2. PROCESSES IN THE GAS, RESULTING TO THERMAL ELECTROMAGNETIC RADIATION

There is interest to consider in more detail processes in the gas, resulting to thermal electromagnetic radiation.

In **Figure 1** some constant volume of gas V which is heated up due to submission of heat Q is shown. It results in radiation by gas of quantums of electromagnetic energy $h\nu$.

Let's carry out the kinetic analysis of processes in the rarefied gas, resulting to electromagnetic radiation.

We shall assume gas the closed thermodynamic system which is taking place in equilibrium state. i.e. how many thermal energy Q gives in system, as much electromagnetic energy from it is radiated. Interaction external of molecules electrons in such gas occurs only at their collision.

Let's consider probability of transition of electron in a molecule of gas from the top energy level on bottom. This probability p_{21}, it is similar to a method to Einstein [3], we shall divide into two parts, on the spontaneous

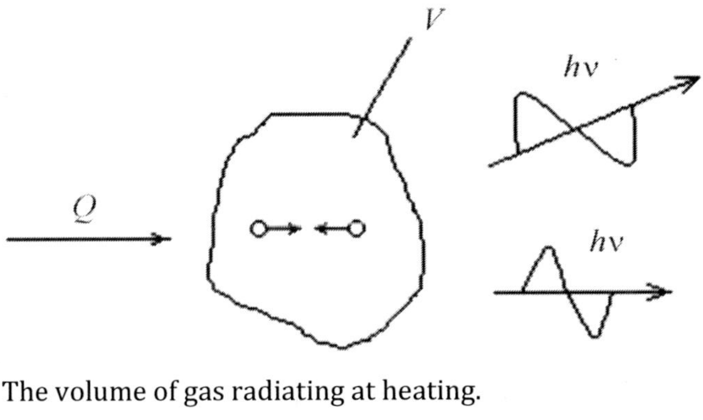

Figure 1. The volume of gas radiating at heating.

p_{21}^{s} and compelled p_{21}^{c} probabilities. We shall consider, that the compelled transition of electrons occurs only due to collision of molecules, **Figure 2**.

$$p_{21} = p_{21}^{s} + p_{21}^{c} \quad .(1)$$

Let's note, that as at spontaneous transition of electrons on the bottom energy level, and at compelled there is radiation of electromagnetic energy quantums $h\nu$.

Using a principle of detailed balance and Boltzmann distribution, we have:

$$\frac{p_{12}}{p_{21}} = \exp\left(-\frac{\Delta W}{kT}\right), \quad (2)$$

where p_{12} is probability of transition of electron on the top energy level at impact of molecules, T-absolute temperature of gas. The value $\Delta W = W_2 - W_1$ takes into account change of the energy condition of molecules at their collision, i.e. redistribution of kinetic energy of molecules and transitions of electrons in molecules from a level on a level.

Substituting (1) in (2), and taking into account, that the probability of the compelled transition at collision of molecules with the bottom energy level on top is equal to probability of the compelled transition at collision of molecules with the top level on bottom $p_{12} = p_{21}^{c}$, we find:

$$p_{21}^{c} = \frac{p_{21}^{s}}{\exp\left(\frac{\Delta W}{kT}\right) - 1}. \tag{3}$$

Change of the energy condition ΔW of molecule generally can take place or due to transfer of kinetic energy from one molecule to another at their collision, or due to supply of energy from the outside, or due to electromagnetic radiation of gas, or all these cases simultaneously took place.

On the other hand at impact of molecules, in them there is a distribution of external electrons on their energy. This distribution, as well as Maxwell distribution, has thermal character. Since it results from collisions of molecules the internal electrons of molecules do not participate in occurrence of the given distribution.

The probability for electron to appear at a level 1, Figure 2, we shall accept equal p_1, and at a level 2 equal p_2. Taking into account the condition of normalization $p_1 + p_2 = 1$ and Boltzmann distribution $\frac{p_2}{p_1} = \exp\left(-\frac{\Delta\varepsilon}{kT}\right)$, we receive:

$$p_2 = \frac{1}{1 + \exp\left(\frac{\Delta\varepsilon}{kT}\right)}. \tag{4}$$

The value $\Delta\varepsilon = \varepsilon_2 - \varepsilon_1$ takes into account only difference of the electrons energy at energies levels 1 and 2.

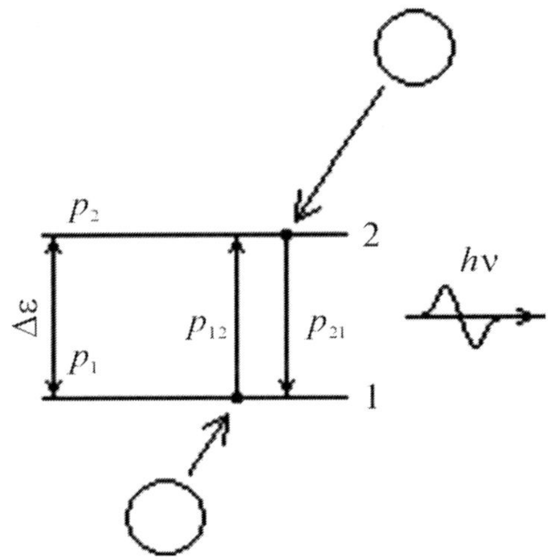

Figure 2. Excitation and the compelled radiation of molecules at their collisions.

In the form of record the received formula is close to Fermi-Dirac [4] distribution, but not identical to it. As against Fermi-Dirac distribution, in the formula (4) value $\Delta\varepsilon$ is not excess of electron energy above Fermi's level, and represents a difference of energies levels of external electrons to which they can transit at collision of molecules. Besides we shall noted, that at the deduction of distribution (4) Pauli principle was not used.

The Maxwell distribution and distributions (3), (4) contradict each other since that there were distributions (3), (4) it is necessary, that impacts of molecules were not elastic, and to Maxwell distribution there correspond absolutely elastic impacts of molecules.

Besides distinction of forms of distributions (3) and (4), concerning as a whole to the same system to rarefied gas pays attention. Hence, distribution of molecules on energy or to velocities in examined gas system cannot be described by the uniform formula.

3. THE QUANTUM KINETIC EQUATION OF RADIATING RAREFIED GAS

To find out how molecules of gas should be distributed on energy in view of quantum processes, we shall consider kinetic Boltzmann equation as [3]:

$$\frac{df}{dt} = \iint_{p_1,\Omega_1} v_{rel}\sigma\left(f_2 f_3 - f f_1\right) dp_1 d\Omega_1 \quad , \qquad (5)$$

where f and f_1 is the density of probabilities of molecules distribution in the velocities (or energy) for two molecules to have velocities before collision v and v_1, and f_2, f_3-density of probabilities for the same molecules to have velocities after collision v_2 and v_3. Besides sizes v_{rel}-relative speed of impacting molecules, s-effective section of interaction of particles, p_1-impulse of examined impacting molecule, Ω_1-a solid angle in which particles scattered, in this case molecule a, after collision, **Figure 3**.

The main lack of Boltzmann Equation (5) is the impossibility with the help of it to describe electromagnetic thermal radiation of gas. It is connected to the assumption of absolutely elastic impacts of gas molecules. We shall notice also, that this equation found before creation of quantum mechanics has no quantum character.

With the purpose of the account of not elastic process of collisions of molecules, finally, resulting in radiation of quantums of electromagnetic energy, and also translation of Boltzmann equation in quantum area, we modify the Equation (5).

Energy of a molecule represents the sum of kinetic energy of its movement and potential energy of excitation due to transition of electrons on higher energies levels

$$W = \frac{mv^2}{2} + \Delta\varepsilon \qquad [5].$$

Let's allocate in all volume of gas the relative share of molecules q (concerning corresponding interval of energy) which have energies W and W_1. As a result of their collisions there are molecules with energies W_2 and W_3, which share too q. The number of collisions of molecules with energies W and W_1 is proportional to product of probabilities ff_1. But after the first collision the share of molecules with these energies will decrease due to occurrence of molecules with energies W_2 and W_3. Therefore the number of collisions will be proportional to reduction of a share of molecules with energies W and W_1 due to increase in a share of molecules with energies W_2 and W_3, i.e. proportional to value $(\theta - f_2 - f_3)$. Hence, Boltzmann kinetic Equation (5) for radiating gas gets a kind:

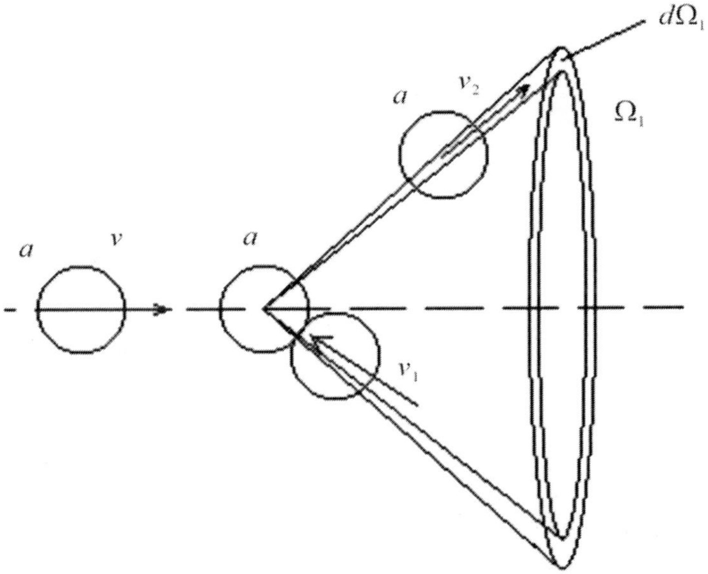

Figure 3. Scattering of molecule a after collision.

$$\frac{df}{dt} = \iint_{p_1,\Omega_1} v_{rel}\sigma\left[f_2 f_3\left(\theta + f + f_1\right) - ff_1\left(\theta - f_2 - f_3\right)\right]d\boldsymbol{p}_1 d\Omega_1. \tag{6}$$

The value $(\theta + f + f_1)$ shows, that the share of molecules with energies W_2 and W_3 grows due to molecules which before collision had energies W and W_1. In the Equation (6) as against (5) integral of collisions it is written down not only on the basis of balance of molecules before and after collision, but also dynamics of change of molecules shares is taken into account during the collisions, resulting to thermal electromagnetic radiation.

Let's find stationary distributions of the molecules f_i, submitting to the Equation (6). Stationary distributions of molecules arise in a case $\frac{df}{dt} = 0$. Hence, from (6) we find:

$$f_2 f_3\left(\theta + f + f_1\right) = ff_1\left(\theta - f_2 - f_3\right), \tag{7}$$

To the functional Equation (7) satisfy functions of distribution of type:

$$f = \frac{\theta}{\exp\left(\frac{W}{kT}\right) - 1}, f_1 = \frac{\theta}{\exp\left(\frac{W_1}{kT}\right) - 1}, \quad (8)$$

$$f_2 = \frac{\theta}{1 + \exp\left(\frac{W_2}{kT}\right)}, f_3 = \frac{\theta}{1 + \exp\left(\frac{W_3}{kT}\right)}. \quad (9)$$

Simply to show, that the Equation (7) will identically be transformed to a kind:

$$\frac{f_2 f_3}{(\theta - f_2)(\theta - f_3)} = \frac{f f_1}{(\theta + f)(\theta + f_1)}, \quad (10)$$

which decision, obviously, Functions (8) and (9) are.

Thus the law of energy conservation is carried out at not elastic collision of molecules:

$$W + W_1 = W_2 + W_3. \quad (11)$$

Substituting $W = \frac{mv^2}{2} + \Delta\varepsilon$, we find:

$$\frac{mv^2}{2} + \Delta\varepsilon + \frac{mv_1^2}{2} + \Delta\varepsilon_1 = \frac{mv_2^2}{2} + \Delta\varepsilon_2 + \frac{mv_3^2}{2} + \Delta\varepsilon_3. \quad (12)$$

In the Formula (12) values $\Delta\varepsilon_i$ characterize change of molecules energy at transition of electrons from level on level. Thus this transition can be both compelled at impact, and spontaneous.

The kind of functions of molecules distribution at collision changes from the form (8) on the form (9) due to excitation of molecules. But after radiation by molecules of electromagnetic energy the kind of their function of distribution comes back to the form (8), **Figure 4**.

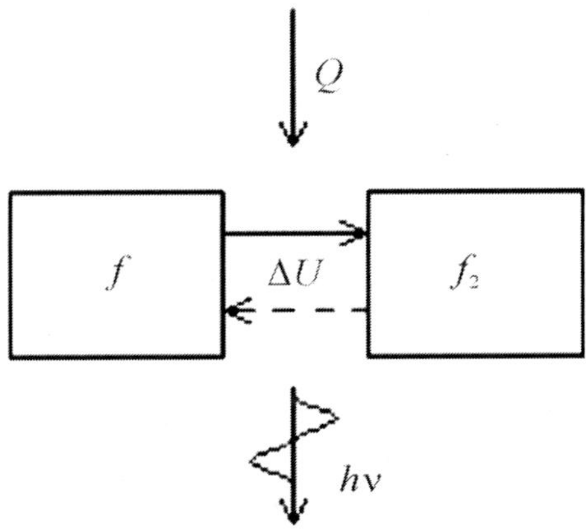

Figure 4. Oscillations of the distribution law of molecules at absorption of heat and radiation of quantums.

In figure it is marked, that is heat Q, submitted to isovolumetric heated up gas goes on change of internal energy of gas ΔU that results in change of the law of distribution of gas molecules at their excitation and to radiation of electromagnetic energy. Thus, not excited molecules have function of distribution (8), and excited-function of distribution (9). At gas simultaneously there are the molecules distributed under forms (8) and (9).

The kinetic Equation (6) has irreversible character on time that reflects the second law of thermodynamics. Really, collisions of molecules with energies W_2 and W_3 will not lead to molecules with energies W and W_1. Mathematically, it is connected by that the left part of the difference in integral of collisions (6) is not symmetric of the right part, as against Boltzmann Equation (5). It is connected by that a fraction of molecules energy after their collision is irreversible lost due to radiation.

Thus, distribution of molecules in the velocities in case of their not elastic impacts cannot be described one function of distribution.

Distribution of molecules on energies in the given problem where it is examined not isolated, but closed isovolumetric molecules system submits to laws (8) and (9). Thus the given system of molecules there is can, both to be heated up from external source, and to emit electromagnetic thermal radiation. But the examined thermodynamic system of molecules is not open since change of it mass is excluded.

Concerning value q in the Equation (6) it is necessary to make the following explanations. In the kinetic Equation (6) interaction only molecules with energies W and W_1 is examined. Before the first collision of molecules, information about resulting system of molecules with energies W_2 and W_3, for example, their relative shares in gas system, no. But at the first collision of molecules with energies W and W_1 already there should be the information on relative shares of molecules with energies W_2 and W_3. Therefore the factor $(\theta - f_2 - f_3)$ is used. Actually this factor before collision of molecules with energies W also W_1 takes into account the information on the condition of molecules system with energies W_2 and W_3. From this point of view probably and a quantum explanation of a factor $(\theta - f_2 - f_3)$.

For the account of Pauli principle it is necessary to use the multiplier $(1 - f_2 - f_3)$. The matter is that at collision of molecules, there is the interaction of their external electronic shells and Pauli's principle starts to act. Due to this multiplier number of molecules transit in the filled conditions for which $f_2 + f_3 = 1$, it is equal to zero. i.e. collisions can result molecules only in yet not occupied conditions [1,6]. In other words, at fulfilment of a condition $f_2 + f_3 = 1$ there can not be a molecule with energies W_2 and W_3. This position should be reflected in the Equation (6).

Thus, quantum consideration shows, that size $\theta = 1$. It also means, that in rarefied gas, by consideration of collisions of molecules with energies W and W_1 other molecules with others energies to consider there is no necessity and a relative share of molecules with these energies at anyone including the first collision, it is necessary to accept to equal one. From the quantum point of view the information on the condition of molecules with energies W_2 and W_3 is determined by Pauli's principle. Therefore classical and quantum explanations of presence of the factor $(1 - f_2 - f_3)$ actually containing the information on the condition of system of interacting molecules after collision, it is equally correctly.

4. DISTRIBUTIONS OF GAS MOLECULES IN THE VELOCITIES IN RADIATING GAS

Taking into account the duality of function of molecules distribution on energies, we shall estimate functions of distribution of molecules in the velocities. First we shall consider a case when the form of distribution corresponds (9), i.e. molecules are in the excited condition. The probability of a finding of a molecule energy W in phase volume $\mathrm{d}\gamma$ can be received under the formula:

$$\mathrm{d}P = f\frac{\mathrm{d}\gamma}{\gamma_1} = f\frac{\mathrm{d}X\mathrm{d}Y\mathrm{d}Z\mathrm{d}p_X\mathrm{d}p_Y\mathrm{d}p_Z}{\gamma_1} = f\frac{\mathrm{d}V 4\pi p^2\mathrm{d}p}{\gamma_1}, \tag{13}$$

where $\mathrm{d}\gamma = \mathrm{d}X\mathrm{d}Y\mathrm{d}Z\mathrm{d}p_X\mathrm{d}p_Y\mathrm{d}p_Z = \mathrm{d}V 4\pi p^2\mathrm{d}p$ is element of phase volume, $\mathrm{d}X\mathrm{d}Y\mathrm{d}Z = \mathrm{d}V$ -the volume, falling one molecule in gas, $\mathrm{d}p_X\mathrm{d}p_Y\mathrm{d}p_Z$ -product of components of molecule impulse, p-the module of molecule impulse, γ_1 -phase volume falling one molecule. Function of distribution according to (9) is $f = \dfrac{1}{1+\exp\left(\dfrac{mv^2/2+\Delta\varepsilon}{kT}\right)}$.

In the Formula (13) integration executed of impulses on angles of spherical layer by thickness $\mathrm{d}p$ that is equivalent to replacement $\mathrm{d}p_X\mathrm{d}p_Y\mathrm{d}p_Z = 4\pi p^2\mathrm{d}p$ [4].

Transit in (13) from the impulse to velocity of molecule v, we shall find:

$$\mathrm{d}P = f\frac{V 4\pi m^3 v^2\mathrm{d}v}{\gamma_1} = \frac{V 4\pi m^3}{\gamma_1}\frac{v^2\mathrm{d}v}{1+\exp\left(\dfrac{mv^2/2+\Delta\varepsilon}{kT}\right)}. \tag{14}$$

In the Formula (14) integration on volume of gas V is carried out.

Taking into account $\mathrm{d}P = \dfrac{\mathrm{d}n}{n}$ where $\mathrm{d}n$ is number of molecules with a speed from v up to $v+\Delta v$, and n the general number of molecules in volume V, from (14) we shall receive:

$$dn = F\frac{v^2 dv}{1+\exp\left(\frac{mv^2/2+\Delta\varepsilon}{kT}\right)}, \quad (15)$$

where the constant $F = \frac{nV4\pi m^3}{\gamma_1}$ can be found from the condition of normalization:

$$n = F\int_0^\infty \frac{v^2 dv}{1+\exp\left(\frac{mv^2/2+\Delta\varepsilon}{kT}\right)}. \quad (16)$$

The integral (16) in elementary functions is not calculated.

The density of probability of molecules distribution in the velocities, using (15), looks like:

$$\frac{dn}{ndv} = \frac{F}{n}\frac{v^2}{1+\exp\left(\frac{mv^2/2+\Delta\varepsilon}{kT}\right)}. \quad (17)$$

The value $\Delta\varepsilon$ is change of molecules energy at transition of electrons from level on level depends on a kind of gas. For rarefied gas in atomic condition the spectrum of radiation is lined. And, on any length of wave at the certain temperature of gas the maximal energy is radiated. For example, in a spectrum of seen light at steam of mercury (in mercury-quartz lamps) the brightest green line with length of a wave $\lambda = 546.1$ nm. Strictly speaking, the value $\Delta\varepsilon$ is necessary to calculate under the formula $\Delta\varepsilon = \sum_i \Delta\varepsilon_i$ where summation executed on all possible energies transitions in a spectrum of gas. For the further analysis we shall assume, that radiation of rarefied gas monochromatic, occurs on one length of the wave corresponding to a maximum of energy of thermal equilibrium radiation of absolutely black body. Energy of electrons transition $\Delta\varepsilon$ at which the maximum of energy of thermal radiation is observed, calculate from a ratio (the Wien's law) $\Delta\varepsilon = h\nu = 4.96\,kT$, [3].

Let's designate in (16) value $\frac{mv^2}{2kT} = X$. In result the Formula (16) it is possible to transform:

$$n = F\frac{1}{2}\left(\frac{2kT}{m}\right)^{\frac{3}{2}}\int_0^{\infty}\frac{\sqrt{X}\mathrm{d}X}{1+\exp(X+4.96)}$$
$$= F\frac{1}{2}\left(\frac{2kT}{m}\right)^{\frac{3}{2}}\times 6.131\times 10^{-3}. \qquad (18)$$

In the Formula (18) and lower all integrals are found by the numerical method.

Hence, value

$$F = 4\pi a\left(\frac{m}{2\pi kT}\right)^{\frac{3}{2}} n \qquad ,(19)$$

where constant a = 144.5. Size a and number p used for comparison of distribution (17) with Maxwell distribution [4]:

$$\frac{\mathrm{d}n}{n\mathrm{d}v} = 4\pi\left(\frac{m}{2\pi kT}\right)^{\frac{3}{2}} v^2 \exp\left(-\frac{mv^2/2}{kT}\right). \qquad (20)$$

Substituting (19) in (17), we find:

$$\frac{\mathrm{d}n}{n\mathrm{d}v} = 4\pi a\left(\frac{m}{2\pi kT}\right)^{\frac{3}{2}} \frac{v^2}{1+\exp\left(\frac{mv^2/2+\Delta\varepsilon}{kT}\right)}. \qquad (21)$$

Distribution (21) actually depends on three variables: $v, \Delta\varepsilon$, and T.

Let's consider now a case of not excited molecules when function of distribution has the form (8).

Similarly (15) we shall write down:

$$\mathrm{d}n = F\frac{v^2\mathrm{d}v}{\exp\left(\frac{mv^2/2+\Delta\varepsilon}{kT}\right)-1}. \qquad (22)$$

As before we accept $\Delta\varepsilon = 4.96\,kT$. Constant F we shall find from the condition of normalization:

$$n = F\int_0^\infty \frac{v^2 \mathrm{d}v}{\exp\left(\frac{mv^2/2 + \Delta\varepsilon}{kT}\right) - 1}. \qquad (23)$$

Hence:

$$F = 4\pi b\left(\frac{m}{2\pi kT}\right)^{\frac{3}{2}} n, \qquad (24)$$

where b = 143.8.

Thus, distribution of not excited molecules looks like:

$$\frac{\mathrm{d}n}{n\mathrm{d}v} = 4\pi b\left(\frac{m}{2\pi kT}\right)^{\frac{3}{2}} \frac{v^2}{\exp\left(\frac{mv^2/2 + \Delta\varepsilon}{kT}\right) - 1}. \qquad (25)$$

In **Figure 5** curves of functions of distribution which are described by Formulas (21), (25)-curve 1 are shown, and Maxwell distribution (20)-curve 2. The temperature of gas (air with equivalent molecular mass 29) is accepted equal T = 300 K.

From figure it is visible, that the difference between distributions (21) and (25) practically is absent. Distinction between them and Maxwell distribution does not exceed 1%. The curve of Maxwell distribution passes a little below the curves plotted under Formulas (21) and (25) only in areas of the most probable velocity of molecules.

Hence, for calculation of speeds of molecules and other statistical parameters of molecular radiating systems with the big accuracy it is possible to use Maxwell distribution.

Comparison of distributions (21) and (25) and Maxwell distributions shows, that Maxwell distribution (20) is some average distribution between (21) and (25).

That distributions of velocities of the excited and not excited molecules (21) and (25) are numerically practically identical, is defined by action of the law of quantity movement conservation of molecules p_i before and after collision which for differentials looks like:

$$\mathbf{d}\boldsymbol{p} + \mathbf{d}\boldsymbol{p}_1 = \mathbf{d}\boldsymbol{p}_2 + \mathbf{d}\boldsymbol{p}_3 \quad (26)$$

Concerning the law of conservation (26) the following

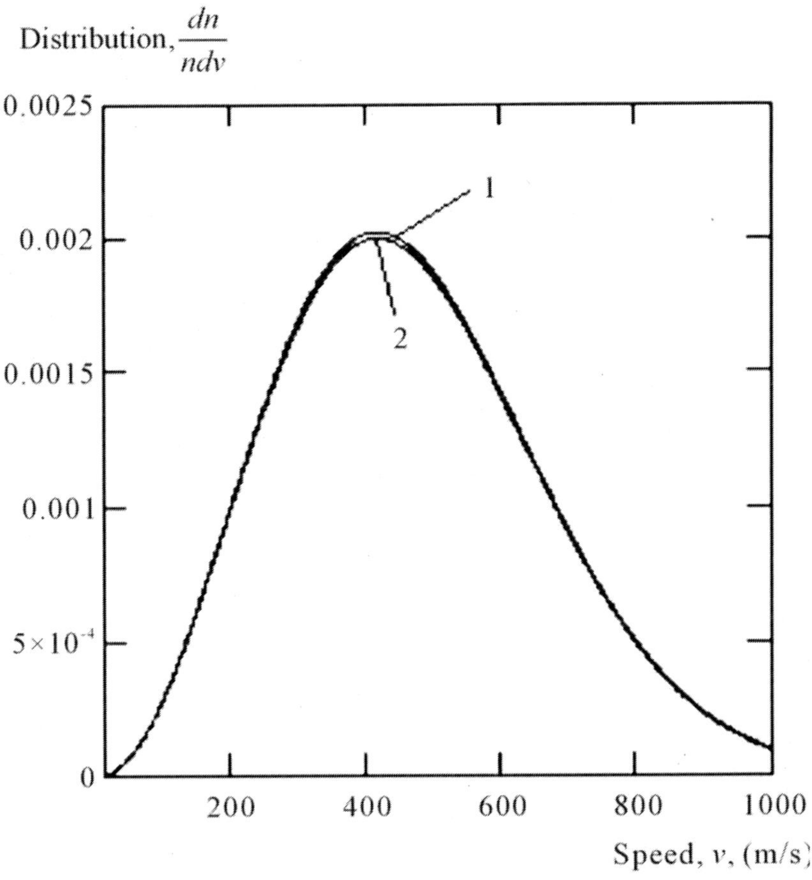

Figure 5. Distribution laws of molecules in the speeds in rarefied radiating gas.

is necessary remark. It is applicable and for a situation when in the condition of balance of system the left part reflects quantity of movement of particles leaving some phase volume at collision, and the right part quantity of movement of the particles passing at collision in this phase volume. The same can be attributed and to the law of conservation of energy (12).

From (26) it is received:

$$(\mathrm{d}p)^2 + (\mathrm{d}p_1)^2 + 2\mathrm{d}\boldsymbol{p}\mathrm{d}\boldsymbol{p}_1 = (\mathrm{d}p_2)^2 + (\mathrm{d}p_3)^2 + 2\mathrm{d}\boldsymbol{p}_2\mathrm{d}\boldsymbol{p}_3 \quad .(27)$$

At interaction of molecules in the condition of balance, the quantity of molecules leaving and passing in the same phase volume is kept. Therefore it is possible to write, that $\mathrm{d}p\mathrm{d}p_1 = \mathrm{d}p_2\mathrm{d}p_3$ [3]. This equality is used at deduction of Boltzmann kinetic Equation (5). Hence, from (27) we have:

$$(\mathrm{d}v)^2 + (\mathrm{d}v_1)^2 = (\mathrm{d}v_2)^2 + (\mathrm{d}v_3)^2 . \quad (28)$$

The ratio (28) shows that change of velocities of molecules 2 and 3 can occur only owing to change of velocities of molecules without index and 1, but not due to radiation by the excited molecules 2 and 3 of electromagnetic quantums. It also determines similarity (**Figure 5**) of distributions (21) and (25), despite of their various forms.

5. ENERGY ASPECTS OF RAREFIED GAS RADIATION

Let's find a share of energy which passes on radiation of gas.

The general energy of n gas molecules can be found under the formula:

$$U_{\pm} = \int_0^{\infty} W \mathrm{d}n_{\pm} , \quad (29)$$

where the top sign attributes to the excited molecules, and bottom to not excited.

Using Formulas (15) and (22), we shall find:

$$U_{\pm} = F_{\pm} \int_0^{\infty} \frac{\left(mv^2/2 + \Delta\varepsilon\right) v^2}{\exp\left(\dfrac{mv^2/2 + \Delta\varepsilon}{kT}\right) \pm 1} \mathrm{d}v . \quad (30)$$

Carrying out separately integration for the excited and not excited molecules, we have:

$$U_{+} = \frac{2a}{\sqrt{\pi}} nkT \int_0^{\infty} \frac{(X + 4.96) X^{1/2}}{\exp(X + 4.96) + 1} \mathrm{d}X = 6.52\, nkT , \quad (31)$$

$$U_{-}=\frac{2b}{\sqrt{\pi}}nkT\int_{0}^{\infty}\frac{(X+4.96)X^{1/2}}{\exp(X+4.96)-1}\mathrm{d}X$$
$$=6.517\ nkT. \tag{32}$$

As before, it is made the substitution $\frac{mv^2}{2kT}=X$. Besides sizes of constants F from Formulas (19) and (24) are used.

From (31) and (32) follows, that energy of radiating gas much more exceeds internal energy of ideal gas U = 1.5 nkT. It is connected by that the most fraction of energy Q passing in gas system transit on excitation of molecules. In condition of balance examined isovolumetric gas system, **Figure 4**, on radiation of gas is spent energy:

$$\Delta U=Q=U_{+}-U_{-}=3.2\times10^{-3}nkT. \tag{33}$$

Let's find, following [7], the energy of thermal radiation falling 1 quantum.

Using Bose-Einstein's distribution for photon gas [4], we shall find volumetric density of energy of radiation:

$$\varepsilon=\frac{2}{(2\pi\hbar)^3}\int_{0}^{\infty}\frac{cp\cdot4\pi p^2\mathrm{d}p}{\exp(ap)-1}=\frac{8\pi c}{(2\pi\hbar)^3}\frac{\pi^4}{15a^4}, \tag{34}$$

where p is the impulse of quantum, also it is designated $a=\frac{c}{kT}$.

Number of quantums in unit of volume equally:

$$n_0=\frac{2}{(2\pi\hbar)^3}\int_{0}^{\infty}\frac{4\pi p^2\mathrm{d}p}{\exp(ap)-1}=\frac{8\pi}{(2\pi\hbar)^3}\frac{2\varsigma(3)}{a^3}, \tag{35}$$

where $\varsigma(3)\approx1.202$ is dzeta-function of Riemann.

The average energy of thermal radiation falling 1 quantum is equal:

$$\varepsilon_1=\frac{\varepsilon}{n_0}=\frac{\pi^4}{30\varsigma(3)}\frac{c}{a}=\frac{\pi^4}{30\varsigma(3)}kT\approx2.7kT. \tag{36}$$

Let's find quantity of simultaneously radiating molecules $N = \frac{\Delta U}{\varepsilon_1} = 1.18 \times 10^{-3} n$. Hence, simultaneously radiate only 0.118% of all gas molecules.

If to consider rarefied gas in the given thermodynamic system as the converter of thermal energy in energy of electromagnetic radiation on the basis of the carried out analysis it is possible to make the following conclusion. Energy of transition from the thermal form, for exampleby heating a vessel with gas, in energy of thermal radiation is very small part of the general internal energy of gas. That gas practically submits to Maxwell distribution testifies to it. If to use conditional efficiency of transformation of thermal energy into energy of radiation for gas system it makes:

$$\eta = \frac{\Delta U}{U_+} 100\% \approx 0.05\% \quad (37)$$

6. CONCLUSIONS

For calculation of the molecules velocities and other statistical parameters of molecular radiating systems with the big accuracy it is possible to use Maxwell distribution.

In spite of the fact that transitions of molecules at their collisions from one form of distribution in another practically do not influence distribution of molecules velocities, these transitions are determining in distribution of quantums of thermal radiation on their energies.

Andrey N. Volobuev, Eugene S. Petrov, Eugene L. Ovchinnikov In ideal rarefied gas simultaneously radiate only 0.118% of all molecules.

REFERENCES

1. E. M. Lifshits and L. P. Pitaevsky, "Physical Kinetics," Fizmatlit, Moscow, 2002.
2. V. P. Kochanov, "The Sated Two-Photon Absorption of Unifrequent Radiation in Gas Mediums," Quantum Electronics, Vol. 42, No. 2, 2012, pp. 111-116.
3. V. G. Levich, "Theoretic Physics Course," Phismatgis, Moscow, 1962.[Citation Time(s):7]
4. A. N. Matveev, "Molecular Physics," Higher School, Moscow, 1981.
5. L. D. Landau and E. M. Lifshits, "Statistical Physics," Science, Moscow, 1976.
6. O. Madelung, "The Theory of a Solid Body," Science, Moscow, 1980.

7. J. B. Zeldovich and I. D. Novikov, “The Theory of Gravitation and Evolution of Stars,” Science, Moscow, 1971

CHAPTER 13

Review on Natural Gas Thermopysical Property Measurement Techniques

Mert Atilhan[1] and Santiago Aparicio[2]

[1] *Qatar University, Department of Chemical Engineering, Doha, Qatar*

[2] *University of Burgos, Department of Chemistry, Burgos, Spain*

1. INTRODUCTION

Since it is impossible to measure the thermodynamics properties of all systems in nature, we must rely upon mathematical models to extrapolate the available experimental data. In order to develop such models, very accurate experimental data are necessary for selected complex mixtures, such as those that exist in natural gas. For this reason researchers must collect the most important and fundamental thermodynamics properties for such systems. Two of the most important thermodynamics properties are the pressure volume (density) and temperature (PρT) surface and the phase equilibrium properties of mixtures. Accurate volumetric property data are used in custody transfer operations for natural gas. Also accurate PρT data are necessary for calculating energy functions. On the other hand, phase equilibria data are needed mostly for design calculations involving separation processes. Additionally, very accurate phase equilibrium knowledge is necessary for natural gas transfer through pipelines to avoid condensation in the pipelines. Atilhan et. al. [1] have shown that even widely used equations of state (EOS) such as Peng-Robinson or Redlich-Kwong (RK) cannot predict the retrograde condensation region for simple natural gas-like mixtures that do not contain heavy fractions.

When natural gas rises from the reservoir to the ocean floor at offshore platforms, the stream temperature can drop quickly (perhaps 5 to 10 °C) until it reaches the surrounding ocean temperature. This rapid temperature drop at high pressure along with moisture in the natural gas stream make conditions favorable for gas hydrate formation in the pipeline. Hydrates can cause several serious problems such as: plugging the pipeline and blowouts [2]. Similar problems are as well common in natural gas compression facilities at offshore and onshore processing plants. Such problems can be avoided by increasing the temperature and insulating the stream that comes from the ocean bed, or by lowering the pressure of the pipeline. Another possible solution is lowering the dew point of water in the stream by adding polar solvents to the line such as methanol or glycols. In order to apply all these methods, accurate knowledge on PρT behavior of the natural gas stream is necessary.

Because of their importance, density measurements essential for both industrial applications and scientific research. Very accurate PρT data is required not only to calculate custody transfer of natural gas in pipelines but also to develop new EOS for industrial and scientific use. Experimental PρT data is employed to calculate thermal properties of fluids required for industrial process design calculations. Loss of accuracy from density predictions directly impacts processes; therefore only exceptionally good density values ensure good thermal properties [3].

Knowledge of temperature, pressure and composition enables determination of the density from an EOS. The equation most widely used in custody transfer of natural gas is the Detailed Characterization Method or AGA8-DC92 EOS developed by American Gas Association (AGA) in 1992 [4]. This EOS was derived using an extensive and reliable experimental PρT database that included real natural gas mixtures as well as high order hydrocarbon mixtures (mostly binary mixtures of natural gas components). AGA8-DC92 EOS has different accuracy regions as shown in Figure 1.

As seen in Figure 1, the maximum uncertainty claimed for the EOS is 0.1% in region 1, 0.3% in region 2, 0.5% in region 3 and 1.0% in region 4. However, the equation is valid only for lean natural gas mixtures over this wide range of conditions, and its ability to describe rich natural gases is untested. The equation cannot perform equilibrium calculations, as it is only valid for gas phase calculations. Also, use of the equation is not recommended near the critical point. The deviations from density measurements are 2 to 2.5% for North Sea natural gas samples at 8 to 17 MPa and 40 to 80 °C [6]. Following this, the Gas Research Institute (GRI) and National Engineering Laboratory (NEL) started a collaborative, joint industry project to extend the range of applicability of AGA8-DC92 EOS for natural gas mixtures to include the gas compositions observed in the North Sea. GERG-2008 EOS is the latest benchmark EOS that addresses the drawbacks of AGA8-DC92 EOS and it has proven to have better uncertainty in PρT predictions [7].

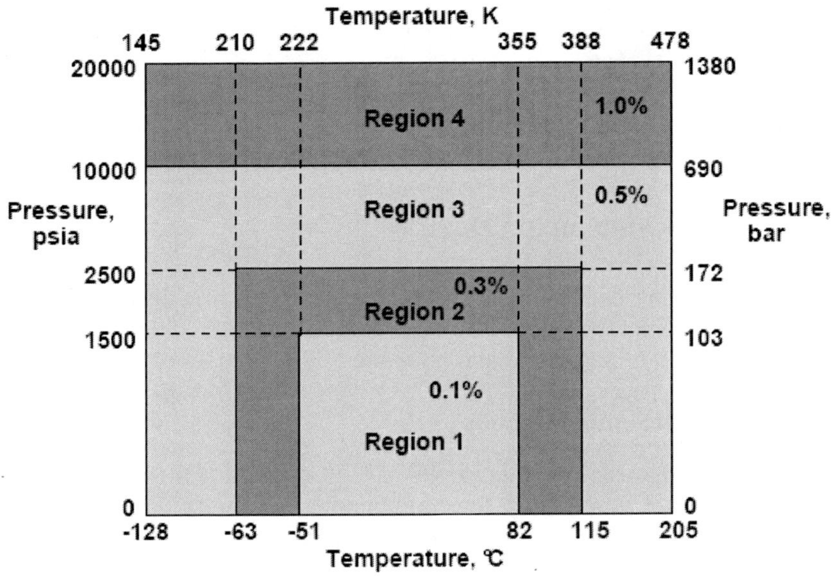

FIGURE 1. AGA8-DC92 EOS uncertainty regions [5].

2. DENSITY MEASUREMENT TECHNIQUES

2.1. Introduction

This section contains reviews of several experimental methods for density measurements, and discusses the relative strengths and weakness of each method. An EOS can describe the thermodynamic state or vapor liquid equilibrium (VLE) of pure fluids and mixtures with accuracy that depends upon the application. The accuracy of an EOS depends upon the experimental data used during development of the equation. Historically, the quality of predictions obtained from EOS has improved dramatically as advanced technologies and new instrumentation have become more common for experimental methods. Among the thermodynamic properties, density is the most directly predicted property using EOS. The measured densities should be approximated by suitable EOS and the measurements should be traceable to the International System of Units [8]. According to Kleinrahm et. al. [9], the following considerations are important when deciding upon a density measurement technique:

i. Large pressure and temperature range for wide operations.

ii. Low total uncertainty and high accuracy of the method for the overall range.

iii. Simplicity in design and ease of maintenance and operation.

iv. Little time required for each data point measurement.

2.2. Density Measuring Devices

Several different density-measurement techniques are described in this section, including: speed of sound methods, vibrating body techniques, continuous weighing method and buoyancy-based densimeters.

2.2.1. Speed Of Sound Methods

Speed of sound measurements can be used to determine the performance of an equation of state for thermodynamics property predictions. By correlating the speed of sound to thermodynamics properties, one can build experimental devices and investigate solid, liquid and gas thermodynamics properties for pure components and mixtures. Densities and isothermal and isentropic compressibility factors result from speed of sound measurements experiments [10]. Based upon a pulse technique described by Daridon et. al. [11], a cylindrical-shaped cell is used to measure ultrasonic waves. In the pulse technique, effects of pressure upon piezo-electric materials are isolated by separating piezo-electric elements from the fluid studied. The speed results from the measurements of the transit time through the sample and the length of passage, which is a function of temperature and pressure. Density comes from:

$$\rho(P,T)=\rho_0(P_0,T_0)+\int_{P_0}^{P}u^{-2}dP+\int_{P_0}^{P}\left(\alpha^2/C_p\right)dP \tag{1}$$

In above equation u is the sound speed, α is the isobaric coefficient of thermal expansion, *Cp* is the isobaric heat capacity and P_0 is the atmospheric pressure. The sum of these terms gives the density with as a function of pressure at different temperatures. The first integral, where can be expressed as a polynomial in pressure with coefficients expressed as polynomials in temperature, can be evaluated along the isotherms considered. By using thermodynamics relations for α and Cp, the second integral can be calculated iteratively. The second integral is a few percent of the first integral. More detailed discussion on numerical evaluations for such measurements appears in [10] and [11].

2.2.2. Vibrating Devices

Vibrating tubes and vibrating forks are common density measuring techniques. These devices measure the fluid density of interest by determining the oscillation frequency of the vibrating element in the fluid. These instruments provide accurate results quickly. However, frequent calibration is necessary for this apparatus to maintain its accuracy [12]. Moreover, when the density of the fluid is vastly different from air or pure water (frequently used as reference fluids because of their well-known thermophysical properties) the uncertainty of the measurements increases as reported by Kuramoto et al. [8].

2.2.2.1. Vibrating wires

In vibrating wire densimeters, a wire carrying a diamagnetic weight is suspended in the fluid to be monitored. The wire is placed in a robust position in a uniform magnetic field provided by permanent magnet in both vertical and horizontal directions. When an alternating current passes through the power source to the wire, interaction starts with the current and the magnetic field. This leads to induced harmonic motion that is orthogonal to the magnetic field and the wire. If the mass, density and the dimension of all the solid components of the system are known, the resonant frequency of the wire can be determined experimentally under vacuum conditions. If the viscosity of the fluid of interest is known, experimental measurement of resonant frequency of wire velocity provides the fluid density [13; 14]. Although the vibrating wire technique is suitable for a wide range of pure fluid and mixture gas density applications, it suffers from problems such as surface tension on the wire, adsorption on the weight, detailed knowledge need of exact dimensions of the wire and the assembly. However, the device is used widely as a primary densimeter device because it has a simple operating principle and allows development of an exact physical model. Density, in principle, can be calculated directly from the theory.

2.2.2.2. Vibrating tubes

Vibrating tube densimeters consist of an assembly that includes two thin walled metallic or glass tubes bent in Y or V shapes as shown in figure 2. A permanent magnet and drive coil reside between these two tubes. Generally, a drive coil and a permanent magnet are placed in the middle of the two tubes. The drive coils and magnet are mounted on the opposite legs of the tubes. Each coil and magnet on the side leg forms a pick-off circuit. Alternatively, attractive and repulsive magnetic fields between the coils and magnets are provided by sending alternating current to the drive coil.

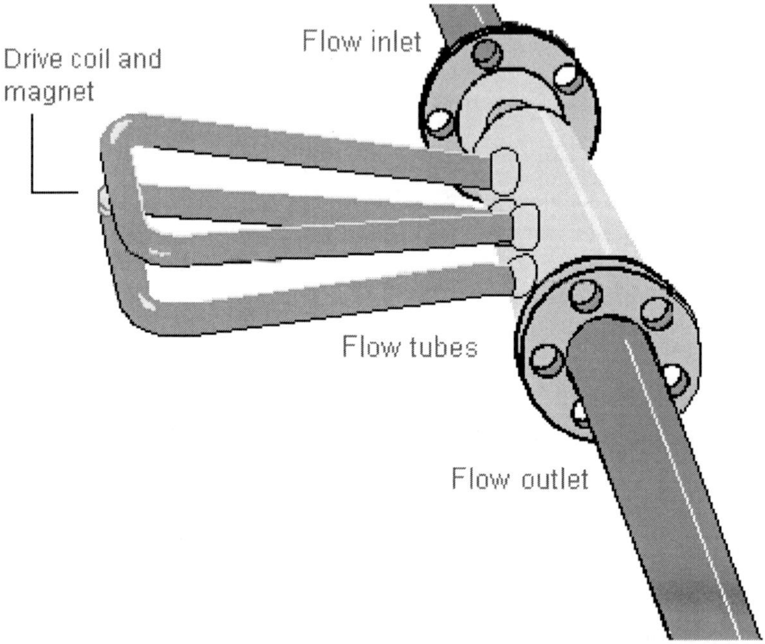

FIGURE 2. Vibrating tube densimeter scheme.

Because the drive coils and the magnets are installed on the opposing side of the tubes, a sine wave generated by the two pick-up circuits represents the motion of one tube relative to the other. The sine waves are in phase if there is no flow in the tube. The density of the fluid is:

$$\rho = K\tau^2 + L \tag{2}$$

where, tube parameters K and L are both pressure and temperature dependent and τ is the period of vibration. Because it is not possible to determine the temperature and pressure dependence of K and L, measurements are performed at the same temperature and pressure conditions with the sample. A reference fluid of well-known properties is used for this reason. Finally, the corresponding density difference equation is:

$$\rho - \rho_r = K\left(\tau^2 - \tau_r^2\right)$$

(3)

where subscript *r* stands for the reference fluid. ρandτ are not exactly linear, and this must be taken into consideration. The vibrating tube densimeters are designed for rapid operation and they perform very precise measurements of density differences. If one assumes linearity between ρ and τ, the highest levels of accuracy expectations are not achieved.

2.2.3. Expansion Devices: Burnett Method

Another well-established and widely-used density measurement device is the Burnett apparatus. Burnett [15] suggested a technique to measure the densities of sample fluids without measuring the mass or volume directly. An expansion device contains two cells. During operation, the sample is charged initially into the first cell and, after pressure and temperature measurement, expanded into the second cell. The ratio of the final volume to the original volume equals the ratio of densities before and after the expansion. Only pressure and temperature are measured before and after expansion of the sample from a single volume (V_A) into the combination of the original volume and a second volume (V_A+V_B). Some of the gas goes through a sequence of isothermal expansions into a chamber, which is evacuated every time the expansion takes place. Both virial coefficient and gas density can be calculated with this method. The ratio of the densities before and after the expansion is calculated for each expansion:

$$\rho_i = \rho_m \pi_\infty^{m-i} \prod_{j=i+1}^{m} \left[\frac{1+\gamma_{ab} P_j}{1+\gamma_a P_{j-1}} \right]$$

(4)

where

$$\pi_\infty = \lim_{P \to 0} \frac{V_a + V_b}{V_a}$$

(5)

In equation 4, γaγaand γabγab are the pressure distortions of the volumes Va and $(Va + Vb)$ respectively, π∞π∞is the zero pressure cell constant, ρmρmis the density at the lowest pressure and subscripts i and superscript m indicate the value after the i-th and m-th (last) expansions, respectively. A serious problem that can affect the Burnett apparatus is adsorption of the sample gas on the inner surfaces of the measuring cell [16]. Also a Burnett apparatus is difficult to automate fully because of frequent valve operations. Because of error accumulation, very high precision is necessary in the pressure measurements, which necessitates use of high-quality dead-weight gauges. The adsorption affect can be minimized by the using two cells with a surface area ratio almost equal to the volume ratio [17]. Eubank et al. have formulated new adsorption correction schemes, based upon the BET adsorption isotherm [18].

2.2.4. Continuously Weighed Pycnometer Method

In the continuously weighed pycnometer method, the mass of the sample is determined by direct weighing of the cell. A typical pycnometer consists of a weight measurement system, constant temperature bath, temperature control system and data acquisition system, a volume bellows cell for changing pressure and density without transferring mass, and a high vacuum system [19]. The major component of this method is a constant volume pycnometer suspended from a digital balance. The pycnometer can be filled and evacuated with an extension tube that enables faster measurements and reduces operator errors. The mass of the pycnometer when empty and when filled with fluid is measured by a digital balance. The density of the fluid being measured at constant temperature and pressure is calculated from the measured mass value of the fluid and the known volume of the pycnometer. One disadvantage of this method is that the long feed tube exposes part of the sample to ambient temperature making it impossible to measure mixture densities when the sample exists as one phase at the cell set point temperature and at room temperature.

2.2.5. Hydrostatic Buoyancy Methods

The hydrostatic buoyancy force technique is based upon Archimedes' Principle. Basically, Archimedes' Principle states "when a solid body is immersed in a fluid, it displaces a volume of fluid the weight of which is equal to the buoyancy force exerted by the fluid on the sinker." This means that the buoyancy force is proportional to the density of the fluid in the measuring cell under pressure. This principle can be applied to determine the gas density of any pure fluid or mixture. Historically, improvements have appeared in the application of buoyancy method based densimeters.

2.2.5.1. Classical methods

In classical hydrostatic buoyancy densimeters, an object (sinker hereafter), usually a sphere or cylinder, is suspended from a commercial digital balance by a thin wire. The fluid is kept in a pressure cell at constant temperature using a temperature control mechanism. The sinker is submerged in the fluid and weight of the sinker is constantly monitored. According to Archimedes' principal, the apparent loss in the true weight of the sinker is equal to the weight of the displaced fluid. Density of the fluid results from:

$$\rho = \frac{m_V - m_a}{V_S(T,P)} \tag{6}$$

In above equation *mv* is the 'true' mass of the sinker in vacuum, *ma* is 'apparent' mass of the sinker in the fluid and *Vs* is the calibrated volume of the sinker, which is a function of temperature and pressure. In such densimeters, several corrections are necessary to reduce the effect of surface tension between the sample liquid and the immersed part of the wire, and the effect of the buoyant force of air on the masses of the analytical balance. Zero shift of balance readings, buoyancy forces on auxiliary devices, adsorption effects and surface tension may reduce the accuracy of such measurements [20].

2.2.5.2. Magnetic suspension devices

To overcome limitations in achievable accuracy, the need for frequent calibration of the apparatus with reference fluids, complexity of operation, limitations on temperature and pressure, Kleinrahm and Wagner [9] introduced an MSD based upon magnetic levitation of the sinker in the measuring cell. The novelty of the magnetic suspension coupling was that it used non-physical-contact force transmission between the sinker in the pressurized cell and the weighing balance at atmospheric pressure, thus allowing a cell design that covered a very wide temperature and pressure range [21]. Then, Kleinrahm and Wagner [20] modified the hydrostatic buoyancy force method by introducing an alternative force transmission method in which they levitated two sinkers through a magnetic suspension coupling. By compensation for surface tension, buoyancy, adsorption effects and shifts in zero-point of the balance, a two-sinker MSD improved the accuracy of the density measurements.

Operation of a two-sinker MSD is rather complex and its advantage is not required for medium or high-density measurements encountered in many practical applications. To extend the instrument range towards higher temperatures and pressures, Wagner et al. [16] have developed a single-sinker densimeter. Although the single sinker design is much simpler than that of the two-sinker densimeter, it is still possible to perform high-accuracy density measurements at relatively low gas densities by applying some of the advantageous features of the two sinker device [22]. The single-sinker densimeter also operates based upon Archimedes' principle and the force transmission comes from levitation of the sinker in the measuring housing of the high-pressure cell. Klimeck *et al.* [23] have concluded that the accuracy of density measurement from a single-sinker densimeter is lower than that from a two-sinker densimeter especially at low densities because it lacks compensation for the adsorption effect. Moreover, the force transmission error has more effect on total density measurement uncertainty than observed in a two-sinker densimeter. Also for small densities, having the load compensation system outside of the measuring cell is less effective than having it inside as with the two-sinker densimeter.

3. VISCOSITY MEASUREMENT TECHNIQUES

3.1. Introduction

Viscosity is a remarkable property for natural gas because of its influence on flow behavior, which is especially important for reservoir conditions. Natural gas viscosity is several orders of magnitude lower than for oil or water, and thus, natural gas mobility in reservoirs is larger than for water or oil. Moreover, gas flow is predominantly laminar in reservoirs, and thus, the influence of viscosity is especially important. The upstream gas industry faces new challenges for precision monitoring of gas supplies, for which accurate and reliable knowledge of the natural gas viscosity is a prerequisite. Davani et al. [24] and Denney [25] analyzed the gas viscosity estimation errors on the gas recovery from a high pressure-high temperature (HPHT) reservoir, showing that a -10 % error in gas viscosity estimation can produce a relative 8.22 % error in estimated cumulative gas production, and a + 10 % error can lead to a relative 5.5 % error in cumulative production. Moreover, uncertainty in gas viscosity data has a direct effect for inflow performance relationship curves [26], Davani et al. [27] showed that a 1 % uncertainty in gas viscosity data leads to a 1 % uncertainty in gas flow rate.

Sanjari et al. [28] reported in recent study that common prediction methods for natural gas viscosity lead to large deviations when applied for high pressure – high temperature ranges, and thus, these methods should be applied with caution for reservoir estimations. This lack of accuracy for current predictive viscosity models, and the large economical impact of viscosity uncertainties, show the need of experimental accurate natural gas viscosity to analyze *i*) the effect of natural gas composition on viscosity for wide pressure – temperature ranges, and *ii*) the predictive performance of current viscosity analyze the predictive performance of current viscosity predictive models and developing more accurate predictive approaches.

Experimental accurate measurement of gas viscosity is very difficult, requires properly designed equipment, especially for high pressure–high temperature conditions, and is very costly both in time and resources. Moreover, possible natural gas mixtures under experimental study is very large considering that natural gas composition depends strongly on the origin, age, and depth of the reservoir [29]. Likewise, conditions of interest, especially the high pressure – high temperature conditions found in many new reservoirs that can be explored with current technologies [30], has a remarkable effect on natural gas viscosity, and discard the use of traditional measurement techniques. The analysis of published viscosity data in the open literature shows its scarcity both in the number of studied systems and the experimental conditions (high pressure data are almost absent) [28,31]. Therefore, systematic studies on natural gas viscosity has to be carried out in wide pressure-temperature ranges and as a function of mixture compositions, for selected mixtures representative of key reservoirs. Our group is involved in a multilaboratory international research project in which natural gas viscosity is measured, and other relevant thermophysical properties, using state-of-the-art equipments [31,32]. As a result of our research, we report in this section a detailed analysis of the available experimental methods for measurement of natural gas viscosity.

3.2. Viscosity Measuring Devices

3.2.1. Rolling Ball Viscometers

Rolling ball viscometers measuring principle is based on the travelling time of a metal or glass ball through a known distance to measure the viscosity of the fluid, Figure 3. The ball of known diameter rolls down a tube of known length, at a known inclination, filled with the fluid under study, under isothermal-isobaric conditions.

The viscosity of the fluid is proportional to the ball travelling time, considering that the fluid flow around the ball is laminar. Sage and Lacey [33] proposed corrections to be applied in case of viscosity measurements under turbulent flow.

The working principle of rolling ball viscometers may be summarized in equation (7):

$$\eta(P,T) = k_1(P,T,\theta) \times t(\theta) \times (\rho_b - \rho) + k_2(P,T,\theta) \tag{7}$$

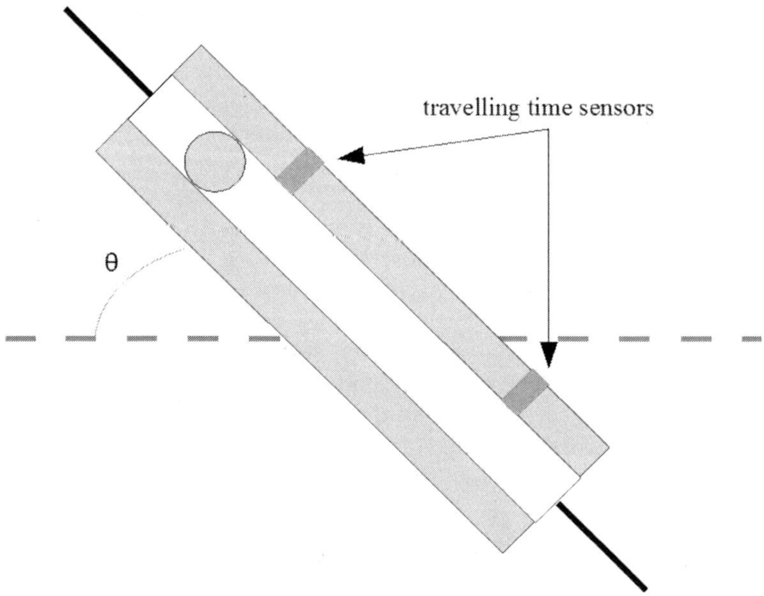

FIGURE 3. Scheme of a rolling ball viscometer.

Where (b and (stand for the ball and fluid density, respectively, *t* for the ball travelling time, and *k1* and *k2* for the calibration constants, which are function of the measurement angle (). Calibration constants are obtained with fluids of known viscosity for the required pressure-temperature ranges. It should be remarked that rolling ball equipments are kinematic viscometers, and thus, and additional instrument is required for density measuring in the studied pressure-temperature ranges.

The use of rolling ball viscometers for gas measurements is very scarce, and high pressure data are almost absent. Sage and Lacey [33] measured viscosity data for methane and two hydrocarbon gases up to 2900 psi, Bicher and Katz [34] measured the viscosities of methane-propane mixtures up to 5000 psi. The uncertainty of viscosity measurements using rolling ball viscometers it is claimed to be ± 2 % when applied for measurements in the liquid state [35], although no detailed uncertainty analysis is available for gas measurements. Dolan et al [36] carried out gas phase measurements for n-butane using a capillary viscometer and showed large difference with rolling ball viscometer data by Sage et al. [37], who claimed a ± 5 % for their gas phase viscosity measurements.

3.2.2. Capillary Tube Viscometers

The working principle of classical capillary viscometers (so-called Rankine viscometers) is based on the introductions of a pellet of mercury into a tube filled with the gas under study, with the mercury pellet completely filling the cross section of the tube. The mercury pellet will reach a steady descending velocity will for any inclination of the tube. The descending pellet will act as a moving piston, which will force the gas through an adjacent capillary, leading to a constant pressure difference across the fine capillary. Gas viscosity is obtained from the measurement of the mercury pellet falling time between two fixed points. Capillary viscometers have been widely studied for gas measurement because of their operational simplicity and the small amounts of required gas. Heath [38] analyzed the performance of capillary viscometers for gas phase measurements, and Kobayashi [39] carried out a detailed analysis of the application of Hagen-Poiseuille's law for gases to the Rankine viscometer. Giddings et al. [40] developed a high-pressure capillary viscometer, which was applied for measurements of methane, propane and their mixtures up to 544.4 atm with a 0.25 % claimed reproducibility. Lee et al. [41] carried out systematic measurements of the viscosity of natural gas mixtures using a capillary viscometer with an estimated accuracy of ±2.00. In a recent work, May et al. [42] obtained reference viscosity data for several gases, including methane, using a modification of capillary viscometer.

3.2.3. Vibrating Wire Viscometers

The falling body and the capillary tube methods for viscosity measurements require hydrodynamic corrections and approximations for ends, edges, and walls. These applied corrections are not always known, and in most cases are sources of important errors. These corrections are avoided by the use of vibrating wire based viscometers. These apparatus are based on the damping of the vibrations of a wire in the fluid under study. An applied external field disrupts the wire immersed in the fluid, leading to periodic oscillations. In the free-decay mode the damping of the oscillations depends on the viscosity and density of the fluid [43]. This technique can be applied in a straightforward manner to low viscosity fluids, in fact most of the available contemporary natural gas viscosity data have been measured using vibrating-wire based approaches. Tough et al. [44] and Trappeniers [45] were the first to apply vibrating wire devices for the measurement of low viscous fluids and gases at high pressures. Assael et al. [46] carried out viscosity measurements of a natural gas mixture with a claimed uncertainty of ± 1 % up to 15 MPa. Langelandsvik et al. [47] measured viscosity of three natural gas mixtures with a ± 1% estimated uncertainty up to 5.0 MPa. Schley al. [48] measured viscosity of methane and two natural gases up to 29 MPa with a ±0.3 % and ±0.5 % estimated uncertainties for methane and gas mixtures, respectively. Likewise, vibrating-wire viscometers are considered as quasi-primary measurement methods [43].

3.2.4. Falling Body Viscometers

The experimental setup for these instruments is very similar to that for rolling body viscometer with the exception that the ball is replaced with a piston. These equipments have been widely for measuring liquid viscosity but measurements for low viscous gases are scarce. In most cases the viscometer arrangement is vertical, which is a serious limitation for gas measurements considering the very short travelling times. Heidaryan et al. [49] used a falling body viscometer for measuring methane viscosity up to 140 MPa.

A modification of traditional falling body viscometers has recently proposed to withdraw the aforementioned disadvantages and to apply this equipment to large pressure temperature ranges. In this equipment a piston is placed inside a cylindrical measuring chamber filled with the fluid (gas or liquid) under study, the piston is driven electromagnetically by two coils located at opposite ends and the time taken by the piston to complete one motion is correlated to the viscosity of the fluid in the measuring chamber by a working equation. A scheme of the equipment installed in our laboratory is showed in Figure 4. The measurement chamber is placed at 45º inclination, and the setup allows measurements from 0.02 to 10000 cp, simply changing the used piston.

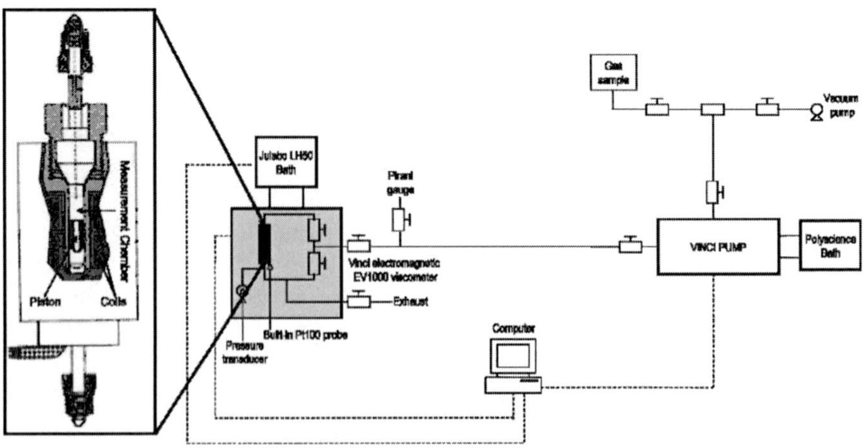

FIGURE 4. Electromagnetic viscosity measuring device [31].

The main advantage of this equipment is the ability to carry out fast measurements in wide pressure-temperature ranges. Nevertheless, Thomas et al. [50] and Viswanathan [51] reported problems for gas viscosity measurements using electromagnetic viscometers, mainly raising from the poorly defined pressure-temperature dependence of the measurement chamber properties. Therefore, we proposed a new calibration method using this equipment to avoid these problems, which was validated against reference viscosity data, leading ± 0.1 % reproducibility and to ± 2.5 % and ± 4.0 % uncertainties, for pressures lower than 30 MPa and higher than 30 MPa, respectively. Nevertheless, we should remark that the uncertainty values are derived from the uncertainties obtained from the calibration fluids. This is a remarkable problem for viscosity measurements, the use of secondary methods requires the knowledge of highly accurate viscosity data for reference fluids, in wide pressure – temperature ranges, which is not currently available [43]. Nevertheless, the use of electromagnetic based viscometer allow to obtain wide collection of viscosity data for gas mixtures, in wide pressure – temperature ranges, with acceptable accuracy and at moderate costs.

4. PHASE EQUILIBRIA MEASUREMENT TECHNIQUES

4.1. Introduction

The prediction of natural gas vapor–liquid equilibria is of primary importance for industrial purposes [52]. This property, together with the knowledge of PVT behavior, is required in all the stages of natural gas production / transportation chain. The vapor–liquid equilibria of natural gas mixtures is commonly analyzed using the pressure–temperature projection, so–called phase envelope, Figure 5, in which a curve separating the two–phase (vapor–liquid equilibria) and single–phase (vapor or liquid) regions is plotted.

In this curve, three points are of remarkable importance: *i*) the critical point, where bubble and dew points curves meet, *ii*) the cricondenbar, the maximum pressure at which vapor and liquid phases may coexist, and *iii*) the cricondentherm, the highest dew point temperature. Therefore the importance of an accurate knowledge of natural gas phase envelopes could be summarized in three main areas: *i*) reservoir engineering, *ii*) custody transfer and *iii*) gas transportation through pipelines [53].

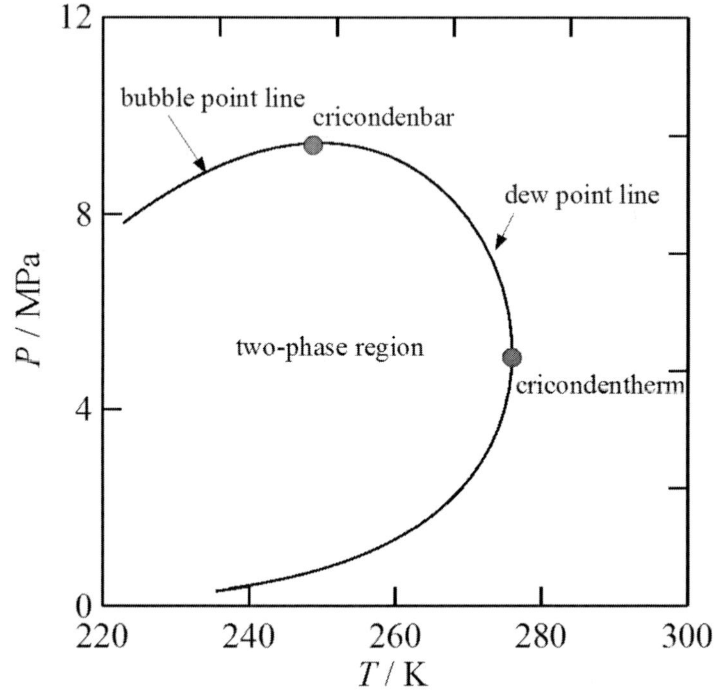

FIGURE 5. Phase envelope of a typical natural gas.

For reservoir engineering purposes, care must be taken to maintain the gas in single phase conditions because during exploration a pressure depletion may lead to an erroneous determination of reservoir composition (and thus to incorrect field–development designs), because dew point curve is crossed and liquid is formed, and during production liquid drop-out may lead to valuable hydrocarbons left behind in the reservoir [54]. If liquid drop-out appears in the wellbore, it may lead to liquid loading concerns. For custody transfer purposes, the inaccuracy in the knowledge of phase envelopes may lead to contractual disputes rising from the contract sales gas dew point specifications between the company that sales the gas and the purchaser. For transportation through pipelines [29], it is essential to maintain the fluid pressure above the cricondenbar to protect the compressors which may be damaged by the presence of liquid drops [55]. Moreover, the sizing of the compressors depends on the value of the predicted cricondenbar, and thus, inaccuracy in the knowledge of this value may lead to an oversizing of the expensive compressors (if the real value is lower than the predicted ones) or to a two–phase flow (if the real value is above the predicted one) along the pipelines. Hence, accurate knowledge of phase envelopes allows reducing the design margins, to optimize the pipelines capacity, and thus, to improve the economic viability of new pipelines projects [56].

Therefore phase envelopes must be known before production / transportation operations are designed for any new gas–field. The most accurate and reliable way to obtain the phase envelope for any natural gas mixtures is its determination through experimental measurements [12] Nevertheless, these measurements require state–of–the–art apparatus, which are very costly, both in time and resources, and, considering that the composition, and thus properties, of natural gases can vary widely depending upon the reservoir from which the fluid comes, it is almost impossible carry out measurements for all the possible mixtures over the wide temperature and pressure ranges required. Thus, the common way to obtain phase envelopes for design purposes in the gas industry is through available theoretical models, using gas chromatography determined compositions, and thus, the reliability of the designs stands on the accuracy of the used models. In the natural gas industry, equations of state (EOS) are the common choice for phase equilibria predictions, and cubic EOS are mainly used because of their simplicity. Multiparametric EOS developed in the last years, such as AGA8–DC92 [4], which are used with high accuracy for density predictions are not applicable for phase equilibrium calculations or for liquid properties, and thus other EOS such as cubic ones or new multiparametric GERG2008 [7] are used to characterize natural gas. Nonetheless, the predictive ability of available models for phase equilibria have to be studied against accurate experimental data to test their reliability, and thus, experimental measurements for selected natural gas mixtures using accurate methods are required [57].

Experimental methods for measurement of vapor-liquid equilibria in high pressure conditions may be classified as *i*) analytical and *ii*) synthetic [58].

4.2. Phase Equilibria Measuring Methods

4.2.1. Analytic Methods

For the analytic method, pressure and temperature are fixed and then phase separation is produced, sampling for each phase under equilibrium is carried out and then composition is determined by chromatographic methods. This technique may carried out using static or dynamic circulation approaches. The experimental results from this type of experimental equipment are usually isothermal or isobaric phase diagrams. Natural gas mixtures usually contain small fractions of heavy components (C_{6+} fractions), which are difficult to measure accurately through chromatographic methods, and thus, this approach is not very useful for the study of natural gas mixtures.

4.2.2. Synthetic Methods

In the synthetic method a mixture of known composition is prepared and its behavior observed as a function of pressure or temperature. The experimental results from this type of equipment are isopleth phase diagrams. The synthetic method may be subdivided into visual and non-visual.

4.2.2.1. Visual synthetic methods

Chilled mirror dew point meters is the simplest and most widely applied method of hydrocarbon dew point measurement in the gas industry. These instruments have a metallic mirror surface inside a high-pressure sample cell. The instruments are also equipped with a glass viewing port through which operator can observe the mirror surface. Mirror dew point meters normally are used for periodic spot check measurements [59].

The principle of the mirror dew point meter is to observe the very first signs of condensate, Figure 6. Therefore, cooling samples for several pressures allows to build the phase envelope, Figure 7.

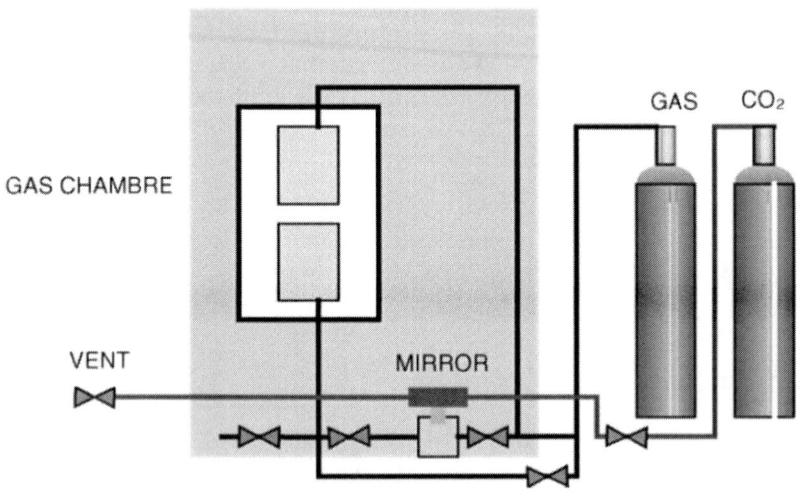

FIGURE 6. Scheme of a chilled mirror equipment for phase envelope measurements.

The main difficulties in making such a measurement lie in the characteristics of hydrocarbon condensates. The natural gas condensates are colorless and have low surface tension. This means that the liquid film that forms as the sample cools through the dew point temperature is almost invisible to operators. At the same time, the decreasing temperature measurement reading must be observed as the mirror temperature drops. To achieve the best sensitivity and repeatability of measurement, the rate of mirror cooling is critical and should be as slow as possible through the region in which the dew point is likely to be found [60]. The subjective nature of such a measurement technique can result in large uncertainty in natural gas dew point measurement [59].

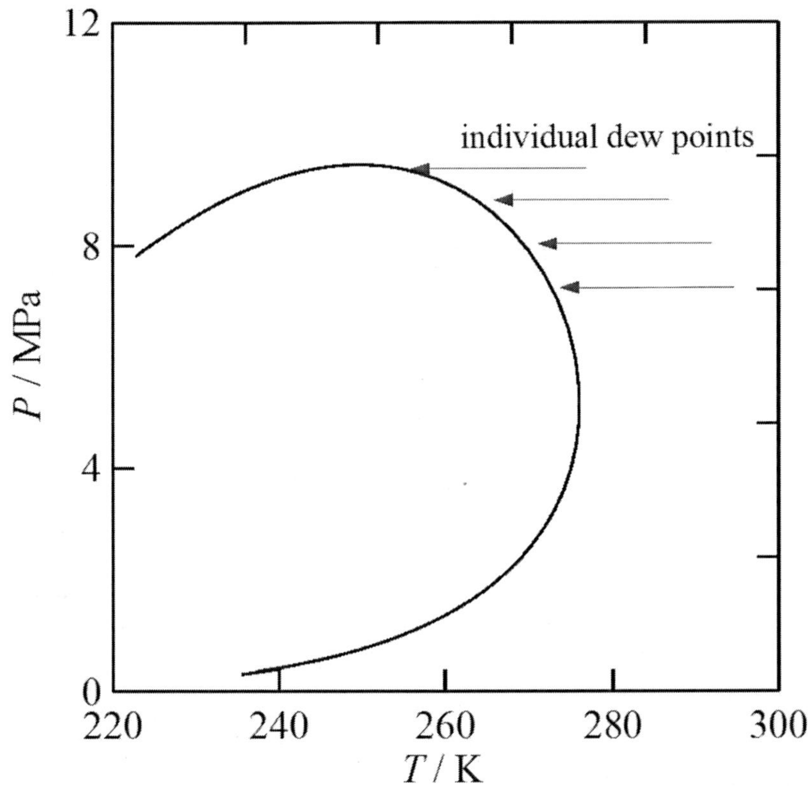

FIGURE 7. Scheme of experimental procedure for measuring phase envelopes using visual chilled mirror method.

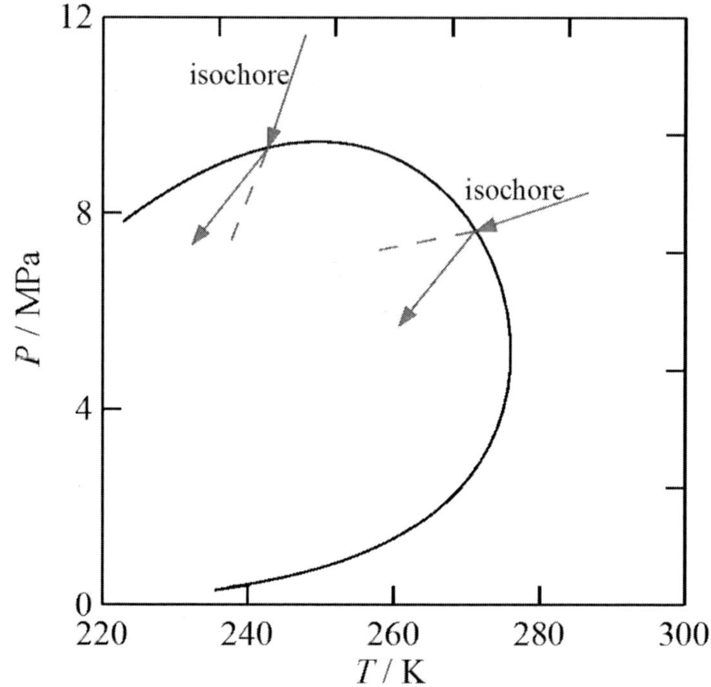

FIGURE 8. Method for determine phase envelope from isochoric measurements.

4.2.2.2. Non-visual synthetic methods

The method usually involves a blind *pVT* cell filled in the single-phase region with a mixture of known composition. Variation of one of the measured quantities allows measurement of a second quantity while the third is held constant. Therefore, measurements are made along either an isobaric or isochoric or isothermal path with the homogenous / heterogeneous boundary (phase envelope) determined from a discontinuity in the slope of the other two variables [61]. Among the non-visual methods, the isochoric technique is probably the most useful for the determination of phase envelopes in natural gas mixtures, its amenability to automation allow the collection of large quantities of accurate data at moderate costs.

The technique for determining phase loops using isochoric method utilizes the change of the slope of an isochore as it crosses the phase boundary, Figure 8.

It should be remarked that the change of slope does not occur at the cricondentherm, which has a collinear isochore [62]. The apparatus cell volume changes slightly with pressure and temperature, and thus, experimental data require an application of the cell distortion equation to correct the results to truly isochores. In a recent work Acosta et al. [63] developed a new computational method for the analysis of isochoric data to obtain phase envelope experimental data points, reporting that the phase boundary temperatures and pressures could be determined within 0.45% and 0.04%, respectively. Our research group has developed a systematic measurement program on selected natural gas-like mixtures using the isochoric method with the cell design reported in Figure 9 [64-66]. The reported results allowed to analyze the predictive ability of the equations of state commonly used in the gas industry for phase envelope predictions showing large deviations, which justify the need of very accurate experimental data for the analysis and prediction of natural gas phase equilibria.

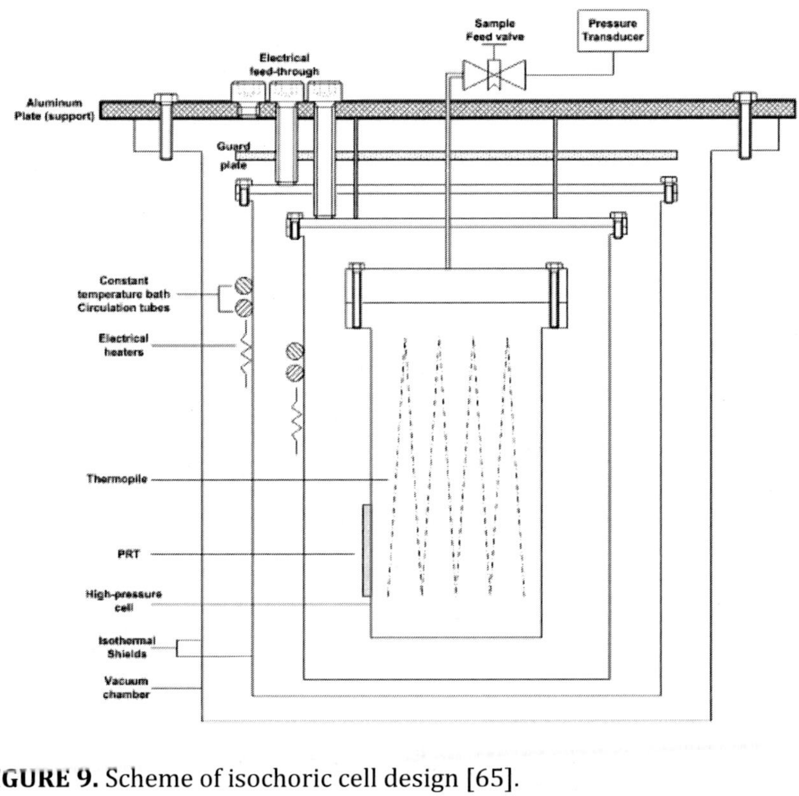

FIGURE 9. Scheme of isochoric cell design [65].

5. CONCLUSIONS

Techniques for the measurement of the most relevant thermophysical properties for the characterization of natural gas mixtures are reviewed showing the state-of-the-art, and the weaknesses and strengths of the current methods. Accurate determination of density, viscosity and phase equilibria are extremely important for the natural gas industry, both for technical and economical reasons, for production, processing and transportation purposes. Likewise, exploration and production using non-conventional reservoirs, including high pressure conditions, requires the accurate measurements in wide pressure-temperature ranges. Moreover, current predictive theoretical models using in the gas industry can not be used out of their tested pressure – temperature ranges, and thus, new accurate data are required to check the models performance and / or develop new theoretical predictive approaches. Therefore, this review could help to guide in the selection of the most suitable experimental approach for the determination of the required properties in wide pressure – temperature ranges.

REFERENCES

1. M. Atilhan, S. Ejaz, S. Aparicio-Martinez, K.R. Hall, Prediction of Natural Gas Mixtures Properties from Cubic and Molecular-Based Equations of State, Sixteenth Symposium on Thermophysical Properties, Boulder, CO, 2006.
2. X. Wang, J.G. Bomba, Hydrate Problems in Deepwater Flowlines, Proceedings of the 13th International Conference on Offshore Mechanics and Arctic Engineering, Houston, 1994.
3. W. Wagner, R. Kleinrahm, . Metrologia, 2004S24S39.
4. American, Gas, Association, Compressibility Factors of Natural Gas and Other Related Hydrocarbon Gases, Compressibility Factors of Natural Gas and Other Related Hydrocarbon Gases, Washington DC, 1992.
5. http://www.4uengineer.com/topic/FlowMeter/AGA8.gif.
6. J.T.R. Watson, B. Millington, Project No DRG001, Report No 110/97, National Engineering Lab (1998).
7. O. Kunz, W. Wagner, for. submitted, to. J. publication, Chem. Eng. Data, 2012
8. N. Kuramoto, K. Fujii, A. Waseda, . Metrologia, 2004S84S94.
9. R. Kleinrahm, W. Wagner, Reports. Progress, the. V. D. I. of, . Journals, 1984
10. A.R.H. Goodwin, K.N.Marsh, W.A. Wakeham, (Eds.), Measurement of the Thermodynamics Properties of Single Phases, 2003.
11. J. L. Daridon, A. Lagrabette, B. Lagourette, J. Chem, . Thermodyn, 1998
12. J. J. Zhou, P. Patil, S. Ejaz, M. Atilhan, J. C. Holste, K. R. Hall, J. Chem, . Thermodyn, 2006
13. V. Majer, A. A. H. Padua, of. Measurement, with. Density, Bodies. Vibrating, in: A.R.H. Goodwin, K.N.Marsh, W.A.Wakeham, (Eds.), Measurement of the Thermodynamic Properties of Single Phases, Experimental Thermodynamics, Amsterdam, 2003
14. A.A.H. Padua, J.M.N.A. Fareleira, J.C.G. Calado, W.A. Wakeham, Review of Scientific Instruments691998199823922399

15. E. S. Burnett, J. Appl, . Mech, 1936A-136.
16. W. Wagner, K. Brachthäuser, R. Kleinrahm, H. W. Losch, J. Int, . Thermophys, 1995
17. R. Tillner-Roth, H. D. Baehr, J. Chem, . Thermodyn, 1992
18. P. T. Eubank, L. L. Joffrion, M. R. Patel, W. Warowny, J. Chem, . Thermodyn, 1988
19. W. R. Lau, C. A. Hwang, H. B. Brugge, G. A. Iglesias-Silva, H. A. Duarte-Garza, W. J. Rogers, K. R. Hall, J. C. Holste, B. E. Gammon, K. N. Marsh, J. Chem, Data. . Eng, 1997
20. R. Kleinrahm, W. Wagner, J. Chem, . Thermodyn, 1986
21. H.W. Lösch, Development and Design of New Magnetic Suspension Balances for Non-Contact Measurements of Vertical Forces, VDI Verlog, Dusseldolf, 1987.
22. W. Wagner, R. Kleinrahm, H. W. Lösch, J. T. R. Watson, Balance. Hydrostatic, with. Densimeters, Suspension. Magnetic, Coupling, in: K.N. Marsh, W.A. Wakeham, (Eds.), Measurements of the Thermodynamic Properties of Single Phases, Elsevier, Amsterdam, 2003
23. J. Klimeck, R. Kleinrahm, W. Wagner, J. Chem, . Thermodyn, 1998
24. E. Davani, K. Ling, C. Teodoriu, W. D. Mc Cain, G. Falcone, accurate. More, viscosity. gas, for. correlation, at. use, conditions. hpht, better. ensures, estimation. reserves, Paper SPE 124734, SPE Annual Technical Conference and Exhibition, New Orleans, LA, 2009
25. D. Denney, J. Pet, . Technol, 2010
26. T. Billiter, J. Lee, Proceedings of the Presentation at the SPE Eastern Regional Meeting, Paper SPE 59759, Canton, OH, 2001.
27. E. Davani, K. Ling, C. Teodoriu, W.D. McCain, G. Falcone, Proceedings of the Latin American and Caribbean Petroleum Engineering Conference, Paper SPE 122827, Cartagena de Indias, Colombia, 2009.
28. E. Sanjari, E. N. Lay, M. Peymani, J. Nat, Chem. . Gas, 2011
29. S. Mokhatab, W.A. Poe, J. G. Speight, Handbook of Natural Gas Transmission and Processing; Gulf Professional Publishing, Burlington, MA, 2006.
30. T. S. Dyman, R. E. Wyman, V. A. Kuuskraa, M. D. Lewan, T. A. Cook, Resour. Nat, . Res, 2003
31. M. Atilhan, S. Aparicio, R. Alcalde, G. A. Iglesias-Silva, M. El -Halwagi, K. R. Hall, J. Chem, Data. . Eng, 2010
32. M. Atilhan, S. Aparicio, G. A. Iglesias-Silva, M. El -Halwagi, K. R. Hall, J. Chem, Data. . Eng, 2010
33. B. H. Sage, W. N. Lacey, Am. Trans, Mining. Inst, Engrs. 1. Met, 1938
34. L. B. Bicher, D. L. Katz, Eng. Ind, . Chem, 1943
35. A. S. Pensado, M. J. P. Comuñas, L. Lugo, J. Fernández, J. Chem, Data. . Eng, 2005
36. J.P. Dolan, K.E. Starling, A.L. Lee, B.E. Eakin, R.T. Ellington, J. Chem. Eng. Data 8 (1963) 396-399.
37. B. H. Sage, W. D. Yale, W. N. Lacey, Eng. Ind, . Chem, 1939
38. H. R. Heath, Phys. Proc, B. . Soc, 1953
39. R. Kobayashi, A Study on the Rankine Gas Viscometer, Research Reports of Faculty of Engineering, Tokyo University, 2001.
40. J. G. Giddings, J. T. F. Kao, R. Kobayashi, J. Chem, . Phys, 1966
41. A. L. Lee, M. H. Gonzalez, B. E. Eakin, J. Petrol, . Technol, 1966
42. E. F. May, R. F. Berg, M. R. Moldover, J. Int, Thermophys, . , 2007
43. C. A. Nieto, F. J. V. de Castro, J. M. N. A. Santos, W. A. Fareleira, J. Wakeham, Eng. Chem, . Data, 2009
44. J. T. Tough, W. D. Mc Cormick, J. G. Dash, Sci. Rev, . Instrum, 1964
45. N. J. Trappeniers, P. S. van der Gulik, H. van den, Chem. Hooff, Lett. . Phys, 1980
46. M. J. Assael, N. K. Dalaouti, V. Vesovic, J. Int, . Thermophys, 2001
47. L. I. Langelandsvik, S. Solvang, M. Rousselet, I. N. Metaxa, M. J. Assael, J. Int, Thermophys, . , 2007
48. P. Schley, M. Jaeschke, C. Kuchenmeister, E. Vogel, J. Int, Thermophys, . , 2004
49. E. Heidaryan, J. Moghadasi, A. Salarabadi, J. Nat, Chem. . Gas, 2010
50. F. B. Thomas, E. Stephani, D. B. Bennion, J. Rushing, J. Can, Technol. . Petrol, 2003

51. A. Viswanathan, of. Viscosities, Gases. Natural, High. at, Pressures, Temperatures. M. High, M.Sc. Thesis, Texas A&M University, College Station, TX, 2007
52. K.S. Pedersen, P.L. Christensen, Phase Behavior of Petroleum Reservoir Fluids.CRC Press, Boca Raton, FL, 2007
53. White Paper on Liquid Hydrocarbon Drop Out inNatural Gas Infrastructure. Natural Gas Council for the Federal Energy Regulatory Commission, NGC+ Liquid Hydrocarbon Dropout Task Group, 2005
54. D. Denney, Condensate. Identifying, With. Banking, Flowmeters. Multiphase-A, Study. Case, J. Petrol. Technol. 39 (2007
55. M. Atilhan, J. Zhou, S. Ejaz, D. Cristancho, J. Holste, K. R. Hall, Behavior. Phase, for. Concerns, Natural. Multicomponent, Mixtures. Gas-Like, Proceedings of the 1st Annual Gas Processing Symposium. H. Alfadala, G. V. Rex, M. M. El-Halwagi, (Eds.), Elsevier, 2009
56. E. Skouras, E. , Thermodynamics in Gas Processing-Phase Envelope Predictions and Process Design. 1st Trondheim Gas Technology Conference, 2009
57. K. Nasrifar, O. Bolland, M. Moshfeghian, Natural. Predicting, Dew. Gas, from. . Points, of. Equations, State, Energy & Fuels 19 (2005
58. U. Deiters, G. Schneider, Phase. Fluid, . Equilibria, 1986
59. A.J. Benton, Hydrocarbon Engineering,7200220025964
60. H.R. Warner, E.E. Leamer, A.P. Spence, R.L. Bone, R.A. Hubbard, J. Bernos, and W. A. Kriel, “Hydrocarbon Dewpoint Determination of Lean Natural Gases,” 80th Annual Convention Presentations, Gas Processors Association, Tulsa, OK (2001).
61. P. T. Eubank, K. R. Hall, J. C. Holste, A. Review, Experimental. of, for. Techniques, Equilibria. Vapor-Liquid, High. at, Phase. Pressures, Equilibria, Properties. Fluid, the. in, Industry. Chemical, 2nd. Proceedings, Conference. H. International, Knapp and S. I. Sandler, eds., Great Neck, NY 675 (1980
62. P. T. Eubank, M. A. Barrufet, conditions. General, collinearity. of, the. at, boundaries. phase, fluid. of, A. I. mixtures, Ch, AIChE J. 331987198718821887
63. P. L. Acosta, D. E. Cristancho, I. D. Mantilla, K. R. Hall, G. A. Iglesias-Silva, Phase. Fluid, 2. Equilibr, 2009
64. M. Atilhan, S. Ejaz, J. Zhou, I. Mantilla, J. Holste, K. R. Hall, J. Chem, Data. . Eng, 2010
65. M. Atilhan, S. Aparicio, S. Ejaz, D. Cristancho, K. R. Hall, J. Chem, Data. . Eng, 2011
66. M. Atilhan, S. Aparicio, S. Ejaz, D. Cristancho, K. R. Hall, J. Chem, Data. . Eng, 2011

CHAPTER 14

Exploring the Effects on Lipid Bilayer Induced by Noble Gases via Molecular Dynamics Simulations

Junlang Chen[1,4], Liang Chen[1,3], Yu Wang[1], Xiaogang Wang[1] & Songwei Zeng[2]

[1] *School of Sciences, Zhejiang A & F University, Lin'an 311300, China.*

[2] *School of Information and Industry, Zhejiang A & F University, Lin'an 311300, China.*

[3] *Shanghai Applied Radiation Institute, Shanghai University, Shanghai 200444, China.*

[4] *Zhejiang Provincial Key Laboratory of Chemical Utilization of Forestry Biomass, Zhejiang A & F University, Lin'an 311300, China.*

ABSTRACT

Noble gases seem to have no significant effect on the anesthetic targets due to their simple, spherical shape. However, xenon has strong narcotic efficacy and can be used clinically, while other noble gases cannot. The mechanism remains unclear. Here, we performed molecular dynamics simulations on phospholipid bilayers with four kinds of noble gases to elucidate the difference of their effects on the membrane. Our results showed that the sequence of effects on membrane exerted by noble gases from weak to strong was Ne, Ar, Kr and Xe, the same order as their relative narcotic potencies as well as their lipid/water partition percentages. Compared with the other three kinds of noble gases, more xenon molecules were distributed between the lipid tails and headgroups, resulting in membrane's lateral expansion and lipid tail disorder. It may contribute to xenon's strong anesthetic potency. The results are well consistent with the membrane mediated mechanism of general anesthesia.

INTRODUCTION

The noble gases (He, Ne, Ar, Kr and Xe), except helium and probably neon, can cause general anesthesia[1]. However, their narcotic potencies vary widely in degree. For example, xenon is an excellent anesthetic and has already been clinically used in surgical operations[2,3,4], though the specific molecular mechanisms of general anesthesia are still under debate.

The mechanisms of general anesthesia are divided into membrane and protein hypotheses[5,6,7,8]. According to the membrane mechanism, the entering of anesthetic in membrane leads to the change of the physical properties of membrane (such as lipid order and lateral pressure), which further affects the function of ion channels embedded within the membrane, and may finally cause general anesthesia[9,10,11]. This hypothesis is supported by Meyer and Overton based on the correlation between anesthetic potency and lipid solubility, which is known as the Meyer-Overton correlation: the anesthetic potency is proportional to the lipid/water partition coefficient[12,13]. Although this rule is true for a wide range of anesthetics including noble gases, it suffers from several weaknesses. For example, stereoisomers of an anesthetic have very different narcotic potency, whereas their oil/gas partition coefficients are similar[14].

Therefore, researchers have resorted to protein hypothesis and focused on the binding of anesthetics to trans-membrane ion channels[15,16,17,18]. Recently, experiments have shown that N-methyl-D-aspartic acid (NMDA) receptor and γ-aminobutyric acid ($GABA_A$) receptor are molecular targets for general anesthesia[19,20,21,22,23]. Also, xenon can inhibit the NMDA receptor while mutations of the NMDA receptor reduce the anesthetic sensitivity[24]. Other anesthetics, however, interact with neurotransmitter receptors very differently compared with xenon, such as isoflurane, which anesthetic target is $GABA_A$ receptor[25]. There is still no consensus in the protein mechanism of general anesthesia.

Since the electrons in outer shells of noble gases are saturated and spherically symmetric, noble gases have apparently minimal capacities to bind with a putative action site. More recently, researchers resume their interests in the membrane mediated mechanism of general anesthesia[26,27]. Yamamoto *et al.* performed molecular dynamics simulation to study the diffusive nature of xenon within the lipid bilayer[28].

They found that xenon molecules are concentrated at the center of the bilayer and lipid tail order increases under extremely high pressure (100 ~ 500bar). They proposed that pressure reveral of general anesthesia originated from the recovery of membrane disorder previously caused by xenon. Booker *et al.* simulated six xenon concentrations (0 ~ 3 xenon molecules per lipid) and reported significant changes in lipid bilayer properties induced by xenon, including bilayer thickness increasing and modulation of lateral pressure profile29. They therefore indicated that these changes may directly affect the gated ion channel function that finally lead to general anesthesia. Through measuring the electric polarizabilities of noble gases in the pure lipid membrane, Sierra-Valdez and Ruiz-Suarez found that the increasing membrane disorder induced by noble gases is He, Ar, Kr and Xe in turn30. They suggested that general anesthesia depends on the ability of certain molecules to increase the general disorder of the membrane. Weinrich and co-worker's experiment demonstrated that inhalation anesthetics could affect the lipid environment of membrane proteins, which supported the membrane-mediated mechanism31.

In this paper, we investigate the interactions of four kinds of noble gases (Ne, Ar, Kr and Xe) with 1-palmitoyl-2-oleoyl phosphatidylethanolamine (POPE) bilayer using molecular dynamics simulations. Noble gases show the wide effects on bilayer structure, including increasing area per lipid and lipid tail diorder. The sequence of changes on membrane from weak to strong is Ne, Ar, Kr and Xe, the same order as their relative narcotic potencies. The results sufficiently support the membrane mediated mechanism of general anesthesia.

METHODS

Molecular dynamics (MD) simulations were performed on POPE bilayers in aqueous solution with or without noble gas molecules. The lipid bilayer was composed of 256 POPE molecules (128 molecules in each leaflet) and was fully hydrated with 40 water molecules per lipid. Four kinds of noble gases (Ne, Ar, Kr and Xe) and five concentrations (0, 1, 1.5, 2 and 3 noble gas molecules per lipid) were simulated. The initial configurations were generated from the pure hydrated system by replacing water molecules with noble gas molecules randomly according to the fixed concentration. We then minimized their energy using the conjugate gradient method and continued to perform 200 ns MD simulations for each system.

All MD simulations were performed in the isothermal-isobaric (NPT) ensemble using the Gromacs package 4.5.532. The force field parameters for POPE lipids were taken from Berger *et al.*33,34. Noble gas atoms were treated as uncharged Lennard-Jones (LJ) spheres and their ver der Waals (vdW) parameters were taken from UFF force field35. The temperature was kept stable at 310 K using the V-rescale thermostat and the pressure was controlled semi-isotropically at 1 bar by a Berendsen barostat36,37. The particle-mesh Ewald method (PME) was used to calculate the long-range electrostatic interactions, whereas the vdW interactions were treated with smooth cutoff at a distance of 10 Å38,39. Water was represented by the simple point-charge (SPC) model40. Bond lengths within the DPPC lipids and water molecules were constrained by the LINCS and the SETTLE algorithms, which allowed an integration time of 2 fs. We applied periodic boundary conditions in all three directions and used the last 20 ns trajectory of each system to analyze the properties of noble gases and POPE bilayers.

To demonstrate that the systems had been well equilibrated at $t = 180$ ns, we compared the noble gas density profiles of the first, middle and last 10 ns trajectories, as shown in Figure S1, and monitored the time evolution of area per lipid and lipid volume in Figure S2, which confirmed that the systems had already reached equilibrium from about $t = 100$ ns.

RESULTS AND DISCUSSION

Noble gas molecules partitioning

At dynamical equilibrium, noble gas molecules (Ne, Ar, Kr and Xe) within the bilayer account for 77.3%, 85.2%, 91.0% and 95.7% of the total, initially localized in the aqueous phase. According to the Meyer-Overton correlation, the greater the lipid/water partitioning coefficient is, the stronger its anesthetic potency is. Therefore, our simulation results are in good agreement with this rule. Although the partitioning percentage presents a little fluctuation because of different gas concentration, the results still obey the correlation after equilibrium.

The partitioning of the noble gases within the bilayer can be more quantitatively illustrated by the noble gas density distributions. Figure 1shows the densities of the z direction of four kinds of noble gases. The peak of each curve is at the center of the bilayer for all systems, indicating that noble gas molecules are preferentially aggregated at the center of the membrane, since there is an interspace between the upper and lower leaflets. This localization of gas molecules is almost the same as previous simulations performed by Stimson *et al.* and Yamamoto *et al.*28,29,41. However, the probability of xenon localizing near the lipid head groups is about 10% higher than those of the other three gases (Ne, Ar and Kr), which is of great importance to explain xenon's strong narcotic efficacy.

Yamamoto proposed that the narcotic function of xenon will be decreased when xenon molecules move to the hydrophobic core of the lipid bilayer and are jammed there under high pressures. Other anesthetics, such as halothane, diethyl ether, enflurane and ethanol, prefer two different positions, namely the center of the membrane and the hydrocarbon region, close to the polar headgroups[5,42,43]. Therefore, we suggest that the narcotic potency of xenon is much stronger because more xenon molecules than other noble gases are distributed between the lipid tails and headgroups. This speculation can be further confirmed at higher noble gas concentration (3 noble gas atoms per lipid). As shown in Fig. 2, about 90% neon atoms are congested in the middle of the membrane and stretch the bilayer at the normal direction, while about 40% xenon atoms diffuse near the hydrocarbon region and expand the membrane at the lateral direction.

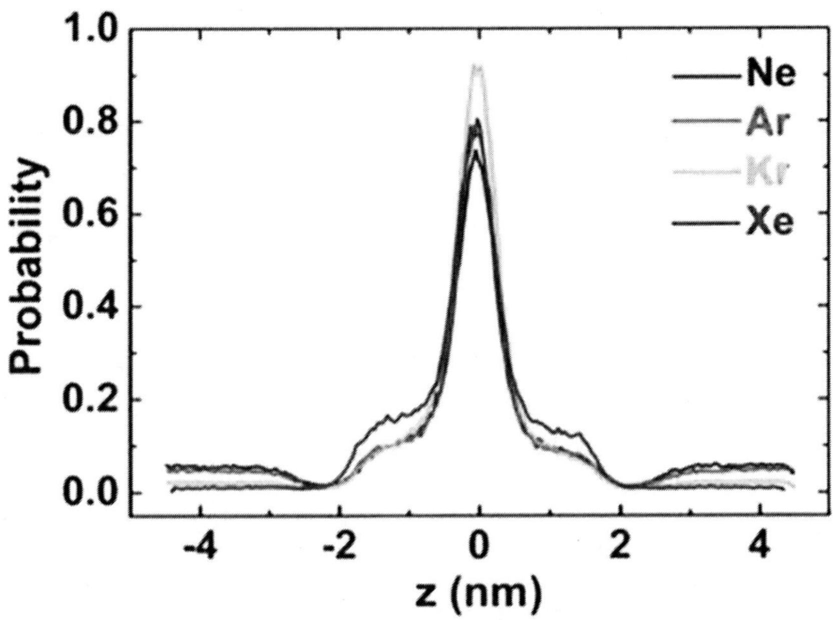

Figure 1: Density of the *z* direction of four kinds of noble gases (Ne, black, Ar, red, Kr, green and Xe blue), where the z axis is normal to the lipid bilayer.

The center of the lipid bilayer is fixed at $z = 0$ nm.

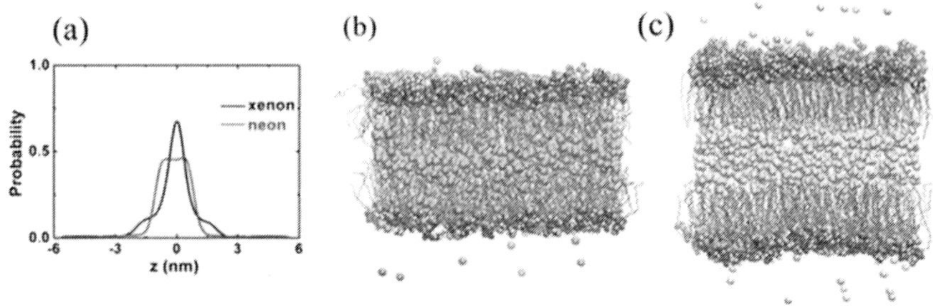

Figure 2: Results of 3 noble gas molecules per lipid system.(**a**) Density distribution. (**b,c**) are final snapshots at $t = 200$ ns. Carbon (cyan), oxygen (red), nitrogen (blue) and phosphorous (tan) in head groups are shown using spheres while the lipid tails are shown as dynamic-bonds. Xenon and neon molecules are depicted as green and yellow balls. Water molecules are omitted for clarity. The images are created by VMD software.

Effects of noble gases on membrane structure

Figure 3 shows area per lipid S, thickness h and volume V in the five independent systems. The area per lipid is defined as the area of the xy-plane of the simulation box divided by the number of lipids per leaflet. The thickness of lipid bilayer h is defined as the distance between the average z-position of the phosphorus atom in the two layers. And the volume per lipid is defined as $V = Sh/2$. In general, the three parameters containing noble gases are bigger than those of pure hydrated membrane. Furthermore, it is interesting to find that the values of S and V rise gradually in the order of Ne, Ar, Kr and Xe, which is in agreement with the sequence of their narcotic potencies from weak to strong. The area per lipid and volume with xenon increase at the most by 12.2% and 9.54%, respectively. However, the thickness h does not hold this trend, especially xenon, as shown in Fig. 3b. For example, the thickness with argon increases 0.87%, while h with xenon decreases 0.33%, compared with pure membrane. The thickness h is mainly determined by two factors: the distributions of gas molecules and lipid tail ordering. As mentioned above, xenon within the membrane leads to its lateral expansion more than normal stretch.

The localization of noble gases in the hydrophobic core of the bilayer has substantial effects on the ordering of the lipid acyl chains. Figure 4shows the two lipid tail (sn-1 and sn-2) deuterium order parameters of five systems. The deuterium order is defined by

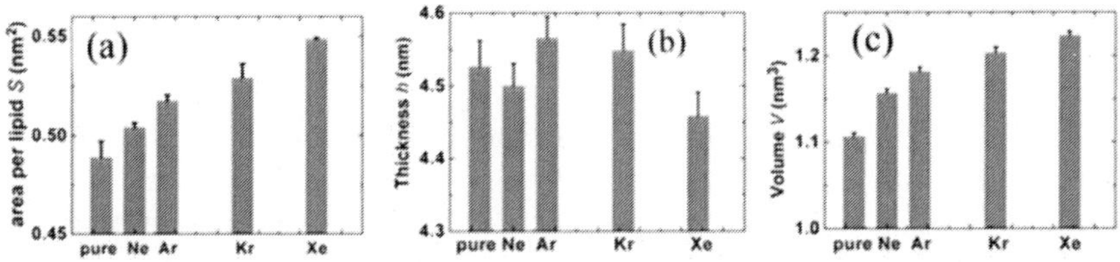

Figure 3: Effects of noble gas molecules on bilayer structural properties. (**a**) Area per lipid, (**b**) thickness and (**c**) volume per lipid. Error bars mean the root mean square deviations.

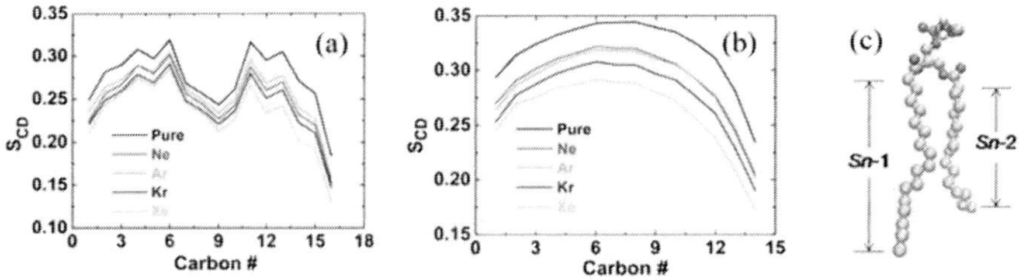

Figure 4. Lipid tail order parameters of (**a**) *sn*-1 chain and (**b**) *sn*-2 chain. (**c**) Atomic structure of POPE (cyan, red, blue and tan balls represent carbon, oxygen, nitrogen and phosphorus atoms, respectively).

$$S_{CD} = \left\langle \frac{3}{2}\cos^2\theta - \frac{1}{2} \right\rangle$$

where θ is the relative angle between the CD bond vector and the bilayer normal and the brackets denote averaging over molecules and simulation time44. The order parameter provides a quantitative measure of the alignment of the lipid tails. The value of S_{CD} with noble gases is lower than that of purely hydrated membrane system because gas molecules diffuse in the lipid bilayer and collide with the lipid tails. Interestingly, it is found that the sequence of effects on lipid tail ordering induced by noble gases is Ne, Ar, Kr and Xe, which is also in agreement with their narcotic potency. This sequence is also in accordance with the previous experiment30. The value of S_{CD} with xenon decreases the most significantly, about 16.2%.

CONCLUSIONS

By extensive MD simulations of POPE membranes in aqueous solution with four kinds of noble gases, we find that the sequence of effects on POPE from weak to strong is Ne, Ar, Kr and Xe, which is in agreement with their relative narcotic potencies. The localization of noble gases within the membrane leads to a relative increase in area, volume per lipid and decrease in lipid tail ordering, which can increase membrane fluidity and depress the membrane gel-liquid crystalline phase transition temperature29.

Due to its deep potential well and wide collision radius, xenon is prone to diffuse between the lipid tails and headgroups than other noble gases, inducing more lateral expansion of the membrane and lipid tail disorder. Of the four noble gases, the distribution of xenon within the membrane is the most similar to that of other anesthetics, such as chloroform and enflurane. Xenon's vigorous effects on membrane structure and localization therein contribute to its strong anesthetic efficacy. These results are in favor of the membrane mediated mechanism of general anesthesia.

ADDITIONAL INFORMATION

How to cite this article: Chen, J. *et al.* Exploring the Effects on Lipid Bilayer Induced by Noble Gases via Molecular Dynamics Simulations. *Sci. Rep.* **5**, 17235; doi: 10.1038/srep17235 (2015).

ACKNOWLEDGEMENTS

The authors thank Prof. Haiping Fang and Yusong Tu for helpful discussion. This work was supported by the National Natural Science Foundation of China (Grant No. 61178016 and 11304284). The authors acknowledge Shanghai Supercomputer Center and Shanghai University HPC ZQ3000&4000 for computational resources.

REFERENCES

1. Brauer, R. W. & Way, R. O. Relative narcotic potencies of hydrogen, helium, nitrogen, and their mixtures. J. Appl. Physiol. 29, 23–31 (1970).
2. Cullen, S. & Gross, E. The anesthetic properties of xenon in animals and human beings with additional observations on krypton. Science 113, 580–582 (1951).

3. Harris, P. & Barnes, R. The uses of helium and xenon in current clinical practice. Anesthesia 63, 284–293 (2008).
4. Brücken, A., Coburn, M., Rex, S., Rossaint, R. & Fries, M. Current developments in xenon research. Importance for anesthesia and intensive care medicine. Anaesthesist 59, 883–895 (2010).
5. Darvas, M., Hoang, P. N., Picaud, S., Sega, M. & Jedlovszky, P. Anesthetic molecules embedded in a lipid membrane: a computer simulation study. Phys. Chem. Chem. Phys. 14, 12956–12969 (2012).
6. Tang, P. & Xu, Y. Large-scale molecular dynamics simulations of general anesthetic effects on the ion channel in the fully hydrated membrane: the implication of molecular mechanisms of general anesthesia. Proc. Natl. Acad. Sci. 99, 16035–16040 (2002).
7. Zhang, M., Zuo, G., Chen, J., Gao, Y. & Fang, H. Aggregated gas molecules: toxic to protein? Sci. Rep. 3, 1660 (2013).
8. Ma, D. et al. Four-alpha-helix bundle with designed anesthetic binding pockets. Part I: structural and dynamical analyses. Biophys. J. 94, 4454–4463 (2008).
9. Cantor, R. S. The influence of membrane lateral pressures on simple geometric models of protein conformational equilibria. Chem. Phys. Lipids 101, 45–56 (1999).
10. Cantor, R. S. The Lateral Pressure Profile in Membranes: A Physical Mechanism of General Anesthesia. Biochemistry 36, 2339–2344 (1997).
11. Cantor, R. S. Lateral Pressures in Cell Membranes: A Mechanism for Modulation of Protein Function. J. Phys. Chem. B 101, 1723–1725 (1997).
12. Meyer, H. Zur theorie der alkoholnarkose, Naunyn-Schmiedebergs Arch. Pharmakol 42, 109–118 (1899).
13. Overton, C. E. & Lipnick, R. L. Studies of Narcosis; Chapman and Hall: London. (1991).
14. Cameron, J. W. The molecular mechanisms of general anaesthesia: dissecting the GABAA receptor. Continuing Education in Anaesthesia, Critical Care & Pain 6, 49–53 (2006).
15. Belelli, D., Lambert, J. J., Peters, J. A., Wafford, K. & Whiting, P. J. The interaction of the general anesthetic etomidate with the g-aminobutyric acid type A receptor is influenced by a single amino acid. Proc. Natl. Acad. Sci. 94, 11031–11036 (1997).
16. Krasowski, M. D. & Harrison, N. L. The actions of ether, alcohol and alkane general anaesthetics on GABAA and glycine receptors and the e ects of TM2 and TM3 mutations. Brit. J. Pharmacol. 129, 731–743 (2000).
17. Jenkins, A. et al. Evidence for a Common Binding Cavity for Three General Anesthetics within the GABAA Receptor. J. Neurosci. 21, 1–4 (2001).
18. Vemparala, S., Domene, C. & Klein, M. L. Computational Studies on the Interactions of Inhalational Anesthetics with Proteins. Accounts. Chem. Res. 43, 103–110 (2009).
19. Franks, N. P., Dickinson, R., Sousa, S. L., Hall, A. C. & Lieb, W. R. How does xenon produce anaesthesia? Nature 396, 324 (1998).
20. Sousa, S. L., Dickinson, R., Lieb, W. R. & Franks, N. P. Contrasting synaptic actions of the inhalational general anesthetics isoflurane and xenon. Anesthesiology 92, 1055–1066 (2000).
21. Yamakura, T. & Harris, R. A. Effects of gaseous anesthetics nitrous oxide and xenon on ligand-gated ion channels. comparison with isoflurane and ethanol. Anesthesiology 93, 1095–1101 (2000).
22. Ogata, J. et al. Effects of anesthetics on mutant N-methyl-D-aspartate receptors expressed in Xenopus oocytes. J. Pharmacol. Exp. Ther. 318, 434–443 (2006).
23. Dickinson, R. et al. Competitive inhibition at the glycine site of the N-methyl-D-aspartate receptor by the anesthetics xenon and isoflurane: evidence from molecular modeling and electrophysiology. Anesthesiology 107, 756–767 (2007).
24. Liu, L. T., Xu, Y. & Tang, P. Mechanistic Insights into Xenon Inhibition of NMDA Receptors from MD Simulations. J. Phys. Chem. B 114, 9010–9016 (2010).

25. Sousa, S. L., Dickinson, R., Lieb, W. R. & Franks, N. P. Contrasting synaptic actions of the inhalational general anesthetics isoflurane and xenon. Anesthesiology 92, 1055–1066 (2000).
26. Marsh, D. Protein modulation of lipids, and vice-versa, in membranes. Biochim Biophys Acta 1778, 1545–1575 (2008).
27. Phillips, R., Ursell, T., Wiggins, P. & Sens, P. Emerging roles for lipids in shaping membrane-protein function. Nature 459, 379–385 (2009).
28. Yamamoto, E. et al. Diffusive nature of xenon anesthetic changes properties of a lipid bilayer: molecular dynamics simulations. J. Phys. Chem. B 116, 8989–8995 (2012).
29. Booker, R. D. & Sum, A. K. Biophysical changes induced by xenon on phospholipid bilayers. Biochim. Biophys. Acta. 1828, 1347–1356 (2013).
30. Sierra-Valdez, F. J. & Ruiz-Suarez, J. C. Noble gases in pure lipid membranes. J. Phys. Chem. B 117, 3167–3172 (2013).
31. Weinrich, M. & Worcester, D. L. Xenon and other volatile anesthetics change domain structure in model lipid raft membranes. J. Phys. Chem. B 117, 16141–16147 (2013).
32. Berendsen, H. J. C., Spoel, D. V. D. & Drunen, R. V. GROMACS: A message-passing parallel molecular dynamics implementation. Comput. Phys. Commun. 91, 43–56 (1995).
33. Berger, O., Edholm, O. & Jähnig, F. Molecular dynamics simulations of a fluid bilayer of dipalmitoylphosphatidylcholine at full hydration, constant pressure, and constant temperature. Biophys. J. 72, 2002–2013 (1997).
34. Ane´zo, C. L., Vries, A. H. D., Holtje, H.-D., Tieleman, D. P. & Marrink, S.-J. Methodological Issues in Lipid Bilayer Simulations. J. Phys. Chem. B 107, 9424–9433 (2003).
35. Rappe, A. K., Casewit, C. J., Colwell, K. S., Goddard, W. A. & Skiff, W. M. UFF, a Full Periodic Table Force Field for Molecular Mechanics and Molecular Dynamics Simulations. J. Am. Chem. Soc. 114, 10024–10035 (1992).
36. Bussi, G., Donadio, D. & Parrinello, M. Canonical sampling through velocity rescaling. J Chem Phys 126, 014101 (2007).
37. Berendsen, H. J. C., Postma, J. P. M., Gunsteren, W. F., DiNola, A. & Haak, J. R. Molecular dynamics with coupling to an external bath. J. Phys. Chem. 81, 3684–3691 (1984).
38. Darden, T., York, D. & Pedersen, L. Particle mesh Ewald: An Nlog(N) method for Ewald sums in large systems. J. Phys. Chem. 98, 10089–10094 (1993).
39. Essmann, U. et al. A smooth particle mesh Ewald method. J. Phys. Chem. 103, 8577–8594 (1995).
40. Jorgensen, W. L., Chandrasekhar, J., Madura, J. D., Impey, R. W. & Klein, M. L. Comparison of simple potential functions for simulating liquid water. J. Phys. Chem. 79, 926–936 (1983).
41. Stimsoni, L. M., Vattulainen, I., Rog, T. & Karttunen, M. Exploring the effect of xenon on biomembranes. Cell. Mol. Biol. Lett. 10, 563–569 (2005).
42. Gurtovenko, A. A. & Anwar, J. Interactin of ethanol with biological membranes: the formation of non-bilayer structures within the membrane interior and their significance. J. Phys. Chem. B 113, 1983–1992 (2009).
43. Fabian, B., Darvas, M., Picaud, S., Sega, M. & Jedlovszky, P. The effect of anaesthetics on the properties of a lipid membrane in the biologically relevant phase: a computer simulation study. Phys. Chem. Chem. Phys. 17, 14750–14760 (2015).
44. Chiu, S. W. et al. Incorporation of Surface Tension into Molecular Dynamics Simulation of an Interface: A Fluid Phase Lipid Bilayer Membrane. Biophys. J 69, 1230–1245 (1995).

CHAPTER 15

Gas Sensing Properties of Conducting Polymer/ Au-Loaded ZnO Nanoparticle Composite Materials at Room Temperature

Viruntachar Kruefu[1]*, Anurat Wisitsoraat[2] , Adisorn Tuantranont[2] and Sukon Phanichphant[3]

[1] Program in Materials Science, Faculty of Science, Maejo University, Chiang Mai 50290, Thailand.

[2] National Electronics and Computer Technology Center, Pathumthani 12120, Thailand.

[3] Materials Science Research Center, Faculty of Science, Chiang Mai University, Chiang Mai 50200, Thailand.

ABSTRACT

In this work, a new poly (3-hexylthiophene):1.00 mol% Au-loaded zinc oxide nanoparticles (P3HT:Au/ZnO NPs) hybrid sensor is developed and systematically studied for ammonia sensing applications. The 1.00 mol% Au/ZnO NPs were synthesized by a one-step flame spray pyrolysis (FSP) process and mixed with P3HT at different mixing ratios (1:1, 2:1, 3:1, 4:1, and 1:2) before drop casting on an Al_2O_3 substrate with interdigitated gold electrodes to form thick film sensors. Particle characterizations by X-ray diffraction (XRD), nitrogen adsorption analysis, and high-resolution transmission electron microscopy (HR-TEM) showed highly crystalline ZnO nanoparticles (5 to 15 nm) loaded with ultrafine Au nanoparticles (1 to 2 nm). Film characterizations by XRD, field-emission scanning electron microscopy (FE-SEM), energy-dispersive X-ray (EDX) spectroscopy, and atomic force microscopy (AFM) revealed the presence of P3HT/ZnO mixed phases and porous nanoparticle structures in the composite thick film. The gas sensing properties of P3HT:1.00 mol% Au/ZnO NPs composite sensors were studied for reducing and oxidizing gases (NH_3, C_2H_5OH, CO, H_2S, NO_2, and H_2O) at room temperature. It was found that the composite film with 4:1 of P3HT:1.00 mol% Au/ZnO NPs exhibited the best NH_3 sensing performances with high response (approximately 32 to 1,000 ppm of NH_3), fast response time (4.2 s), and high selectivity at room temperature. Plausible mechanisms explaining the enhanced NH_3 response by composite films were discussed.

Keywords: P3HT Au-loaded ZnO Composite films NH_3 sensor Flame spray pyrolysis

BACKGROUND

Gas sensors for ammonia (NH_3) detection at low concentration are of great scientific importance in environmental monitoring, medical diagnosis, and various chemical/agricultural industries, since ammonia is very harmful to humans and the environment[1, 2, 3, 4, 5]. Several semiconducting metal oxides are highly promising for NH_3detection due to their excellent response[6, 7, 8]. However, they suffer from some inconvenience including high operating temperatures (200°C to 400°C)[6, 7, 8, 9, 10, 11]. High operating temperature results in high power consumption and complicated sensor design/fabrication[12]. Thus, ammonia sensors operable at room temperature with long life time are of great interest.

Conducting polymers, such as polypyrrole (PPy), polyaniline (Pani), polythiophene (PTh), and their derivatives, have demonstrated gas sensing capability at low or even room temperature[13, 14]. However, they are still not practically useful due to comparatively low response, lack of specificity, and relatively poor stability. A summary of gas sensing properties of NH_3 gas sensor-based conducting polymers as well as their hybrids prepared by various methods is shown in Table 1. Firstly, polyaniline (PANI) nanowires (NWs) prepared by photolithography process are shown to exhibit a high sensor response of approximately 3 at 500 ppm NH_3 in synthetic air[14]. In addition, surface acoustic wave (SAW) NH_3 gas sensors based on PPy prepared by layer-by-layer (LBL) self-assembly method are investigated for NH_3 sensing with different numbers of layer. The sensor with two layers of PPy shows the best performance relative to those with other numbers of PPy layers[15]. Additionally, NH_3 gas sensors based on organic thin-film transistors (OTFTs) made from spin-coated poly (3-hexylthiophene) (P3HT) on a thermally grown SiO_2/Si wafer exhibit a sensor response of 0.31 to 100 ppm NH_3 at room temperature[16]. Among these, P3HT is particularly promising for gas sensing applications due to its selective room-temperature response toward some gases especially ammonia and NO_2[16, 17, 18] and its relatively high stability. P3HT is known to have high oxidation potential making it highly stable in doped/undoped states under ambient conditions at room temperature and has specific chemical interactions with some gases[17].

Table 1. Summary of nh $_3$ sensing properties of a conducting polymer and metal or metal oxide/conducting polymer sensor

Authors/reference	Method	Materials	NH_3 concentration (ppm)	NH_3 sensing performances
Chen et al. [15]	Layer-by-layer (LBL) self-assembly method	Polypyrrole (PPy) and Pt-doped two-layer PPy thin films	100	Response: approximately 3 to 100 ppm NH_3 at room temperature
Jeong et al. [16]	Spin coating	P3HT thin-film transistors	10 to 100	Response: 0.31 to 100 ppm NH_3 at room temperature
Saxena et al. [27]	Drop casting	P3HT:ZnO nanowire thin films	4	Response: <1% to 4 ppm NH_3 at room temperature
Chougule et al. [13]	Low-frequency AC spin coating	CSA (30 wt.%) doped PPy-ZnO hybrid films	100	Response: approximately 11 to 100 ppm NH_3 at room temperature
Baratto [18]	Drop casting	Hybrid poly (3-hexylthiophene)-ZnO nanocomposite thin films	25	Response: small response to 25 ppm NH_3 at room temperature
Tuan et al. [14]	A standard photolithography technique	Polyaniline (PANI) nanowires (NWs)	25 to 500	Response: 2.9 to 500 ppm NH_3 at room temperature
Tai et al. [21]	*In situ* self-assembly	Polyaniline/titanium dioxide (PANI/TiO_2) nanocomposite thin films	23 to 141	Response: approximately 9 to 140 ppm NH_3, response time 2 s, and recovery time 20 to 60 s at room temperature
Huang et al. [26]	Spin coating	Graphene oxide (RGO)-polyaniline (PANI) hybrids	50	Response: approximately 10.4 to 50 ppm NH_3 at room temperature
Dhingra et al. [23]	Dipping	Zinc oxide/polyaniline (ZnO/PANI) hybrid	300	Response: approximately 23 to 300 ppm NH_3 at room temperature
This work	Drop casting	P3HT:1.00 mol% Au/ZnO NPs (4:1)	50 to 1,000	Response: approximately 32 to 1,000 ppm NH_3 at room temperature

The advantages of organic materials can be further exploited by their combinations with metal oxides[13, 18, 19, 20, 21, 22, 23] and metals[15, 19, 24, 25]. For instance, the hybrid of graphene oxide (GO)-polyaniline (PANI) nanoparticles prepared by spin coating is shown to exhibit higher response to 50 ppm of NH_3 than those of bare PANI nanofiber and bare GO sensors by a factor of 3.4 and 10.4, respectively[26]. In addition, P3HT/ZnO NWs and polypyrrole-zinc oxide (PPy-ZnO) composites are reported for sensitive detection of NH_3[13, 27]. In contrast, another report of P3HT-ZnO NW thin films demonstrates high sensitivity for NO_2 or H_2S and a moderate sensitivity for CO[27], while the response to NH_3 was very low ($S < 1\%$) at room temperature. Furthermore, PPy-ZnO hybrid films are doped with camphor sulfonic acid (30 wt.%) and exhibit high selectivity to NO_2, high sensitivity at low NO_2concentration (80% to 100 ppm), fast response time (120 s), and good stability but relatively sluggish response to reducing gases (H_2S, NH_3, C_2H_5OH, and CH_3OH) at room temperature[13]. Moreover, novel P3HT-ZnO nanocomposite hybrid thin films show a high relative response of 2.2 to 200 ppb of NO_2 but virtually no response to CO or C_2H_5OH and very small response to NH_3 at room temperature[18]. Besides, zinc oxide/ polyaniline (ZnO/PANI) hybrid structures are confirmed to exhibit much higher sensitivity to NH_3 gas at room temperature than bare ZnO[23, 28].

It can be observed that ZnO nanostructures are among the most widely employed metal oxides in polymer-based hybrid gas sensors, which should be due to its observed gas sensing enhancement, abundance, low cost, high stability, high electron mobility, low crystallization temperature, and ease of fabrication. However, mechanisms for gas sensing enhancement provided by ZnO nanostructures are not yet well understood. Nevertheless, it is widely observed that sensing properties of the hybrid sensors are related to surface characteristics of ZnO, which significantly depend on fabrication processes[29]. Most reported work mostly employs chemical-route and chemical vapor deposition (CVD) methods, which suffer from either poor reproducibility or high cost. Alternative low-cost, effective, and reliable methods for mass production of metal oxide nanostructured components in composite are still needed. Flame spray pyrolysis (FSP) is one of the most promising routes for the formation of single and multi-component functional nanoparticles with well-controlled diameter at low cost and high production rate. FSP has been applied to prepare metal oxide-supported nanoparticles and heterogeneous catalysts. However, FSP-made materials have not been employed in polymer-metal oxide hybrid sensors. It is thus interesting to apply them in this sensor system.

Gold (Au) is another effective means to improve sensing performance of polymer-based gas sensors via catalytic effects, which may be attained at low or room temperature. For instance, Pd incorporation in PANI considerably improved the response to methanol[19]. Similarly, Pt loading in PPy gas-sensitive films considerably improved NH_3responses of the PPy sensor[15]. Au is another effective catalyst for gas sensing[30]. Au loading on metal oxides particularly ZnO nanostructures has demonstrated to enhance response toward C_2H_5OH[31] and CO[30]. However, Au is relatively much less employed in polymer-based hybrid gas sensors. Its effect on gas sensing of a polymer-based hybrid sensor should thus be investigated.

Furthermore, the combination of noble metal catalyst, metal oxide, and polymer is expected to offer superior room-temperature gas sensors. To date, there has been development of noble metal/metal oxide/polymer composite gas sensors. In this work, we propose a practical implementation of this approach by blending a P3HT conductive polymer with Au-loaded ZnO nanoparticles (NPs) prepared by FSP. The novel hybrid materials are structurally characterized and tested for ammonia detection. In addition, the effects of ZnO and gold loading on gas sensing properties of P3HT sensing films are systematically analyzed by comparing the performances of P3HT with and without unloaded and 1.00 mol% Au-loaded ZnO NPs.

METHODS

Synthesis and characterization of nanoparticles

The 1.00 mol% Au-loaded ZnO nanoparticles (Au/ZnO NP_S) were successfully synthesized by the FSP process schematically illustrated in Figure 1. The precursor solution for FSP was prepared from zinc naphthenate (Sigma-Aldrich, St. Louis, MO, USA; 8 wt.% Zn) and gold (III)-chloride hydrate (Sigma-Aldrich; ≥49% Au) diluted in ethanol (Carlo Erba Reagenti SpA, Rodano, Italy; 98.5%). The precursor solution was injected at 5 mL min^{-1} through the reactor nozzle and dispersed with 5.0 L min^{-1} of oxygen into a fine spray (5/5 flame) while maintaining a constant pressure drop of 1.5 bar across the nozzle tip. A premixed flame fueled by 1.19 L min^{-1} of methane and 2.46 L $min^{-1}$of oxygen was ignited and maintained to support the combustion of the spray. The flames have yellowish orange color with a height of approximately 10 to 11 cm for both unloaded ZnO and 1.00 mol% Au/ZnO as shown in Figure 1.

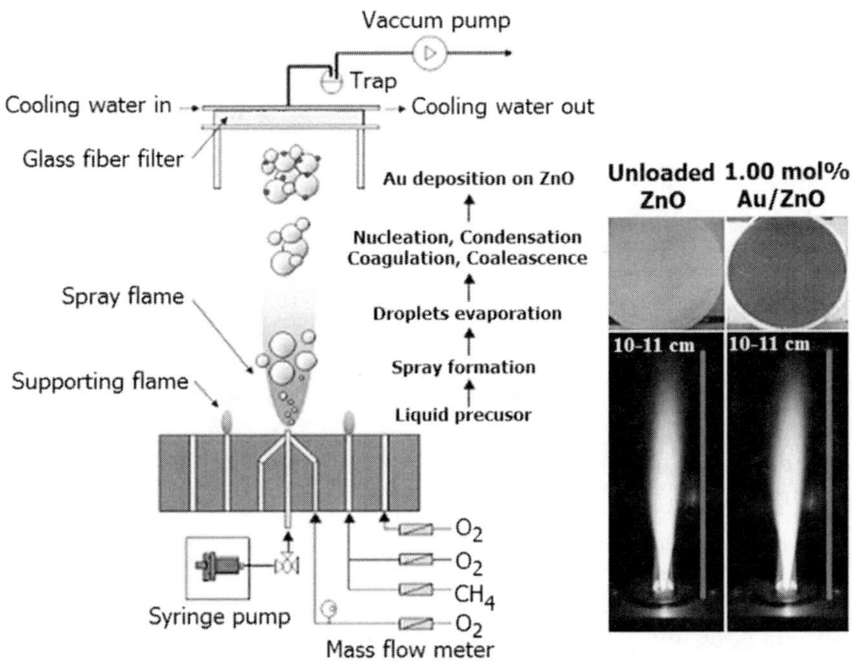

Figure 1. The experimental setup for flame-made unloaded zno and 1.00 mol% au/zno nps.

Upon evaporation and combustion of precursor droplets, particles are formed by nucleation, condensation, coagulation, coalescence, and Au deposition on a ZnO support. Finally, the nanoparticles were collected from glass microfiber filters (Whatmann GF/D, 25.7 cm in diameter) placed above the flame with an aid of a vacuum pump. X-ray diffraction (TTRAXIII diffractometer, Rigaku Corporation, Tokyo, Japan) was employed to confirm the phase and crystallinity of obtained nanoparticles using CuKα radiation at $2\theta = 20°$ to $80°$ with a step size of 0.06° and a scanning speed of 0.72°/min. Brunauer-Emmett-Teller (BET) analysis by nitrogen absorption (Micromeritics Tristar 3000, Micromeritics Instrument Co., Norcross, GA, USA) at liquid nitrogen temperature (77.4 K) was performed to obtain the specific surface area of the nanoparticles. The average particle size (d_{BET}) in nanometers is equal to $6/(SSA_{BET} \times \rho_{av})$, where SSA_{BET} is the specific surface area (m^2/g) and ρ_{av} is the average density of the flame-made 1.00 mol% Au/ZnO NPs with $\rho_{ZnO} = 5.606$ g cm^{-3}[32, 33] and $\rho_{Au} = 19.32$ g cm^{-3}[24], which took into account their weight content. High-resolution transmission electron microscopy (HR-TEM) was employed to examine the morphology and size of nanoparticles. The elemental composition of nanoparticles was analyzed by energy-dispersive X-ray spectroscopy (EDX) in mapping mode to confirm Au content in the resultant powders.

Sensor fabrication and sensing film characterization

Composite sensors were prepared by blending P3HT (Rieke Metals, Inc., Lincoln, NE, USA; M_w 48,000 g mol^{-1}) solution with 1.00 mol% Au/ZnO NP colloidal solution and drop casting onto prefabricated Cr/Au interdigitated electrodes. Cr (50 nm thick) and Au (200 nm thick) layers were deposited by DC sputtering in argon gas at a pressure of 3×10^{-3} mbar on an alumina substrate (0.40 cm × 0.55 cm × 0.04 cm). The interdigit spacing, width, and length were 100 μm, 100 μm, and 0.24 cm, respectively. P3HT solution was prepared by dissolving 30 mg of P3HT in 0.50 mL of chlorobenzene, and Au/ZnO NP colloidal solution was made by dispersing 5 to 25 mg of ZnO nanoparticles (unloaded ZnO and 1.00 mol% Au/ZnO) in 0.50 mL of 1-butanol.

To prepared hybrid films with various compositions, 1.00 mol% Au/ZnO NP colloidal solution was added to the stirred P3HT solution with five different mixing ratios (1:1, 2:1, 3:1, 4:1, and 1:2). The blended solution was drop casted on the interdigitated electrode and then baked at 150°C for 3 min in an oven. The active area of these sensing devices is 0.12 ± 0.04 cm^2. After completion, the crystalline phase of composite films was characterized by X-ray diffraction (XRD). The surface morphologies, elemental analysis, and cross section of the sensing layers were verified by field-emission scanning electron microscopy (FE-SEM) equipped with an EDX analysis system.

Finally, the devices were transferred to a stainless steel chamber for gas sensing measurement at room temperature.

Electrical and sensing test

P3HT and P3HT:1.00 mol% Au/ZnO NPs sensors were then tested by the standard flow through method in a stainless steel chamber at room temperature (25°C). The sensing experiment was carried out by measuring the reversible change of electrical resistance of sensors taken through a 6517 Keithley resistance meter (Keithley Instruments Inc., Cleveland, OH, USA) under a DC applied voltage of 10 V. A constant flux of synthetic dry air of 1 L/min as gas carrier was flowed to mix with the desired concentration of pollutants dispersed in synthetic air, and gas flow rates were precisely manipulated using a computer-controlled multi-channel mass flow controller. The background relative humidity (RH) under a flux of dry air was measured to be around 10%. The NH_3 pollutant source is a calibrated ammonia vapor balanced in dry air at 4,000 ppm (Linde Co. Ltd, Bangkok, Thailand). Ammonia (NH_3) vapor concentration was varied from 25 to 1,000 ppm. The gas sample exposure time and the dry air restoring time were fixed at 10 and 25 min, respectively. For selectivity performance, the sensors were also tested toward C_2H_5OH, CO, H_2S, and NO_2 at 1,000 ppm. The effect of humidity was also tested at 80% RH.

RESULTS AND DISCUSSION

Particles and sensing film properties

The XRD pattern of 1.00 mol% Au/ZnO NPs as shown in Figure 2a reveals that the nanoparticle is highly crystalline and has the hexagonal structure of ZnO according to JCPDS no. 89–1397. Au peaks are also found in these patterns and well matched with a face-centered cubic phase of Au (JCPDS file no. 89–3697[34]). The XRD patterns of P3HT and P3HT:1.00 mol% Au/ZnO NPs composite sensing films coated on Au/Al_2O_3 substrates in Figure 2b indicate the presence of the P3HT monoclinic crystal (the JCPDS no. 48–2040), the hexagonal ZnO phase of the NPs, a fcc phase of Au (JCPDS file no. 89–3697[34]), and a corundum phase of Al_2O_3 (JCPDS file no. 88–0826[35]). It can be seen that Au peaks of the hybrid film are relatively pronounced compared with those of 1.00 mol% Au/ZnO NPs. These observed Au peaks are mainly attributed to the diffraction from the interdigitated Au electrode, which almost completely overrides the very weak diffraction from Au loaded on ZnO NPs.

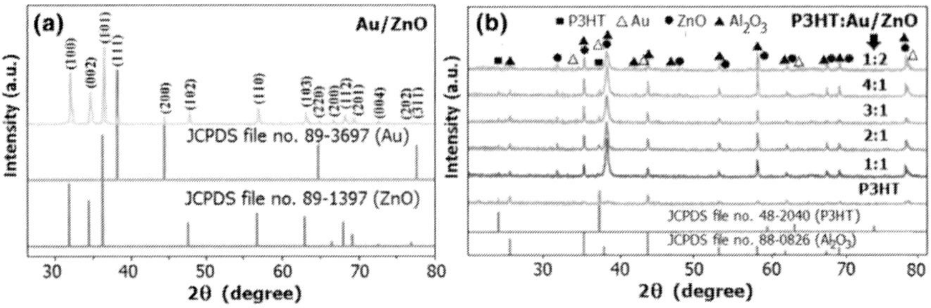

Figure 2. XRD PATTERNS. (A) 1.00 mol% Au/ZnO NPs. **(B)** Sensing films of P3HT:1.00 mol% Au/ZnO NPs in difference ratio.

The specific surface area of the unloaded ZnO and 1.00 mol% Au/ZnO NPs was measured by nitrogen absorption using BET analysis. It was found that the specific surface area (SSA_{BET}) of unloaded ZnO and 1.00 mol% Au/ZnO NPs is about 86.3 and 100 $m^2 g^{-1}$, respectively. The corresponding BET equivalent particle diameters (d_{BET}) of unloaded ZnO and 1.00 mol% Au/ZnO NPs are calculated to be about 10 and 9 nm, respectively. Thus, 1.00 mol% Au loading on ZnO NPs increases the specific surface area by 15% and reduces the particle diameter by about 10%. HR-TEM images of unloaded ZnO and 1.00 mol% Au/ZnO NPs in Figure 3 show spherical nanoparticles along with a few nanorods having a size in the range of 5 to 15 nm. For Au-loaded ZnO (Figure 3b), smaller spherical NPs with an average diameter of approximately 1.5 nm are clearly observed on the surface of ZnO as the darker spots as indicated in the figure. These NPs are confirmed to be Au NPs on ZnO support by EDX analysis in mapping mode (data not shown). The observed particle diameters by HR-TEM are in the same range as BET data. The observed smaller Au nanoparticle diameter of approximately 1.5 explains the result that the average BET nanoparticle diameter becomes smaller with Au loading as the average particle size will be reduced by the contribution of smaller particles.

Figure 4 shows FE-SEM images of P3HT and P3HT:1.00 mol% Au/ZnO NPs composite sensing films with the ratios of 4:1, 2:1, and 1:2 deposited on Al_2O_3 substrates with interdigitated Au electrodes. It can be seen that the surface of P3HT film is smooth and conformal with Au lines of the interdigitated electrode (Figure 4a). Upon mixing with 1.00 mol% Au/ZnO NPs, the surface becomes a relatively rough covering with fine white spots of NPs. The distribution of these spots on the Au interdigitated electrode surface is quite uniform, and the density of white spots increases accordingly with increasing content of NPs (Figure 4b, c, d). The results confirm the homogenous dispersion of 1.00 mol% Au/ZnO NPs in the P3HT matrix and its conformal coating on the substrate. In addition, the specific surface area of the composite film should be increased with increasing content of 1.00 mol% Au/ZnO NPs.

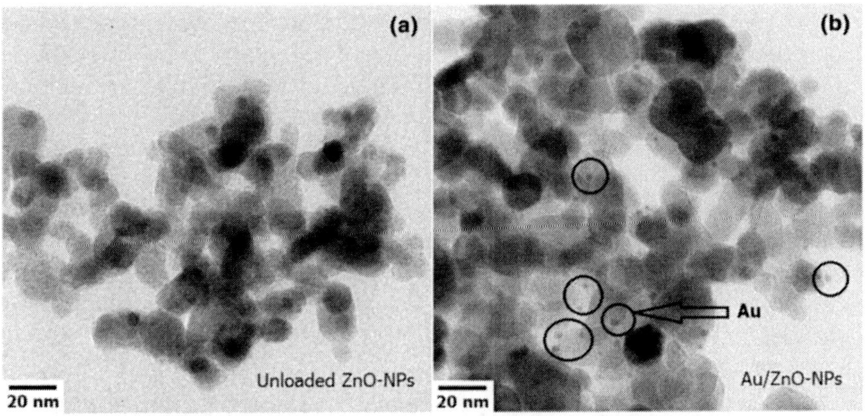

Figure 3. HR-TEM BRIGHT-FIELD IMAGE. (A) Unloaded ZnO. **(B)** 1.00 mol% Au/ZnO NPs.

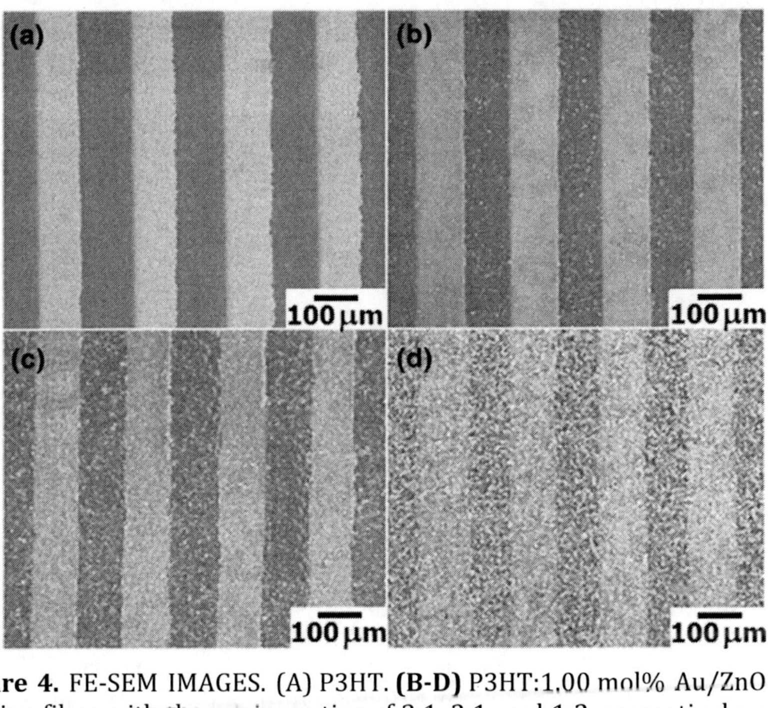

Figure 4. FE-SEM IMAGES. (A) P3HT. **(B-D)** P3HT:1.00 mol% Au/ZnO NPs sensing films with the mixing ratios of 3:1, 2:1, and 1:2, respectively, on an Al_2O_3 substrate with interdigitated Au electrodes.

The cross-sectional FE-SEM images along with EDX analyses of P3HT and P3HT:1.00 mol% Au/ZnO NPs (4:1) composite sensing films on an Al_2O_3 substrate with interdigitated Au electrodes after sensing test at room temperature in dry air are illustrated in Figure 5. It can be seen that the P3HT film is a smooth and solid layer (Figure 5a, b, c), while the composite film demonstrates porous asperities of the nanoparticle-polymer mixture (Figure 5d, e, f). The thicknesses of P3HT and composite films are estimated in the same range of 6 to 8 μm. The elemental composition on the surface and across P3HT and P3HT:1.00 mol% Au/ZnO NP layers is demonstrated in the EDX spectra and line scan profiles (Figure 5b, c and5e, f, respectively). It confirms that the P3HT film contains only oxygen (O), carbon (C), and sulfur (S) and the P3HT:1.00 mol% Au/ZnO NP layer has one additional element of zinc (Zn) while the gold (Au) loaded element cannot be observed due to its very low content. In addition, the line scan profiles indicate that elemental compositions through the films are quite uniform.

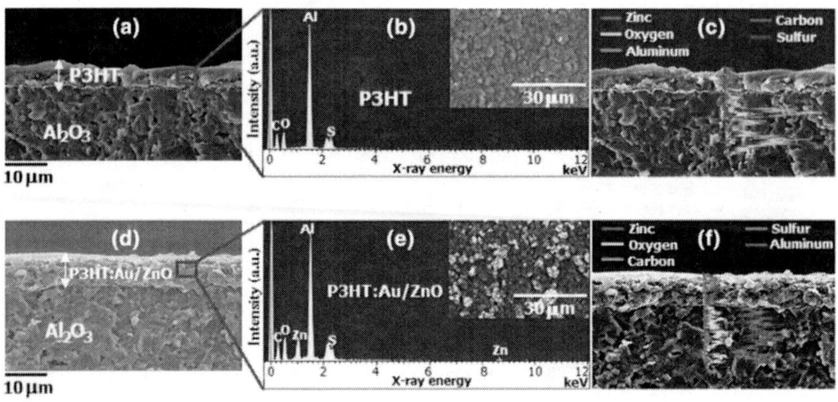

Figure 5. FE-SEM MICROGRAPHS OF THE CROSS-SECTIONAL STRUCTURE. (A) P3HT. **(D)** P3HT:1.00 mol% Au/ZnO NPs sensing films on an alumina substrate. **(B, E)** Corresponding EDX. **(C, F)** Corresponding line scan profiles.

Atomic force microscopy (AFM) was employed to quantitatively investigate the morphology of P3HT and P3HT:1.00 mol% Au/ZnO NPs (4:1) composite sensing films drop casted on the Al_2O_3 substrate (Figure 6). The results indicate that the film surfaces are quite uniform, containing only tiny defects within a scan area of 20 μm × 20 μm. The average surface roughness of P3HT and the P3HT:1.00 mol% Au/ZnO NPs film is calculated from AFM data to be 130.1 and 135.2 nm, respectively. In addition, the composite film exhibits a relatively sharp granular morphology with a uniform grain size of approximately 80 to 100 nm, suggesting the presence of a nanosized grain structure in the composite sensing film due to the addition of 1.00 mol% Au/ZnO NPs.

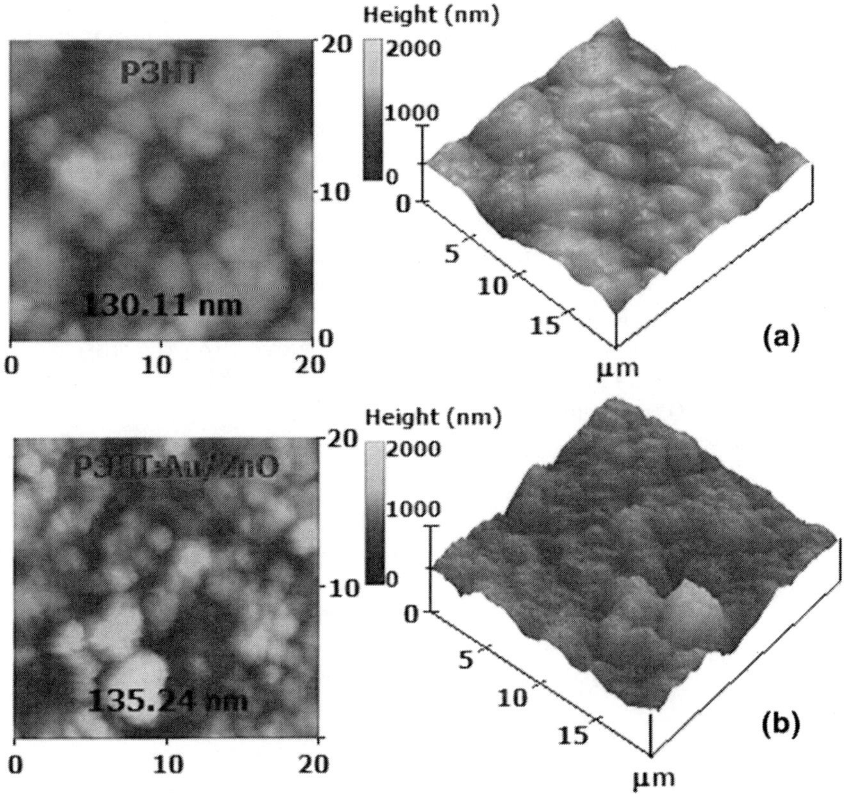

Figure 6. AFM MORPHOLOGY. (A) P3HT. **(B)** P3HT:1.00 mol% Au/ZnO NPs (4:1) hybridized sensing film drop casted on an Al_2O_3substrate.

Gas sensing properties

The dynamic changes in resistance of sensors with different mixing ratios of P3HT:1.00 mol% Au/ZnO NPs (1:0, 1:1, 2:1, 3:1, 4:1, 1:2, and 0:1) are shown in Figure 7. It is seen that all sensors exhibit an increase of resistance during NH_3 exposure, indicating a p-type-like gas sensing behavior. In addition, it is observed that the baseline resistance monotonically increases with increasing content of 1.00 mol% Au/ZnO NPs in accordance with the typical combination of materials' resistances. Furthermore, P3HT exhibits a moderate NH_3 response, while 1.00 mol% Au/ZnO NPs give very low response to NH_3 at room temperature. Moreover, the addition of 1.00 mol% Au/ZnO NPs into P3HT at a mixing ratio up to 1:1 leads to significant enhancement in the NH_3 response compared with the P3HT sensor. However, the response rapidly degrades when the amount of 1.00 mol% Au/ZnO NPs exceeds that of P3HT (1:2). From calculated changes of resistance, it is found that the sensor with 4:1 of P3HT:1.00 mol% Au/ZnO NPs exhibits the highest value, indicating that it is the optimal P3HT:1.00 mol% Au/ZnO NPs composite sensor. Since the optimal mixing ratio of the Au/ZnO NPs and P3HT of 1:4 is at the lowest border of the investigated range, it is possible that the actual optimal concentration will be at a lower concentration value and further detailed investigation should be conducted to refine the result. The obtained optimal performances of P3HT:Au/ZnO sensors are superior to other reports presented in Table 1 with a relatively high response magnitude of 32 and wide concentration range of 1,000 ppm. However, the response at lower concentration may be lower than some work such as ZnO/ PANI hybrid[23] and PANI/TiO_2 nanocomposite thin films[21].

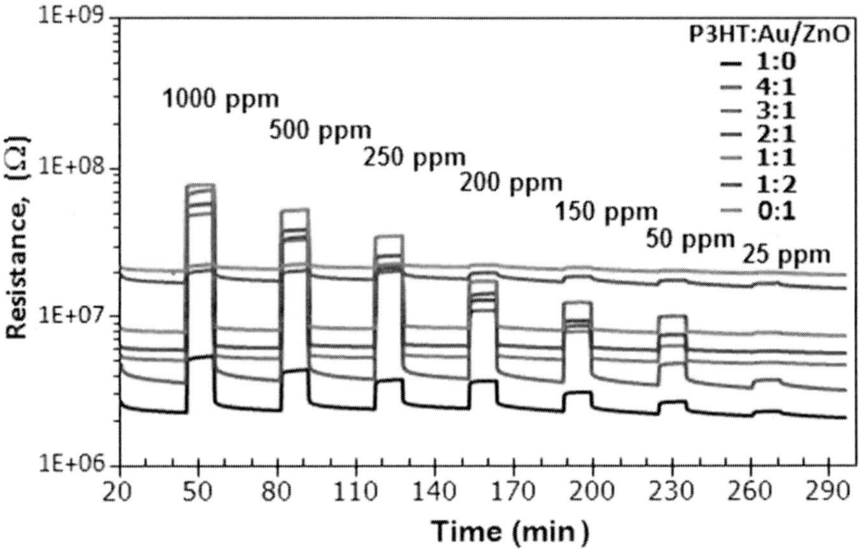

Figure 7 . CHANGE IN RESISTANCE. The resistance of sensors with difference ratio of P3HT:1.00 mol% Au/ZnO NPs (1:0, 1:1, 2:1, 3:1, 4:1, 1:2, and 0:1) toward 25 to 1,000 ppm NH_3 at room temperature.

The sensor characteristics are then analyzed in terms of sensor response and response time. The sensor response (S) is determined from the electrical resistance change of P3HT:1.00 mol% Au/ZnO NPs sensors upon exposure to target gas using the following relation: $S = R_{gas}/R_{air}$, where R_{gas} and R_{air} are the stable electrical resistance of a sensor upon exposure to NH_3 and the initial resistance in air, respectively. The response time is defined as the time needed for a sensor to attain 90% of maximum change in resistance upon exposure to a test gas. The calculated sensor response and response time of optimal sensors with 4:1 of P3HT:1.00 mol% Au/ZnO NPs are shown in Figure 8. Apparently, the sensor response to NH_3 gas monotonically increases upon exposure with increasing NH_3concentration from 25 to 1,000 ppm. At 1,000 ppm, the composite sensor prepared with the 4:1 ratio exhibits the highest NH_3 response of 32 and a short response time of 4.2 s. Thus, 1.00 mol% Au/ZnO NPs:P3HT composite sensors offer excellent NH_3 sensing performances with high response, short response time, and room-temperature operation. It should be noted that *y*-error bars of all data correspond to the statistical spread from five sensors of each composition with five evaluations. The statistical results show that fabricated composite sensors offer good repeatability and reproducibility with maximum variation of less than 20%.

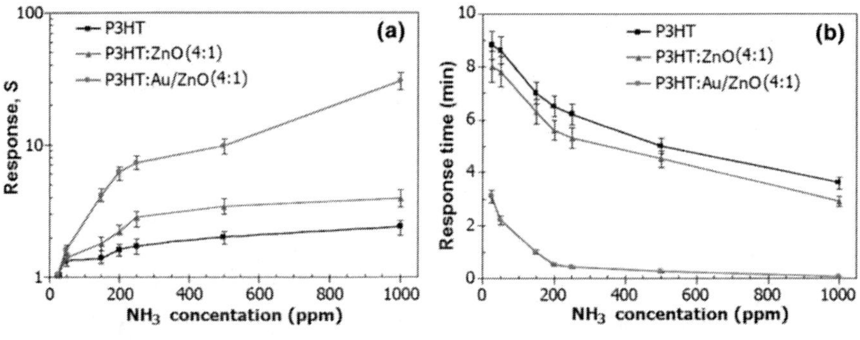

Figure 8. SENSOR RESPONSE AND RESPONSE TIME. (A) Sensor response. **(B)** Response time versus NH_3 concentration (25 to 1,000 ppm) of P3HT:unloaded ZnO NPs (4:1), P3HT:1.00 mol% Au/ZnO NPs (4:1), and pure P3HT sensors at room temperature.

The enhanced gas sensing response of the P3HT:1.00 mol% Au/ZnO NPs (4:1) composite sensor may be attributed to the high specific surface area of P3HT surface-coated on granular 1.00 mol% Au/ZnO, which enhances gas adsorption and interaction at the interface[13, 21, 36]. In order to distinguish the roles of ZnO and gold nanoparticles, the NH_3 sensing performances of P3HT:ZnO loaded with Au (4:1) are compared with those of P3HT:unloaded ZnO (4:1) and pure P3HT as also demonstrated in Figure 8. It can be seen that the response of the P3HT sensor is only slightly improved by the addition of unloaded ZnO at the mixing ratio of 4:1, while Au addition by loading on ZnO NPs leads to significant increase of NH_3 response by almost an order of magnitude. In addition, the response time is also substantially reduced to a few minutes or seconds, while ZnO addition does not notably decrease the response time. Thus, Au plays a much more important role than ZnO NPs in enhancing NH_3 response of the composite sensor. Moreover, it was found from our preliminary study that NH_3 response of the P3HT:Au-loaded ZnO film increased monotonically as Au loading level increased from 0 to 1.00 mol%. Thus, if Au content increased further, the NH_3 response should increase to an optimal point and then reduce due to particle aggregation. Further study will be conducted to determine the ultimate optimal Au loading level of the P3HT:flame-made Au-loaded ZnO film for NH_3 sensing and fully reported elsewhere.

The gas sensing mechanism for the composite sensors may be explained on the basis of interactions between the sensing film and adsorbed gas. For pure P3HT, it has been proposed that NH_3 can adsorb and donate a lone pair of its electrons to the pentagonal sulfur ring in the P3HT structure[22]. Electrons will recombine with existing holes in the p-type P3HT, leading to a resistance increase in agreement with the observed NH_3 response. By adding unloaded ZnO NPs, the response is enhanced by a factor of approximately 1.5. This could reasonably be explained by the increase of specific surface area for gas interaction of the composite film by ZnO NPs. From the FE-SEM image in Figure 5, ZnO NP addition results in considerable increase of film porosity and hence the surface area. However, the magnitude of this increase cannot be quantitatively confirmed as BET measurement, which is currently the only quantitative analysis method of specific surface area, of the film surface is presently not available because the standard BET method requires a large amount of material (>0.3 g) which is difficult to prepare in the form of film and the presence of substrate will considerably complicate nitrogen adsorption evaluation. Thus, the specific surface area of the composite film can only be inferred from BET data of the corresponding powder and SEM images.

With Au loading, the response is enhanced much more drastically than ZnO NPs and the response also increases with increasing Au loading level from 0 to 1.00 mol%. Considering the effect of surface area change, the BET specific surface area of ZnO NPs is found to increase from 86.3 to 100 $m^2 g^{-1}$ with 1.00 mol% Au loading (see the 'Particles and sensing film properties' section). This corresponds to the 15.9% increase, and the influence of specific surface area alone cannot explain the observed large response enhancement by Au loading on ZnO. From the results, Au loading on ZnO increases not only the response magnitude but also the response rate substantially. Thus, the most plausible mechanism for such enhancement should be the catalytic effect of Au on ZnO NPs. Figure 9 depicts our proposed model for the catalytic effect of Au/ZnO NPs based on a P3HT-ammonia interaction mechanism reported recently[17]. In this model, it is assumed that Au/ZnO NPs located around sulfur atoms in the pentagonal rings of P3HT catalyze the reaction, causing more NH_3 molecules to give lone-pair electrons and form the weak binding. The probability for Au/ZnO NPs should be high since gold and sulfur have rather strong binding affinity. To obtain effective catalyst activity, Au NPs should be uniformly dispersed throughout the P3HT matrix. Thus, Au plays the main role in enhancing NH_3 interaction and response with P3HT, while the role of ZnO NPs is the supports that help formation and dispersion of ultrafine Au nanoparticles. However, when only 1.00 mol% Au/ZnO is used, there is no response since Au catalyzes the reaction between NH_3 and P3HT.

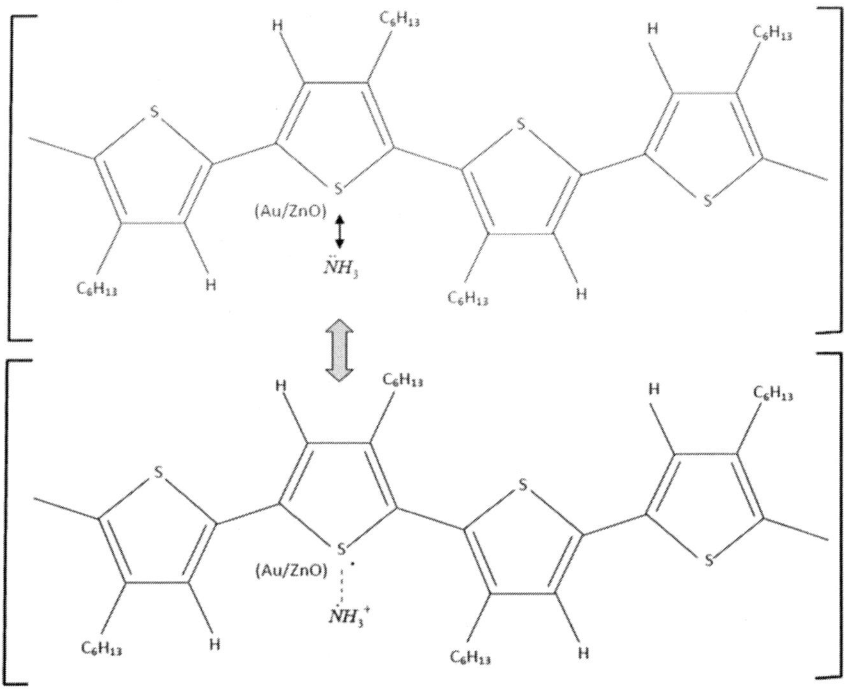

Figure 9. Proposed model for catalytic effect of au/zno nps in p3ht:au/zno sensors on nh $_3$ sensing.

For the effect of composite composition, the results show that 4:1 of P3HT:1.00 mol% Au/ZnO NPs, which is the composite with the lowest 1.00 mol% Au/ZnO NP content, offers the highest NH_3 sensing enhancement and the enhancement decreases with increasing Au/ZnO NP content. A plausible explanation is that 1.00 mol% Au/ZnO NPs are well dispersed in the P3HT matrix at this low concentration, yielding a homogeneous distribution of Au/ZnO NPs throughout the layer and enabling effective catalytic interaction with NH_3 gas. In addition, the well-dispersed structure should be highly porous and exhibit large surface area for gas interaction. As the content of 1.00 mol% Au/ZnO NPs increases, the Au/ZnO NPs may begin to agglomerate resulting in less homogenous formation of Au/ZnO NPs and less porous composite structures, which lead to low NH_3 sensing response. Based on the current results, NH_3 sensing properties of the composite film may be further improved by optimizing the structure/composition of the Au loading material as well as metal oxide support to maximize the catalytic effect and by adding intercalating nanomaterials with different dimensionalities (i.e., 2D graphene, 1D metal oxide nanowire, 1D carbon nanotubes, etc.) to reduce particle agglomeration and increase effective surface area. Moreover, new catalysts based on the composite of Au and other catalytic materials should be explored to further improve the catalytic effect.

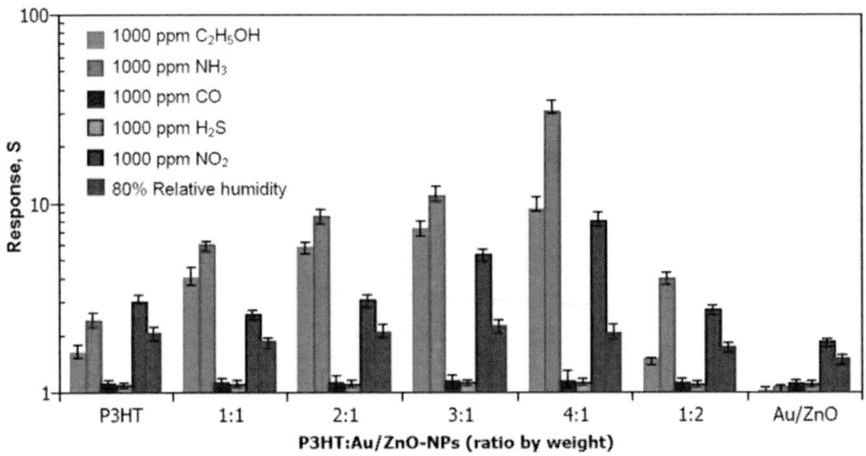

Figure 10. RELATIVE RESPONSE. The relative response to NH_3 (1,000 ppm), C_2H_5OH (1,000 ppm), CO (1,000 ppm), H_2S (1,000 ppm), NO_2 (1,000 ppm), and H_2O (80% RH) of sensors with difference ratio of P3HT:1.00 mol% Au/ZnO NPs (1:0, 1:1, 2:1, 3:1, 4:1, 1:2, and 0:1).

Selectivity can be defined as the ability of a sensor to respond to a target gas in the presence of other interfering gases[12]. The NH_3 sensing selectivity of composite sensors is characterized toward various reducing and oxidizing gases including ethanol (C_2H_5OH), carbon monoxide (CO), hydrogen sulfide (H_2S), and nitrogen dioxide (NO_2) at 1,000 ppm and room temperature as shown in Figure 10. In addition, the effect of water vapor is included at 80% RH. It is evident that the composite sensor of P3HT:1.00 mol% Au/ZnO NPs (4:1) exhibits a relatively high response of 32 to 1,000 ppm of NH_3 while the response to 1,000 ppm of C_2H_5OH and NO_2 is relatively low (approximately 9 and approximately 8, respectively), and those of 1,000 ppm of CO and

1,000 ppm of H_2S are almost negligible. Additionally, the optimal sensor exhibits a quite low response of approximately 2.2 to a high relative humidity of 80%. For P3HT and other composite combinations, the response to 1,000 ppm of NH_3 is not much higher than that to C_2H_5OH, NO_2, and humidity. The results indicate that P3HT:1.00 mol% Au/ZnO NPs also has better selectivity to NH_3 against C_2H_5OH, CO, H_2S, NO_2, and humidity than other sensors. Therefore, the composite sensor of P3HT:1.00 mol% Au/ZnO NPs (4:1) can be used for selective detection of NH_3.

Lastly, the stability of P3HT-based sensors has been evaluated by monitoring the response change over 30 days. It was found that the pure P3HT sensor had an average response reduction of around 4.8%/day, while P3HT with 1.00 mol% Au/ZnO NPs and unloaded ZnO NPs at different ratios exhibits slightly lower average response reduction in the range of 4.2% to 4.6%/day. It is not conclusive whether ZnO NPs help improve the stability of P3HT sensors. Nevertheless, it is seen that the ZnO NPs:P3HT sensor has fair medium-term stability, which is relatively high compared with other conductive polymers.

CONCLUSIONS

In conclusion, novel composite P3HT:1.00 mol% Au/ZnO NPs films have been systemically studied for NH_3 sensing applications at room temperature. The unloaded ZnO and ZnO NPs loaded with Au (1.00 mol%) were produced by a single-step FSP technique. The particle analyses using XRD, HR-TEM, and BET indicated that ZnO NPs were highly crystalline with a typical hexagonal structure of ZnO, and ultrafine Au NPs with 1 to 2 nm in diameter were formed around ZnO NPs. Composite P3HT:1.00 mol% Au/ZnO NPs films with different compositions were prepared by solution mixing and casting. Film characterizations by XRD and FE-SEM confirmed the presence of P3HT/ZnO phases and porous nanoparticle structures in the composite thick film. The gas sensing results showed that the inclusion of 1.00 mol% Au/ZnO NPs at a low content provided significant NH_3 sensing enhancement. In particular, the P3HT:1.00 mol% Au/ZnO NPs composite film with the ratio of 4:1 exhibited the best NH_3 sensing performances with a high sensor response of approximately 32 and short response time within a minute to 1,000 ppm of NH_3 at a room temperature. In addition, the optimal composite film exhibited higher NH_3 selectivity against C_2H_5OH, CO, H_2S, NO_2, and H_2O than other composites as well as P3HT and 1.00 mol% Au/ZnO NPs. The observed composite gas sensing behaviors were explained based on the increased specific surface area by porous blended nanoparticle structure and catalytic effect of Au/ZnO NPs. From overall results, the P3HT:1.00 mol% Au/ZnO NPs composite sensor is a highly promising candidate for the efficient detection of NH_3 at room temperature.

ACKNOWLEDGEMENTS

The authors gratefully acknowledge the financial support from the Thailand Research Fund (TRF), the Office of the Higher Education Commission and Maejo University, Thailand (MRG5580067); Program in Materials Science, Faculty of Science, Maejo University, Thailand; the National Research Council of Thailand; the National Research University under the Office of Higher Education Commission; Materials Science Research Center, Faculty of Science, Chiang Mai University, Thailand; and National Electronics and Computer Technology Center (NECTEC), Pathumthani, Thailand.

COMPETING INTERESTS

The authors declare that they have no competing interests.

AUTHORS' CONTRIBUTIONS

VK carried out the experiments, acquired the original data, participated in the sequence alignment, and drafted the manuscript. AW and SP have made substantial contributions to the conception and design for this article. AT read the final manuscript. All the authors read and approved the final manuscript.

REFERENCES

1. Narasimhan LR, Goodman W, Kumar C, Patel N: Correlation of breath ammonia with blood urea nitrogen and creatinine during hemodialysis. *Proc Natl Acad Sci U S A* 2001, 98: 4617–4621.
2. de la Hoz RE, Schueter DP, Rom WN: Chronic lung disease secondary to ammonia inhalation injury: a report on three cases. *Am J Ind Med* 1996, 29: 209–214.
3. Leung CM, Foo CL: Mass ammonia inhalation burns-experience in the management of patients. *Ann Acad Med Singapore*1992, 21: 624–629.
4. Michaels RA: Emergency planning and acute toxic potency of inhaled ammonia. *Environ Health Perspect* 1999, 107: 617–627.
5. Close LG, Catlin FI, Cohn AM: Acute and chronic effects of ammonia burns on the respiratory track. *Arch Otolaryngol* 1980, 106: 151–158.

6. Shimizu K, Chinzei I, Nishiyama H, Kakimoto S, Sugaya S, Matsutani W, Stasuma A: Doped-vanadium oxides as sensing materials for high temperature operative selective ammonia gas sensors. *Sens Actuators B Chem* 2009, 141: 410–416.
7. Sun F, Hu M, Sun P, Zheng J, Liu B: NH_3 sensing characteristics of nano-WO_3 thin films deposited on porous silicon. *J Nanosci Nanotechnol* 2010, 10: 7739–7742.
8. Jimenez I, Centeno MA, Scotti R, Morazzoni F, Arbiol J, Cornet A, Morante JR: NH_3 interaction with chromium-doped WO_3nanocrystalline powders for gas sensing application. *J Mater Chem* 2004, 15: 2412–2420.
9. Balamurugan C, Bhuvanalogini G, Subramania A: Development of nanocrystalline $CrNbO_4$ based p-type semiconducting gas sensor for LPG, ethanol and ammonia. *Sens Actuators B Chem* 2012, 168: 165–171.
10. Tulliani JM, Cavalieri A, Musso S, Sardella E, Geobaldo F: Room temperature ammonia sensors based on zinc oxide and functionalized graphite and multi-walled carbon nanotubes. *Sens Actuators B Chem* 2011, 152: 144–154.
11. Kolmakov A, Moskovits M: Chemical sensing and catalysis by one-dimensional metal-oxide nanostructures. *Annu Rev Mater Res* 2004, 34: 151–180.
12. Patil LA, Sonawane LS, Patil DG: Room temperature ammonia gas sensing using MnO_2-modified ZnO thick film resistors. *J Model Phys* 2011, 2: 1215–1221.
13. Chougule MA, Sen S, Patil VB: Polypyrrole–ZnO hybrid sensor: effect of camphor sulfonic acid doping on physical and gas sensing properties. *Synth Met* 2012, 162: 1598–1603.
14. Tuan CV, Tuan MA, Hieu NV, Trung T: Electrochemical synthesis of polyaniline nanowires on Pt interdigitated. *Curr Appl Phys* 2012, 12: 1011–1016.
15. Chen X, Li DM, Liang SF, Zhan S, Liu M: Gas sensing properties of surface acoustic wave NH_3 gas sensor based on Pt doped polypyrrole sensitive film. *Sens Actuators B Chem* 2013, 177: 364–369.
16. Jeong JW, Lee YD, Kim YM, Park YW, Choi JH, Park TH, Soo CD, Won SM, Han IK, Ju BK: The response characteristics of a gas sensor based on poly-3-hexylithiophene thin-film transistors. *Sens Actuators B Chem* 2010, 146: 40–45.
17. Tiwari S, Kumar S, Joshi L, Chakrabarti P, Takashimac W, Kaneto K, Prakash R: Poly-3-hexylthiophene based organic field-effect transistor: detection of low concentration of ammonia. *Sens Actuators B Chem* 2012, 171–172: 962–968.
18. Baratto C: *Proceeding of the 5th IEEE Conference on Sensors: IEEE Sensors 2006, 22–25 October 2006*. Daegu, South Korea; 2006.
19. Athawale AA, Bhagwat SV, Katre PP: Nanocomposite of Pd-polyaniline as a selective methanol sensor. *Sens Actuators B Chem* 2006, 114: 263–267.
20. Kruefu V, Liewhiran C, Wisitsoraat A, Phanichphant S: Selectivity of flame-spray-made Nb/ZnO thick films towards NO_2 gas. *Sens Actuators B Chem* 2011, 156: 360–367.
21. Tai H, Juang Y, Xie G, Yu J, Chen X, Ying Z: Influence of polymerization temperature on NH_3 response of PANI/TiO_2 thin film gas sensor. *Sens Actuators B Chem* 2008, 129: 319–326.
22. Ai X, Anderson N, Guo JC, Kowalik J, Tolbert LM, Lian TQ: Ultrafast photoinduced charge separation dynamics in polythiophene/SnO_2 nanocomposites. *J Appl Chem B* 2006, 110: 25496–25503.
23. Dhingra M, Kumar Shrivastava S, Kumra PS, Annapoorni S: Impact of interfacial interactions on optical and ammonia sensing in zinc oxide/polyaniline structures. *Bull Mater Sci* 2013, 36: 647–652.
24. Tsai TH, Lin KC, Chen SM: Electrochemical synthesis of poly (3,4-ethylenedioxythiophene) and gold nanocomposite and its application for hypochlorite sensor. *Int J Electrochem Sci* 2011, 6: 2672–2687.
25. Chang SJ, Weng WY, Hsu CL, Hsueh TJ: High sensitivity of a ZnO nanowire-based ammonia gas sensor with Pt nano-particles. *Nano Commun Netw* 2010, 1: 283–288.
26. Huang X, Hu N, Gao R, Yu Y, Wang Y, Yang Z, Kong E, Wei H, Zhang Y: Reduced graphene oxide-polyaniline hybrid: preparation, characterization and its applications for ammonia gas sensing. *J Mater Chem* 2012, 22: 22488–22495.

27. Saxena V, Aswal DK, Kaur M, Koiry SP, Gupta SK, Yakhmi JV: Enhanced NO_2 selectivity of hybrid poly (3-hexylthiophen): ZnO-nanowire thin films. *Appl Phys Lett* 2007, 90: 043516–1–043516–3.
28. Lima JPH: *Proceeding of the International Conference on Advanced Materials: Brazil-MRS, 20–25 September 2009*. Rio de Janeiro, Brazil; 2009.
29. Wang H, Xie C, Zhang W, Cai S, Gui Z, Hazard J: Comparison of dye degradation efficiency using ZnO powders with various size scales. *J Hazard Mater* 2007, 141: 645–652.
30. Chang SJ, Hsueh TJ, Chen IC, Huang BR: Highly sensitive ZnO nanowire CO sensors with the adsorption of Au nanoparticles. *Nanotechnology* 2008, 19: 1–5.
31. Wongrat E, Pimpang P, Choopun S: Comparative study of ethanol sensor based on gold nanoparticles: ZnO nanostructure and gold: ZnO nanostructure. *Appl Surf Sci* 2009, 256: 968–971.
32. Yu HF, Qian DW: Characterization and photocatalytic kinetics of the ZnO powder prepared using a polyol process. *Part Sci Technol* 2013, 31: 482–487.
33. John R, Rajakumari R: Synthesis and characterization of rare earth ion doped nano ZnO. *Nano Micro Lett* 2012, 4: 65–72.
34. Hua Q, Shi F, Chen K, Chang S, Ma Y, Jiang Z, Pan G, Huang W: Cu_2O-Au nanocomposites with novel structures and remarkable chemisorption capacity and photocatalytic activity. *Nano Res* 2011, 4: 948–962.
35. Lee JS, Kim HS, Park NK, Lee TJ, Kang M: Low temperature synthesis of α-alumina from aluminum hydroxide hydrothermally synthesized using [Al $(C_2O_4)_x(OH)_y$] complexes. *Chem Eng J* 2013, 230: 351–360.
36. Pawar SG, Patil SL, Chougule MA, Raut BT, Godase PR, Mulik RN, Sen S, Patil VB: New method for fabrication of CSA doped PANi (TiO_2) thin-film ammonia sensor. *IEEE Sens J* 2011, 11: 2980–2985.

CHAPTER 16

Effect of Tubular Chiralities and Diameters of Single Carbon Nanotubes on Gas Sensing Behavior: A DFT Analysis

A. A. EL-Barbary[1,2*], Kh. M. Eid[1,3], M. A. Kamel[1], H. M. Osman[1], G. H. Ismail[1,2]

[1]Physics Department, Faculty of Education, Ain Shams University, Cairo, Egypt

[2]Physics Department, Faculty of Science, Jazan University, Jazan, KSA

[3]Bukairiayh for Science, Qassim University, Buraydah, KSA

ABSTRACT

Using density functional theory, the adsorption of CO, CO_2, NO and CO_2 gas molecules on different chiralities and diameters of single carbon nanotubes is investigated in terms of energetic, electronic properties and surface reactivity. We found that the adsorption of CO and CO_2 gas molecules is dependent on the chiralities and diameters of CNTs and it is vice versa for NO and NO_2 gas molecules. Also, the electronic character of CNTs is not affected by the adsorption of CO and CO_2 gas molecules while it is strongly affected by NO and NO_2 gas molecules. In addition, it is found that the dipole moments of zig-zag CNTs are always higher than the arm-chair CNTs. Therefore, we conclude that the zig-zag carbon nanotubes are more preferred as gas sensors than the arm-chair carbon nanotubes, especially for detecting NO and NO_2 gas molecules.

Keywords: Carbon Nanotubes, DFT, Gas Sensors

1. INTRODUCTION

Monitoring of combustible gas alarms, gas leak detection, and environmental pollution is of great concern in public security. Advances in nanotechnology give great promise for achieving new sensing materials. Since the discovery of carbon nanotubes in 1991, the single-walled carbon nanotubes (SWCNTs) have been intensively investigated as nanoscale gas sensors because of their great surface areas to bulk ratio and their abilities to modulate electrical properties upon adsorption of various kinds of gas molecules [1] -[17] . The emission of carbon and nitrogen oxides (CO, CO_2, NO and NO_2) results from the combustion of fossil fuels, contributing to both smog and acid precipitation, and affecting both terrestrial and aquatic ecosystems [18] . Although many efforts have been made to use catalysts to reduce the amount of carbon or nitrogen oxides in the air [19] -[25] , an efficient method of sensing and removing carbon and nitrogen oxides is still required.

Because carbon and nitrogen oxides are the most dangerous air pollutants, toxic and global warming gases, our work is concentrated on investigating the effect of tubular chiralities and diameters of single carbon nanotubes on gas sensing behavior for CO, CO_2, NO and NO_2 gas molecules, applying the first principle calculations.

2. COMPUTATIONAL METHODS

All calculations were performed with the density functional theory as implemented within G03W package [26] - [29] , using B3LYP exchange-functional and applying basis set 6 - 31g (d,p). Pure carbon nanotubes $(5,0)$ and $(9,0)$, $(5,5)$ and $(6,6)$ are fully optimized with spin average as well as the adsorption of CO, CO_2, NO and NO_2 gas molecules.

The obtained diameters [30] and the adsorption energies of gas molecules on CNTs (E_{ads}) [31] are calculated from the following relations:

$$D = 0.73\left(n^2 + nm + m^2\right)^{1/2}$$

where n and m are integral numbers, the composition of chiral vector.

$$E_{ads} = E_{(nanotube+gas\ molecules)} - E_{nanotube} - E_{gas\ molecules}$$

where $E_{(nanotube+gas\ molecules)}$ is the total energy of nanotube and gas molecules, $E_{nanotube}$ is the energy of the carbon nanotube, and $E_{gas\ molecules}$ is the energy of gas molecules.

3. RESULTS AND DISCUSSION

We will investigate the adsorption of gas molecules, CO, CO_2, NO and NO_2 on four carbon nanotubes with different charilities and diameters $(5,0)$ CNT, $(9,0)$ CNT, $(5,5)$ CNT and $(6,6)$ CNT as shown in Figure 1 and Table 1.

3.1. Adsorption of CO, CO_2, NO and NO_2 Gas Molecules on CNTs

We have adsorbed CO and CO_2 gas molecules vertically on different three positions of $(5,0)$, $(9,0)$, $(5,5)$ and $(6,6)$ CNTs: above a carbon atom (carbon site), above a bond between two carbon atoms (bond site) and above a center of a hexagon ring (vacant site). The calculated adsorption energies of CO and CO_2 gas molecules

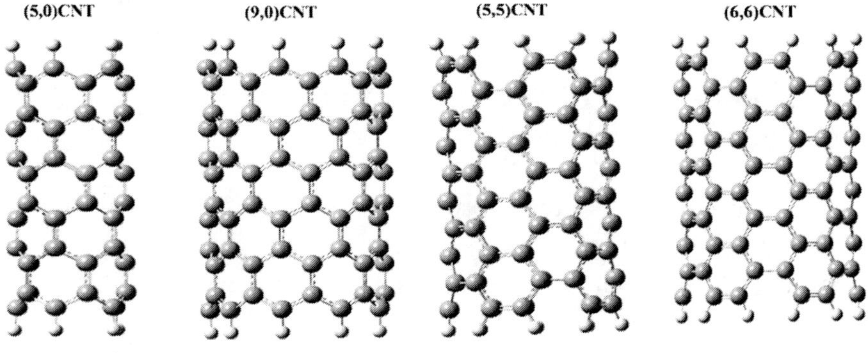

Figure 1. The fully optimized structures of $(5,0)$, $(9,0)$, $(5,5)$ and $(6,6)$ CNTs. Carbon atoms (gray) and hydrogen atoms (white).

are listed in Table 2. It is found that the best position and adsorption energy for CO gas molecule is above the bond site on $(9,0)$ CNT with adsorption energy of −0.43 eV, however for CO_2 gas molecule is above the vacant site on $(9,0)$ CNT with adsorption energy of −0.26 eV. Therefore, one can conclude that the best CNT gas sensor for CO and CO_2 gas molecules is the $(9,0)$ CNT.

Also, we have adsorbed NO and NO_2 gas molecules vertically on different three positions of $(5,0)$, $(9,0)$, $(5,5)$ and $(6,6)$ CNTs: above a carbon site, above a bond site and above a vacant site. The calculated adsorption energies of NO and NO_2 gas molecules are listed in Table 3. It is found that the best adsorption energies of NO gas molecule are on the $(9,0)$ CNT above a bond site, then above a carbon site and after that above a vacant site with adsorption energies of −1.65 eV, −1.55 eV and −1.34 eV, respectively. However, for NO_2 gas molecule is found to be above the bond site on $(9,0)$ CNT with adsorption energy of −1.75 eV. Also, it is noticed that the vacant site is always preferred for NO_2 gas adsorption on all the studied CNTs except for $(9,0)$ CNT. Therefore, one can conclude that all CNTs can be used as gas sensors for NO and NO_2 gas molecules.

From Table 2, Table 3, one can investigate the effect of the chiralities and the diameters on the CNT gas sensors behavior. It is clear that the adsorption of CO and CO_2 gas molecules is dependent on the chiralities and the diameters of CNTs. The adsorption of CO and CO_2 gas molecules is enhanced with increasing the diameter of the zig-zag CNTs. However, the adsorption of NO and NO_2 gas molecules is independent on the chiralities and the diameters of CNTs.

3.2. Energy Gaps of Adsorbed CO, CO_2, NO and NO_2 Gas Molecules on CNTs

From Table 4, it is clear that the adsorption of CO and CO_2 gas molecules on CNTs does not affect the elec

Table 1. The configuration structures and diameters of the studied CNTs.

System	Configuration Structures	Diameters/Å
(5,0) CNT	$C_{60}H_{10}$	3.65
(9,0) CNT	$C_{108}H_{18}$	6.57
(5,5) CNT	$C_{100}H_{20}$	6.32
(6,6) CNT	$C_{120}H_{24}$	7.59

Table 2. The calculated adsorption energies (E_{ads}) of CO and CO_2 above a carbon site, a bond site and a vacant site of pristine $(5,0)$, $(9,0)$, $(5,5)$ and $(6,6)$ CNTs. All energies are given by eV.

	CO			CO_2		
System	Carbon site	Bond site	Vacant site	Carbon site	Bond Site	Vacant Site
(5,0) CNT	0.04	−0.00	−0.04	−0.00	0.00	0.00
(9,0) CNT	0.03	−0.43	−0.24	−0.23	−0.00	−0.26
(5,5) CNT	0.04	0.04	0.04	0.13	0.003	2.19
(6,6) CNT	−0.01	−0.01	−0.03	−0.00	−0.00	−0.16

Table 3. The calculated adsorption energies (E_{ads}) of NO and NO_2 above a carbon site, a bond site and a vacant site of pristine $(5,0)$, $(9,0)$, $(5,5)$ and $(6,6)$ CNTs. All energies are given by eV.

	NO			NO_2		
System	Carbon site	Bond site	Vacant site	Carbon Site	Bond site	Vacant site
(5,0) CNT	−0.35	−0.34	0.29	0.07	−0.36	−1.04
(9,0) CNT	−1.55	−1.65	−1.34	−1.19	−1.75	−1.40
(5,5) CNT	0.06	0.22	0.07	2.26	0.92	−1.32
(6,6) CNT	−0.01	0.25	0.07	0.06	0.35	−1.34

tronic character of the CNTs. Also, the band gaps of pristine CNTs and the adsorbed CO and CO_2gas molecules on CNTs are so close.

From Table 5, the adsorption of NO and NO_2 gas molecules on CNTs is strongly affected the electronic character of the $(9,0)$ and $(5,0)$CNTs. However, there is not any change of the electronic character for $(5,5)$ and $(6,6)$ CNTs. The band gap of pristine $(5,0)$ CNT is increased from 0.70 eV to 1.61 eV and to 1.37 eV when NO and NO_2 gas molecules are adsorbed on it, respectively. Also, The band gap of pristine $(9,0)$ CNT is increased from 0.25 eV to 1.34 eV and to 1.25 eV when NO and NO_2 gas molecules are adsorbed on it, respectively. One can conclude that the electronic character of$(5,0)$, $(9,0)$, $(5,5)$ and $(6,6)$ CNTs is not affected by the adsorption of CO and CO_2 gas molecules. The adsorption of NO and NO_2 gas molecules on CNTs is only strongly affected the electronic character of the $(9,0)$ and $(5,0)$CNTs, however the $(5,5)$ and $(6,6)$ CNTs are not affected at all.

3.3. HOMO-LUMO Orbitals of Adsorbing CO, CO_2, NO and NO_2 Gas Molecules on CNTs

Our calculated band gaps show that the adsorption of CO and CO_2 gas molecules on CNTs is not affected the band gaps of the pristine CNTs, however the adsorption of NO and NO_2 gas molecules is strongly affected the band gaps. To explain that the molecular orbitals of adsorbing CO, CO_2, NO and NO_2 gas molecules on $(5,0)$, $(9,0)$, $(5,5)$ and $(6,6)$ CNTs are investigated, see Figure 2, Figure 3. The band gaps of the pristine CNTs are calculated and are listed in Table 4. The HOMO and LUMO energy orbitals for pristine $(5,0)$, $(9,0)$, $(5,5)$ and $(6,6)$ CNTs are found to be (−4.13 eV, −3.42 eV), (−3.93 eV, −3.68 eV), (−4.62 eV, −2.82 eV) and (−4.70 eV, −2.82 eV), respectively. Comparing the HOMO-LUMO energies of the pristine CNTs with ones after the adsorption of CO and CO_2 gas molecules, it is clear that the energy values are so close. Also, it is noticed that there is not any contribution from the gas molecules at the molecular orbitals and the electron density of HOMO and LUMO is distributed over all the carbon atoms of CNTs except for $(9,0)$ CNT is located at the terminals of the tube, see Figure 2. Comparing the HOMO-LUMO energies of the pristine CNTs with ones after the adsorption of NO and NO_2 gas molecules, it is clear that the energy values are so close in case of $(5,5)$ and $(6,6)$ CNTs and are quite far in case of $(5,0)$ and $(9,0)$ CNTs. The HOMO energy levels in case of $(5,0)$ and $(9,0)$ CNTs after adsorbing NO and NO_2 gas molecular are getting deep (lower) in energy however the LUMO energy levels are getting higher in energy. Results in increasing the band gap from 0.70 eV to 1.81 eV in

Table 4. The calculated energy gaps (E_g) of CO and CO_2 above a carbon site, a bond site and a vacant site of pristine $(5,0)$, $(9,0)$, $(5,5)$ and $(6,6)$ CNTs. All energies are given by eV.

		CO			CO_2		
System	Pristine CNTs	Carbon site	Bond site	Vacant site	Carbon Site	Bond site	Vacant Site
(5,0) CNT	0.70	0.61	0.70	0.70	0.61	0.61	0.60
(9,0) CNT	0.25	0.25	0.25	0.25	0.25	0.24	0.25
(5,5) CNT	1.79	1.80	1.79	1.80	1.80	1.79	1.75
(6,6) CNT	1.90	1.92	1.92	1.92	1.92	1.92	1.92

Table 5. The calculated energy gaps (E_g) of NO and NO_2 above a carbon site, a bond site and a vacant site of pristine $(5,0)$, $(9,0)$, $(5,5)$ and $(6,6)$ CNTs. All energies are given by eV.

	NO			NO_2		
system	Carbon site	Bond Site	Vacant Site	Carbon site	Bond Site	Vacant site
(5,0) CNT	1.61	0.99	0.94	0.61	0.99	1.37
(9,0) CNT	1.34	1.24	1.14	1.25	1.25	1.14
(5,5) CNT	1.79	1.81	1.79	1.72	1.76	1.79
(6,6) CNT	1.92	1.90	1.92	1.94	1.88	1.91

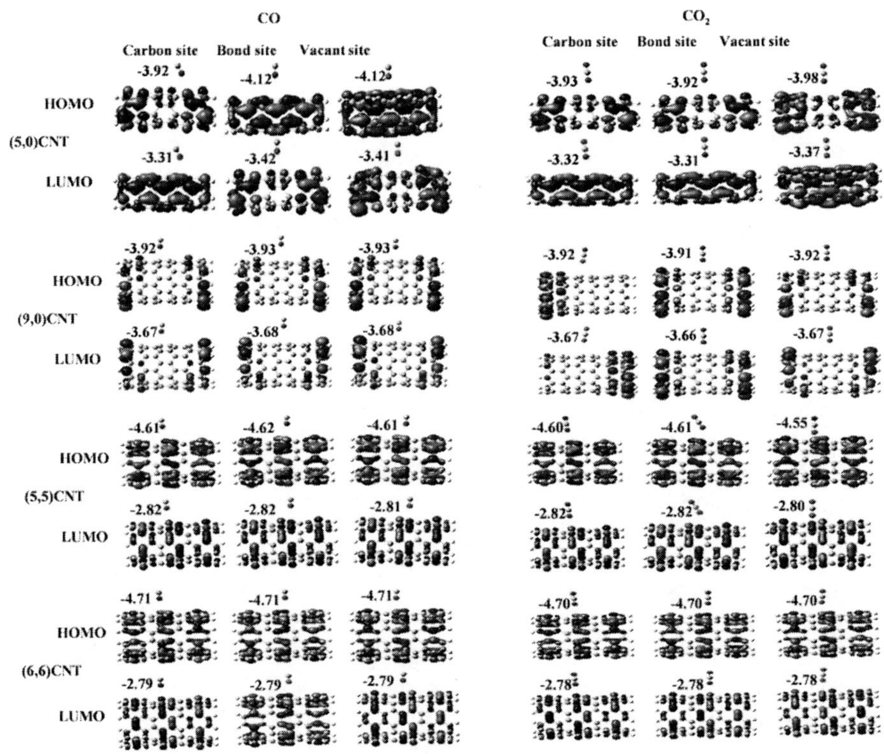

Figure 2. HOMO and LUMO molecular orbitals of adsorbing CO and CO_2gas molecules on the pristine $(5,0)$, $(9,0)$, $(5,5)$ and $(6,6)$ CNTs. Energies of HOMO and LUMO are listed above the molecular orbitals and are given by eV.

case of $(5,0)$ CNT and from 0.25 eV to 1.34 eV in case of $(9,0)$ CNT. Also, it is noticed that there is representation from the NO gas molecule at LUMO of $(9,0)$ and $(6,6)$ CNTs, see Figure 3.

3.4. The Reactivity of CNT Surfaces before and after Adsorbing Gas Molecules

Our calculated band gaps and molecular orbitals show that the adsorption of CO and CO_2 gas molecules on CNTs is not affected neither the band gaps nor the molecular orbitals of the pristine CNTs but the adsorption of NO and NO_2 gas molecules is strongly affected both of the band gaps and the molecular orbitals of $(5,0)$ and $(9,0)$ CNTs. To clear that the reactivity of CNT surfaces before and after adsorbing CO, CO_2, NO and NO_2 gas molecules on $(5,0)$, $(9,0)$, $(5,5)$ and $(6,6)$ CNTs are studied, see Table 6, Table 7. The surface reactivity of the pristine CNTs is calculated and is listed in **Table 6**. The dipole moments of pristine $(5,0)$, $(9,0)$, $(5,5)$ and $(6,6)$ CNTs are found to be 0.54 Debye, 0.20 Debye 0.00 Debye and 0.00 Debye, respectively.

Comparing the dipole moments of the pristine CNTs with ones that are adsorbed the CO and CO_2gas molecules, it is clear that the dipole moment values are so close in case of the adsorption of the CO gas molecule but they are higher in case of the adsorption of the CO_2 gas molecule, seeTable 6. Also, it is noticed that the highest dipole moments after the adsorption of the CO_2 gas molecule are 0.74 Debye (when CO_2 is adsorbed above the bond site of $(9,0)$ CNT) and 0.77 Debye (when CO_2 is adsorbed above the vacant site of $(5,5)$ CNT), re spectively. Comparing the dipole moments of the pristine CNTs with ones that are adsorbed the NO and NO_2 gas molecules, it is found that the dipole moments are getting higher. When the NO and NO_2 gas molecules are adsorbed on the vacant sites of CNTs, their dipole moments are either quite close to or are lower than the dipole moments of pristine CNTs, except in case of adsorbing NO_2 on $(5,0)$ CNT, the dipole moment is increased. Also, all the calculated dipole moments of adsorbing NO and NO_2 gas molecules on the carbon sites of CNTs

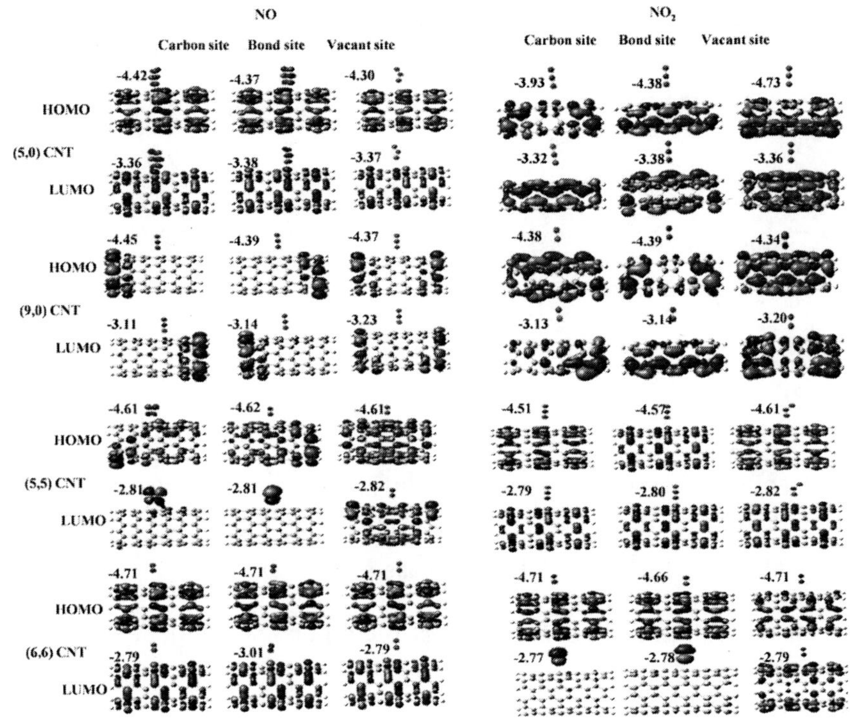

Figure 3. HOMO and LUMO molecular orbitals of adsorbing NO and NO_2gas molecules on the pristine $(5,0)$, $(9,0)$, $(5,5)$ and $(6,6)$ CNTs. Energies of HOMO and LUMO are listed above their molecular orbitals and are given by eV.

Table 6. The calculated dipole moments of pristine and after adsorbing CO and CO_2 gas molecules above a carbon site, a bond site and a vacant site of $(5,0)$, $(9,0)$, $(5,5)$ and $(6,6)$ CNTs. All dipole moments are given by Debye.

		CO			CO_2		
system	Pristine CNTs	Carbon site	Bond site	Vacant Site	Carbon Site	Bond Site	Vacant Site
(5,0) CNT	0.54	0.54	0.56	0.49	0.46	0.47	0.62
(9,0) CNT	0.20	0.40	0.21	0.10	0.23	0.74	0.30
(5,5) CNT	0.00	0.04	0.12	0.06	0.21	0.25	0.77
(6,6) CNT	0.00	0.06	0.05	0.06	0.23	0.23	0.26

Table 7. The calculated dipole moments of pristine and after adsorbing NO and NO_2 gas molecules above a carbon site, a bond site and a vacant site of $(5,0)$, $(9,0)$, $(5,5)$ and $(6,6)$ CNTs. All dipole moments are given by Debye.

	NO			NO_2		
system	Carbon site	Bond Site	Vacant Site	Carbon Site	Bond Site	Vacant site
(5,0) CNT	5.58	1.12	0.36	0.28	0.97	1.47
(9,0) CNT	0.46	0.48	0.11	0.69	0.11	0.07
(5,5) CNT	0.14	0.11	0.11	1.48	0.58	0.13
(6,6) CNT	0.15	0.31	0.17	0.20	0.53	0.17

are increased, except in case of adsorbing NO_2 on $(5,0)$ CNT, the dipole moment is decreased. In case of adsorbing NO and NO_2 gas molecules on the bond sites of CNTs the dipole moments are also increased, except in case of adsorbing NO_2 on $(9,0)$ CNT is decreased, see Table 7.

From Table 6, Table 7, it is clear that the dipole moments of zig-zag $(5,0)$ and $(9,0)$ CNTs are always higher than the arm-chair $(5,5)$ and $(6,6)$ CNTs. Also, it is noticed that the dipole moment of adsorbing NO gas molecule on the bond site of $(5,0)$ CNT is increased by ten times comparing with the dipole moment of pristine $(5,0)$ CNT.

4. CONCLUSION

A. A. EL-Barbary, Kh. M. Eid,, M. A. Kamel, H. M. Osman, G. H. Ismail, (d, p). Three different adsorption sites (above a carbon site, a bond site and a vacant site) are applied on CNTs. It is found that the adsorption of CO and CO_2 gas molecules is dependent on the chiralities and the diameters of CNTs and it is enhanced with increasing the diameter of the zig-zag CNTs. However, the adsorption of NO and NO_2 gas molecules is independent on the chiralities and the diameters of CNTs. Also, the electronic character of $(5,0)$, $(9,0)$, $(5,5)$ and $(6,6)$ CNTs is not affected by the adsorption of CO and CO_2 gas molecules. While, the adsorption of NO and NO_2 gas molecules on CNTs is only strongly affected by the electronic character of the $(9,0)$ and $(5,0)$ CNTs but the $(5,5)$ and $(6,6)$ CNTs are not affected at all. It is found that the dipole moments of zig-zag $(5,0)$ and $(9,0)$ CNTs are always higher than the arm-chair $(5,5)$ and $(6,6)$ CNTs. Also, it is noticed that the dipole moment of adsorbing NO gas molecule on the bond site of $(5,0)$ CNT is increased by ten times compared with the dipole moment of pristine $(5,0)$ CNT. Therefore, these findings prove that the zig-zag carbon nanotubes are better than the arm-chair carbon nanotubes as gas sensors, especially for NO and NO_2 gas molecules.

REFERENCES

1. Kong, J., Franklin, N.R., Zhou, C., Chapline, M.G., Peng, S. and Cho, K. (2000) Nanotube Molecular Wires as Chemical Sensors. Science, 287, 622.
2. Snow, E.S., Perkins, F.K., Houser, E.J., Badescu, S.C. and Reinecke, T.L. (2005) Chemical Detection with a Single-Walled Carbon Nanotube Capacitor. Science, 307, 1942-1945.
3. Baei, M.T., Soltani, A.R., Moradi, A.V. and Lemeski, E.T. (2011) Adsorption Properties of NO on (6, 0), (7, 0), and (8, 0) Zigzag Single-Walled Boron Nitride Nanotubes: A Computational Study. Computational and Theoretical Chemistry, 970, 30-35.
4. Breza, M. (2006) Model Studies of $SOCl_2$ Adsorption on Carbon Nanotubes. Journal of Molecular Structure: THEOCHEM, 767, 159-163.
5. Zhao, J., Buldum, A., Han, J. and Lu, J.P. (2002) Gas Molecule Adsorption in Carbon Nanotubes and Nanotube Bundles. Nanotechnology, 13, 195-200.
6. Ricca, A. and Bauschlicher Jr., C.W. (2006) The Adsorption of NO on (9, 0) and (10, 0) Carbon Nanotubes. Chemical Physics, 323, 511-518.
7. Abbas Rafati, A., Majid Hashemianzadeh, S. and Bolboli Nojini, Z. (2008) Electronic Properties of Adsorption Nitrogen Monoxide on Inside and Outside of the Armchair Single Wall Carbon Nanotubes: A Density Functional Theory Calculations. The Journal of Physical Chemistry C, 112, 3597-3604.
8. Azizi, K., Majid Hashemianzadeh, S. and Bahramifar, Sh. (2011) Density Functional Theory Study of Carbon Monoxide Adsorption on the Inside and Outside of the Armchair Single-Walled Carbon Nanotubes. Current Applied Physics, 11, 776-782.
9. Ricca, A., Bauschlicher Jr., C.W. (2006) The Physisorption of CH_4 on Graphite and on a (9, 0) Carbon Nanotube. Chemical Physics, 324, 455-458.
10. Santucci, S., Picozzi, S., Di Gregorio, F., Lozzi, L., Cantalini, C., Valentini, L., Kenny, J.M. and Delley, B. (2003) NO_2 and CO Gas Adsorption on Carbon Nanotubes: Experiment and Theory. The Journal of Chemical Physics, 119, 10904-10910.
11. Zanolli, Z. and Charlier, J.C. (2009) Defective Carbon Nanotubes for Single-Molecule Sensing. Physical Review B, 80, 155447.
12. Tang, S. and Cao, Z. (2009) Defect-Induced Chemisorption of Nitrogen Oxides on (10, 0) Single-Walled Carbon Nanotubes: Insights from Density Functional Calculations. The Journal of Chemical Physics, 131, 114706.
13. García-Lastra, J.M., Mowbray, D.J., Thygesen, K.S., Rubio, A. and Jacobsen, K.W. (2010) Modeling Nanoscale Gas Sensors under Realistic Conditions: Computational Screening of Metal-Doped Carbon Nanotubes. Physical Review B, 81, 245429.
14. Denis, P.A. (2008) Methane Adsorption Inside and Outside Pristine and N-Doped Single Wall Carbon Nanotubes. Chemical Physics, 353, 79-86.
15. Yeung, C.S., Liu, L.V. and Wang, Y.A. (2008) Adsorption of Small Gas Molecules onto Pt-Doped Single-Walled Carbon Nanotubes. The Journal of Physical Chemistry C, 112, 7401-7411.

16. Zhao, J.X. and Ding, Y.H. (2008) Theoretical Study of the Interactions of Carbon Monoxide with Rh-Decorated (8, 0) Single-Walled Carbon Nanotubes, Materials Chemistry and Physics, 110, 411-416.
17. An, W. and Turner, C.H. (2009) Electronic Structure Calculations of Gas Adsorption on Boron Doped Carbon Nanotubes Sensitized with Tungsten. Chemical Physics, 482, 274-280.
18. Sayago, I., Santos, H., Horrillo, M.C., Aleixandre, M., Fernández, M.J., Terrado, E., Tacchini, I., Aroz, R., Maser, W.K., Benito, A.M., Martínez, M.T., Gutiérrez, J. and Munoz, E. (2008) Carbon Nanotube Networks as Gas Sensors for NO_2 Detection. Talanta, 77, 758-764.
19. Li, X.M., Tian, W.Q., Dong, Q., Huang, X.R., Sun, C.C. and Jiang, L. (2011) Substitutional Doping of BN Nanotube by Transition Metal: A Density Functional Theory Simulation. Computational and Theoretical Chemistry, 964, 199-206.
20. Chen, H.L., Wu, S.Y., Chen, H.T., Chang, J.G., Ju, S.P., Tsai, C. and Hsu, L.C. (2010) Theoretical Study on Adsorption and Dissociation of NO_2 Molecule on Fe(1 1 1) Surface. Langmuir, 26, 7157-7164.
21. Wickham, D.T., Banse, B.A. and Koel, B.E. (1991) Adsorption of Nitrogen Dioxide and Nitric Oxide on Pd(1 1 1). Surface Science, 243, 83-95.
22. Jirsak, T., Kuhn, M. and Rodriguez, J.A. (2000) Chemistry of NO_2 on Mo(1 1 0): Decomposition Reactions and Formation of MoO_2. Surface Science, 457, 254-266.
23. Huang, W., Jiang, Z., Jiao, J., Tan, D., Zhai, R. and Bao, X. (2002) Decomposition of NO on Pt(1 1 0): Formation of a New Oxygen Adsorption State. Surface Science, 506, 287-292.
24. Hellman, A., Panas, I. and Grönbeck, H. (2008) NO_2 Dissociation on Ag(1 1 1) Revisited by Theory. Journal of Chemical Physics, 128, 104704-104709.
25. Yen, M.Y. and Ho, J.J. (2010) Density-Functional Study for the NO_x (x = 1, 2) Dissociation Mechanism on the Cu(1 1 1) Surface. Chemical Physics, 373, 300-306.
26. Frisch, M.J., Trucks, G.W., Schlegel, H.B., Scuseria, G.E., Robb, M.A., Cheeseman, J.R., Zakrzewski, V.G., Montgomery, J.A., Stratmann, R.E., Burant, J.C., Dapprich, S., Millam, J.M., Daniels, A.D., Kudin, K.N., Strain, M.C., Farkas, O., Tomasi, J., Barone, V., Cossi, M., Cammi, R., Mennucci, B., Pomelli, C., Adamo, C., Clifford, S., Ochterski, J., Petersson, G.A., Ayala, P.Y., Cui, Q., Morokuma, K., Malick, D.K., Rabuck, A.D., Raghavachari, K., Foresman, J.B., Cioslowski, J., Ortiz, J.V., Stefanov, B.B., Liu, G., Liashenko, A., Piskorz, P., Komaromi, I., Gomperts, R., Martin, R.L., Fox, D.J., Keith, T., Al-Lamham, M.A., Peng, C.Y., Nanayakkara, A., Gonzalez, C., Challacombe, M., Gill, P.M.W., Johnson, B.G., Chen, W., Wong, M.W., Andres, J.L., Head-Gordon, M., Replogle, E.S. and Pople, J.A. (2004) Gaussian. Wallingford CT, Inc., Wallingford.
27. EL-Barbary, A.A., Lebda, H.I. and Kamel, M.A. (2009) The High Conductivity of Defect Fullerene C_{40} Cage. Computational Materials Science, 46, 128.
28. El-Barbary, A.A., Eid, K.M., Kamel, M.A. and Hassan, M.M. (2013) Band Gap Engineering in Short Heteronanotube Segments via Monovacancy Defects. Computational Materials Science, 69, 87-94.
29. EL-Barbary, A.A., Ismail, G.H. and Babeer, A.M. (2013) Effect of Monovacancy Defects on Adsorbing of CO, CO_2, NO and NO_2 on Carbon Nanotubes: First Principle Calculations. Journal of Surface Engineered Materials and Advanced Technology, 3, 287-294.
30. Nalwa, H. (2002) Nanostructured Materials and Nanotechnology. Academic Press, San Diego.

CHAPTER 17

Effects of Band Nonparabolicity and Band Offset on the Electron Gas Properties in InAs/AlSb Quantum Well

Gafur Gulyamov[1], Bahrom Toshmirza O'g'li Abdulazizov[2], Baymatov Paziljon Jamoldinovich[2]

[1]*Namangan Engineering-Pedagogical Institute, Namangan , Uzbekistan*

[2]*Namangan State University , Namangan , Uzbekistan*

ABSTRACT

One-band effective mass model is used to simulation of electron gas properties in quantum well. We calculate of dispersion curves for first three subbands. Calculation results of Fermi energy, effective mass at Fermi level as function of electron concentration are presented. The obtained results are good agreement with the experimental dates.

Keywords: Quantum Well, In-Plane Dispersion, InAs, AlSb, Two Dimentional Electron Gas, Effective Mass, Cyclotron Mass

1. INTRODUCTION

In semiconductors, InAs and InSb of the conduction band are characterized by a strong nonparabolicity and recently intensively studied heterostructures based on them [1] - [3] . Nonparabolicity of the conduction band and the nature of the spin splitting of the electron in the quantum well (QW) are studied by the cyclotron resonance [4] - [7] .

In [8] has been investigated InAs/AlSb based QW with well width L = 15 nm, where two dimensional (2D) electron concentration ranges from 2.7 × 10^{11} to 8 × 10^{12} cm^{-2}. In this work has been found increase of the effective mass of almost 2 times.

The purpose of this work—the calculation of: 1) subbands dispersion curves, 2) the density of states of 2D electron gas and 3) concentration dependence of effective mass in Fermi level for InAs/AlAs QW with width L = 15 nm.

It is shown that an abrupt change in the density of states leads to a peculiar change in the concentration dependence of effective mass.

2. THE IN-PLANE DISPERSION

Consider a single QW with width L (area A—InAs), concluded between barriers with height V (area B—AlAs). The energy is measured from the bottom of the band of the bulk InAs.

In the one band effective mass approximation, the solution of the three-dimensional Schrödinger equation can be represented as $\psi = e^{i(k_x x + k_y y)}\phi(z)$. Then for the area A and B, respectively, we can write the following one-dimensional equations

$$\frac{\partial^2 \phi_A(z)}{\partial z^2} + q^2 \phi_A(z) = 0, \quad q = \sqrt{\frac{2m_A(E)}{\hbar^2}E - k^2} \tag{1}$$

$$\frac{\partial^2 \phi_B(z)}{\partial z^2} + \chi^2 \phi_B(z) = 0, \quad \chi = \sqrt{\frac{2m_B(E)}{\hbar^2}(V - E) + k^2} \tag{2}$$

Here $k^2 = k_x^2 + k_y^2$ and $k -$ is in plane wave vector, $m_{A,B}(E)$ —energy-dependent effective mass of the electrons in the material A or B. Solving Equations (1) and (2), using the boundary condition

$$\phi_A(0) = \phi_B(0), \quad \frac{1}{m_A(E)} \frac{d\phi_A(z)}{dz}\bigg|_{z=0} = \frac{1}{m_B(E)} \frac{d\phi_B(z)}{dz}\bigg|_{z=0} \tag{3}$$

$$\phi_A(L) = \phi_B(L), \quad \frac{1}{m_A(E)} \frac{d\phi_A(z)}{dz}\bigg|_{z=L} = \frac{1}{m_B(E)} \frac{d\phi_B(z)}{dz}\bigg|_{z=L} \tag{4}$$

we find the dispersion equation

$$E = E_{\|} + E_0 \left[\pi \cdot n - 2\, Arc\sin\left(\sqrt{\frac{\gamma^2 (E - E_{\|})}{\gamma(\gamma - 1)E + \gamma\, V + (1 - \gamma^2) E_{\|}}} \right) \right]^2 \tag{5}$$

here $E_{\|} = \hbar^2 k^2 / 2m_A(E)$, $E_0 = \hbar^2 / 2m_A(E) L^2$, $\gamma = m_B(E)/m_A(E)$.

Nonparabolicity of conduction band well takes into account by formulas

$$\frac{m_0}{m_A(E)} = 1 + \frac{E_{PA}}{3} \left[\frac{2}{E + E_{gA}} + \frac{1}{E + E_{gA} + \Delta_A} \right] \tag{6}$$

$$\frac{m_0}{m_B(E)} = 1 + \frac{E_{PB}}{3} \left[\frac{2}{E - V + E_{gB}} + \frac{1}{E - V + E_{gB} + \Delta_B} \right] \tag{7}$$

where, m_0—the free electron mass, E_P—the Kane parameter, E_g—the band gap, Δ— the spin-orbital splitting of valence band, V—conduction band offset. Band parameters of InAs and AlSb are shown in Table 1.

To describe the statistics of electrons, Equation (5) is non convenient because it is not solvable with respect to E or k. Therefore, we replace Equation (5) is by simple approximation

$$E \frac{m_A(E)}{m_A(0)} \approx \frac{\hbar^2 k^2}{2m_A(0)} + E_n \frac{m_A(E_n)}{m_A(0)} \tag{8}$$

Table 1. Band parameters of InAs and AlSb.

	InAs (*A*)	*AlSb* (*B*)
E_g, eV	0.42	2.37
Δ, eV	0.38	0.75
E_P, eV	21.2	20.85
$m(0)$, $[m_0]$	0.023	0.11
V, eV	0	1.35

where, E_n—is bottom of n-th subbands. Now, approximation (8) is the best solution of (5). However, values of E_n in (8) now are obtained from Equation (5) at k = 0 by use numeric method.

For InAs/AlSb QW with L = 15 nm, we have: $E_1 = 0.0454$ eV, $E_2 = 0.158$ eV, $E_3 = 0.304$ eV, and for case L = 6 nm we have: $E_1 = 0.163$ eV, $E_2 = 0.509$ eV, $E_3 = 0.903$ eV.

Calculated dispersion curves from Equation (5) and approximation (8) are compared in Figure 1(a), Figure 1(b).

From Figure 1(a), Figure 1(b) follows that, the approximation (8) is sufficiently accurate and/or (8) is the best solution of (5) in a wide range of width QW. It is convenient when studying the statistics of electrons, kinetic, optical, or other characteristics of the 2D electron gas. Inconveniences approximation (8) is such that $E_n\left(n=1,2,3,\cdots\right)$ depends on L, V and on other parameters of materials A,B. Therefore every time when changing these parameters, the value of E_n is recalculated from Equation (5) at k = 0.

3. THE FERMI ENERGY AND THERMODYNAMIC DOS

The total electron concentration is

$$n_s = N_0 \sum_{n=1} \int_{E_n}^{E_F} \mathrm{d}\left(\frac{\hbar^2 k^2}{2m_A(0)}\right) \tag{9}$$

where

$$N_0 = \frac{m_A(0)}{\pi\hbar^2} = \frac{m_A(0)\cdot 0.413\times 10^{15}}{\text{эВ}\cdot\text{см}^2}.$$

According (8) we have

$$n_S = \sum_{n=1} n_S^{(n)} \quad , \quad n_s^{(n)} = N_0\left[E_F\frac{m_A(E_F)}{m_A(0)} - E_n\frac{m_A(E_n)}{m_A(0)}\right]\theta(E_F - E_n) \tag{10}$$

where $n_s^{(n)}$ —is n-th subband concentration.

In Equation (10), the terms in the sum should be positive. The negative terms in $n_s^{(n)}$ excluded by the Heaviside function. It establishes a link between the Fermi energy E_F and full 2D electron concentration n_s. They also determine the concentration of electrons in separate subbands $n_s^{(n)}$ for a given n_s.

From (10), we can estimate the critical concentrations of n_{c1}, in which the Fermi level comes to bottom of the second subband $E_F = E_2$. In the structure of InAs/AlSb QW with the well width L = 15 nm can be found

$$n_{c1} = N_0\left[E_2\frac{m_A(E_2)}{m_A(0)} - E_1\frac{m_A(E_1)}{m_A(0)}\right] = 1.52\times 10^{12}\ \text{cm}^{-2} \tag{11}$$

This estimation is close to experimental measured date $n_{c1,\text{exp}} = 1.2\times 10^{12}\ \text{cm}^{-2}$ [8]. Similarly, the critical concentration of n_{c2}, in which the Fermi level comes to bottom of the third subband $E_F = E_3$, is $n_{c2} = 6.87\times 10^{12}\ \text{cm}^{-2}$.

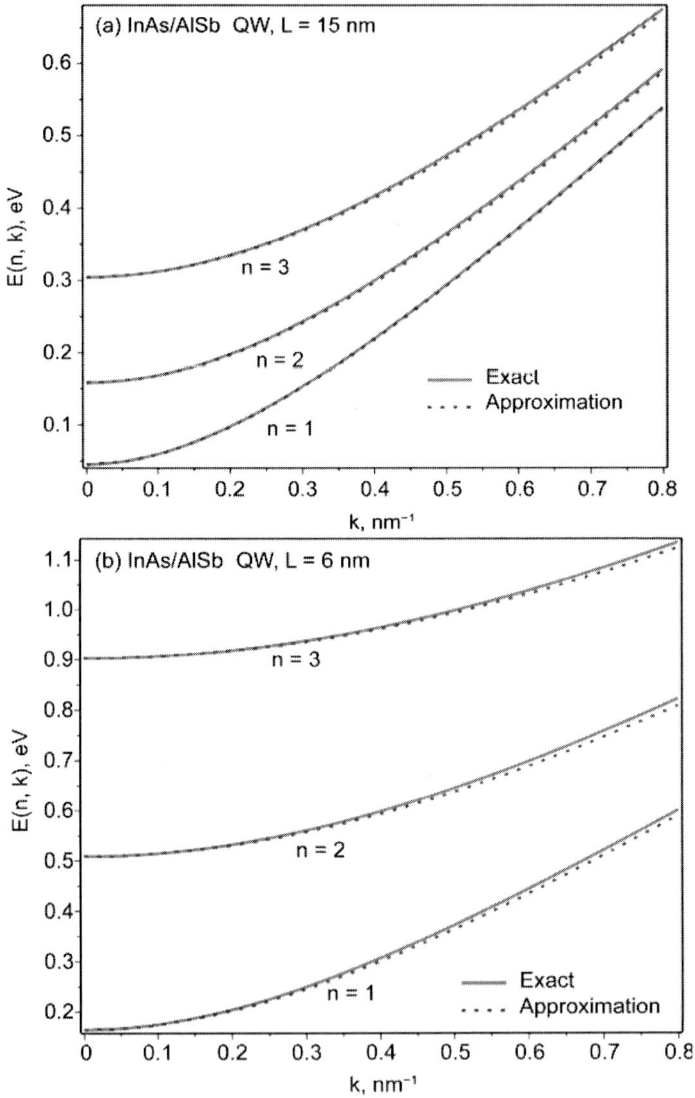

Figure 1. The dispersion curves of the first three subbands (n = 1, 2, 3) in InAs/AlSb QW: the full (red) line—according to Equation (5), the dotted (blue) line—approximation (8); (a) L = 15 nm, (b) L = 6 nm.

The dependence $E_F(n_s)$ is shown in Figure 2. It is obtained from Equations (10) by changing the Fermi energy in the range of $E_F = 0 \div 0.33$ eV. The graph shows that, depending on $E_F(n_s)$ there exist a fractures—slowing of increase the Fermi's energy. They are caused by abrupt changes (by jumps), the density of states at the critical points: $n_s = n_{c1}$, $E_F = E_2$, $n_s = n_{c2}$, $E_F = E_3$.

These fractures occur at the critical concentrations of $n_{c1}, n_{c2} \cdots$, where E_F intersects the Fermi level of the bottom of the next subband.

The thermodynamically DOS of electron gas at Fermi level $N_T(E_F) = \mathrm{d}n_S/\mathrm{d}E_F$ is shown in Figure 3.

4. THE CYCLOTRON MASS

According approximation (8), the electron effective mass at the Fermi level (cyclotron mass) $m_c = \hbar^2 k (\mathrm{d}E/\mathrm{d}k)^{-1}\big|_{E=E_F}$ is

$$m_c = \left(m_A(E) + E\frac{\mathrm{d}m_A(E)}{\mathrm{d}E} \right)_{E=E_F} \tag{12}$$

The dependence $m_c(n_s)$ is shown in Figure 4.

This dependence can be obtained from Equations (10) and (12) by changing the Fermi energy in the range $E_F = 0 - 0.33\ \mathrm{eV}$.

This figure shows also the dependence of experimentally measured value of the effective masses (cyclotron mass) m_c at the Fermi level of the total concentration of 2D n_s [8] . Fracture in the $m_c(n_s)$ occur at critical points: $n_s = n_{c1}$, $E_F = E_2$ and $n_s = n_{c2}$, $E_F = E_3$.

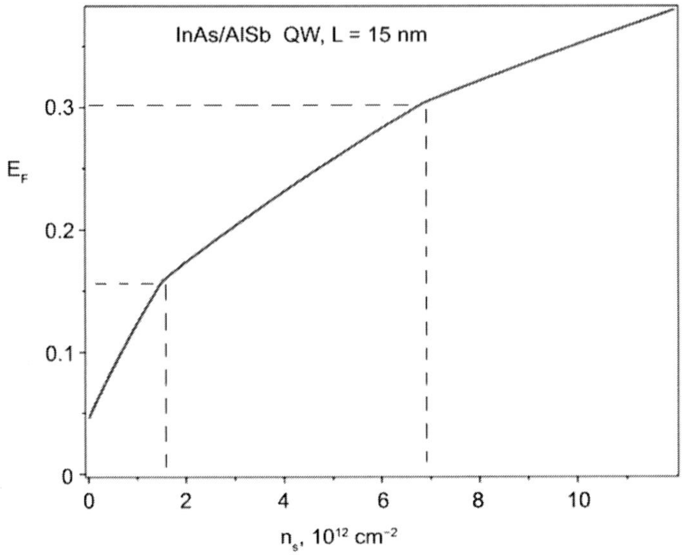

Figure 2. The dependence of the Fermi energy on the 2D concentration in InAs/AlSb QW with L = 15 nm, V = 1.35 eV. The critical points $n_s = n_{c1}$, $E_F = E_2$ and $n_s = n_{c2}$, $E_F = E_3$ shown by the shaded lines.

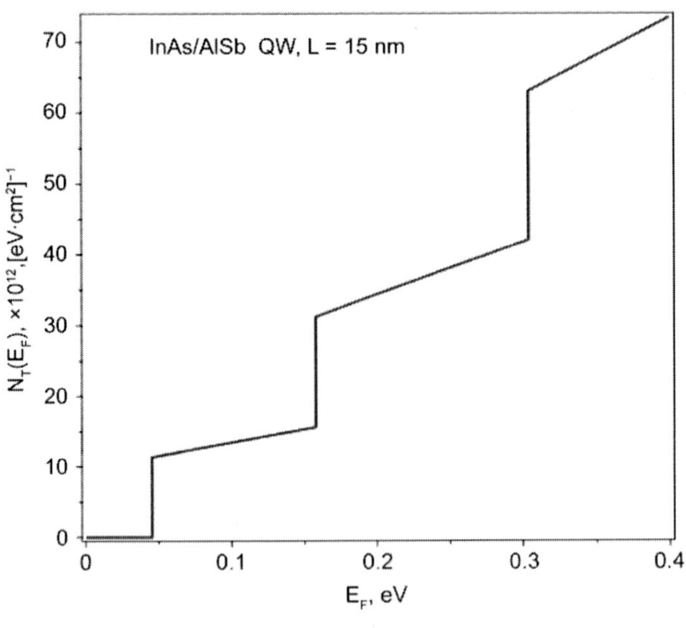

Figure 3. The thermodynamically DOS of electron gas at Fermi level in InAs/AlSb QW, L = 15 nm, V = 1.35 eV.

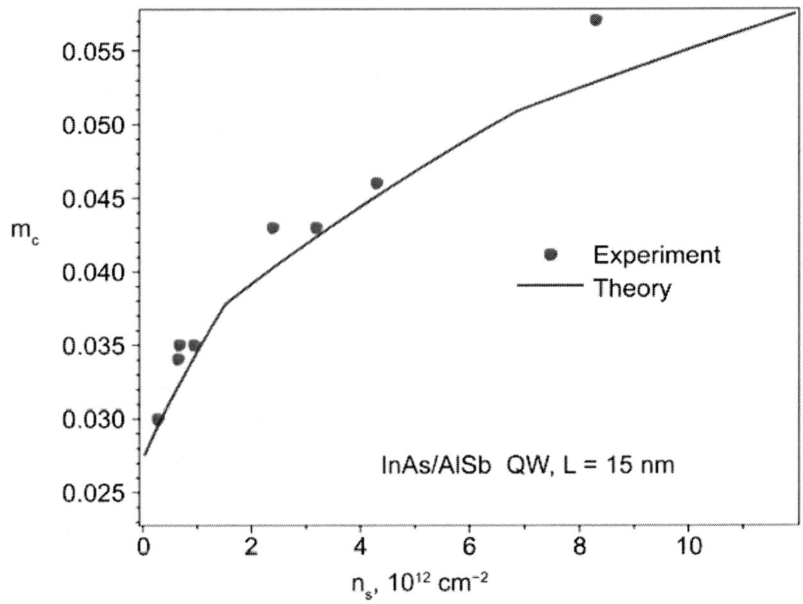

Figure 4. The dependence of the effective mass on the total concentration for InAs/AlSb QW, L = 15 нм, V = 1.35 eV.

5. CONCLUSION

In this study are provided useful approximation (8) of subband dispersions and simplified Equation (10) to calculate the statistics of a degenerate electron gas in heterostructured InAs/AlSb QW, which satisfactorily describes the experimental results [8] . They are also useful to study of calculation of the transport, optical and magnetic properties of electron gas in a Kane type 2D system. The above description of the algorithms can be applied to other QW heterostructures based on semiconductor group A_3B_5.

ACKNOWLEDGEMENTS

This work was supported by the Scientific and Technical program Republic of of Uzbekistan (Grant F2-OT-O-15494).

CITE THIS PAPER

Gulyamov, G., Abdulazizov, B.T. and Jamoldinovich, B.P. (2016) Effects of Band Nonparabolicity and Band Offset on the Electron Gas Properties in InAs/ AlSb Quantum Well. Journal of Modern Physics, 7, 1644-1650.

REFERENCES

1. Vasilyev, Yu.B., Gouider, F., Nachtwei, G. and Buckle, P.D. (2010) Semiconductors, 44, 1511-1514.
2. Spirin, K.E., Kalinin, K.P., Krishtopenko, S.S., Maremyanin, K.V., Gavrilenko, V.I. and Sadofyev, Yu.G. (2012) Semiconductors, 46, 1396-1401.
3. Barate, D., Teissier, R., Wang, Y. and Baranov, A.N. (2005) Applied Physics Letters, 87, 051103.
4. Yang, M.J., Lin-Chung, P.J., Wagner, R.J., Waterman, J.R., Moore, W.J. and Shanabrook, B.V. (1993) Semiconductor Science and Technology, 8, S129.
5. Yang, M.J., Lin-Chung, P.J., Shanabrook, B.V., Waterman, J.R., Wagner, R.J. and Moore, W.J. (1993) Physical Review B, 47, 1691.
6. Gauer, C., Scriba, J., Wixforth, A., Kotthaus, J.P., Bolognesi, C.R., Nguyen, C., Brar, B. and Kroemer, H. (1994) Semiconductor Science and Technology, 9, 1580.
7. Warburton, R.J., Gauer, C., Wixforth, A., Kotthaus, J.P., Brar, B. and Kroemer, H. (1996) Physical Review B, 53, 7903.
8. Aleshkin, V.Ya., Gavrilenko, V.I., Ikonnikov, A.V., Sadofyev, Yu.G., Bird, J.P., Jonhson, S.R. and Zhang, Y.-H. (2005) Semiconductors, 39, 62-66.

CHAPTER 18

Effect of Gas Properties on The Dynamics of The Electrical Slope Asymmetry Effect in Capacitive Plasmas: Comparison of Ar, H_2 And CF_4

B Bruneau[1], T Lafleur[2], T Gans[3], D O'Connell[3], A Greb[3], I Korolov[4], A Derzsi[4], Z Donkó[4], S Brandt[5], E Schüngel[5], J Schulze[5], P Diomede[6,7], D J Economou[6], S Longo[8], E Johnson[1] and J-P Booth[2]

[1] *LPICM-CNRS, Ecole Polytechnique, 91128 Palaiseau, France*

[2] *LPP, Ecole Polytechnique-CNRS-Uni Paris-Sud-UPMC, 91128 Palaiseau, France*

[3] *York Plasma Institute, Department of Physics, University of York, Heslington, York, YO10 5DD, UK*

[4] *Institute for Solid State Physics and Optics, Wigner Research Centre for Physics, Hungarian Academy of Sciences, Konkoly-Thege Miklós str. 29-33, H-1121 Budapest, Hungary*

[5] *Department of Physics, West Virginia University, Morgantown, West Virginia 26506-6315, USA*

[6] *Plasma Processing Laboratory, Department of Chemical & Biomolecular Engineering, University of Houston, Houston, TX 77204-4004, USA*

[7] *FOM Institute DIFFER—Dutch Institute for Fundamental Energy Research, P. O. Box 6336, 5600 H H Eindhoven, The Netherlands*

[8] *Dipartimento di Chimica, Universita' degli studi di Bari, via Orabora 4, 70126 Bari, Italy*

ABSTRACT

Tailored voltage excitation waveforms provide an efficient control of the ion energy (through the electrical asymmetry effect) in capacitive plasmas by varying the 'amplitude' asymmetry of the waveform. In this work, the effect of a 'slope' asymmetry of the waveform is investigated by using sawtooth-like waveforms, through which the sheath dynamic can be manipulated. A remarkably different discharge dynamic is found for Ar, H_2, and CF_4 gases, which is explained by the different dominant electron heating mechanisms and plasma chemistries. In comparison to Argon we find that the electrical asymmetry can even be reversed by using an electronegative gas such as CF_4. Phase resolved optical emission spectroscopy measurements, probing the spatiotemporal distribution of the excitation rate show excellent agreement with the results of particle-in-cell simulations, confirming the high degree of correlation between the excitation rates with the dominant heating mechanisms in the various gases. It is shown that, depending on the gas used, sawtooth-like voltage waveforms may cause a strong asymmetry.

Keywords: tailored voltage waveforms, electrical asymmetry effect, sawtooth waveforms, electron heating, PIC modeling, PROES

Going beyond classic dual-frequency capacitive plasmas between two parallel electrodes [1–6], discharges excited by tailored voltage waveforms (TVWs), comprising a fundamental signal frequency in the radio frequency (RF) domain combined with higher harmonics with controlled phases and amplitudes, have been investigated in recent years [7–15]. Amplitude asymmetry of the waveform, i.e. the difference between its maximum and its minimum value, and the associated dynamics occurring in these plasmas have previously been studied for the separate control of ion energy and flux due to the electrical asymmetry effect (EAE) [16–19].

Recently, a new type of temporally asymmetric TVW was proposed, called 'sawtooth' waveforms [20–22], comprising a fast rise and a slow fall, or vice versa. These waveforms present a slope asymmetry but no amplitude asymmetry. In argon this was shown to induce different sheath motions in front of the two electrodes, leading to strong ionization close to the electrode where the sheath expansion is fast, with only weak ionization close to the electrode where the sheath expansion is slow [22]. However, in molecular, electronegative plasmas significantly different physics phenomena can occur. For example, ionization mechanisms, other than sheath expansion, have been observed [23–25]. This letter compares sawtooth-excited discharges operated in gases with distinctly different properties, namely argon (Ar, electropositive), hydrogen (H_2, light molecular gas), and carbon tetrafluoride (CF_4, electronegative). For each gas, experimental data are compared to specific particle-in-cell (PIC) simulations. For Ar we used the code developed by Lafleur [26], for H_2 that of the Bari University group [27, 28], and for CF_4 the code from the Wigner Research Centre for Physics [18].

The experimental setup is displayed in figure 1. A reactor comprising 20 cm diameter parallel electrodes with an inter-electrode gap of 2.5 cm was made geometrically symmetric by adding a thick Teflon ring (inner diameter 10 cm), fitted with a 2.5 × 2.5 × 10 cm borosilicate glass window to allow optical access. The voltage waveform is generated as presented in detail in [8], without any impedance matching. An impedance-matched generator has been proposed by Franek *et al* [29], but was not employed here as a high coupled power was not required. After an iterative waveform optimization (using the method of Patterson *et al* [30]) the desired waveform (shown in blue for a sawtooth with $n = 5$ frequencies) can be generated at the powered electrode. The gate of the intensified CCD camera (Andor iStar) used for phase resolved optical emission spectroscopy (PROES) measurements is synchronized to the RF generation system via a delay generator. A 2 ns gate was used, providing 91 images within one RF-cycle for a fundamental frequency of 5.5 MHz. The respective emission lines (750.4 nm for Ar, 656 nm (H-alpha) for H_2, and 703.7 nm (F atom) for CF_4,) were selected using interference filters. In all cases, the electron impact excitation rate was deduced from the measured emission by deconvoluting the radiative lifetime of the emitting state [31].

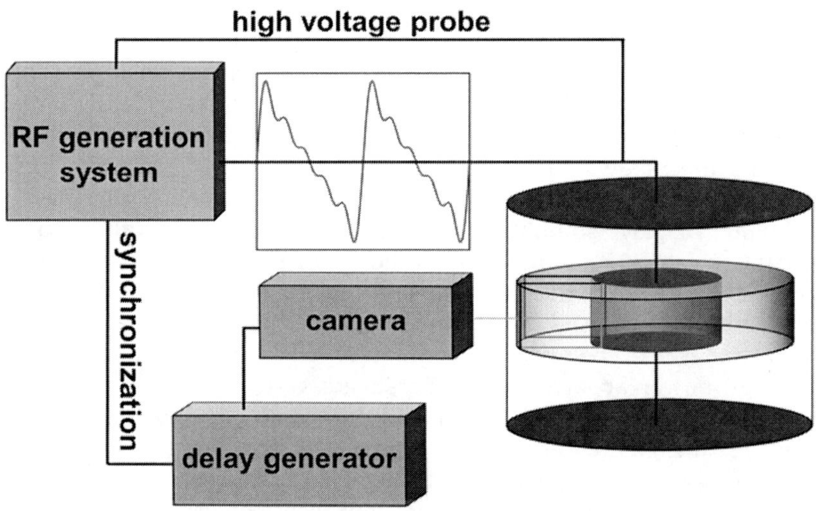

FIGURE 1. Schematic of the experimental setup with RF generation system, delay generator and ultrafast camera for PROES measurements.

Since in any experiment the voltage waveforms delivered have a finite frequency bandwidth, the voltage waveforms used are approximations of an ideal sawtooth, where a waveform consisting of a slow decrease and an infinitely fast increase is referred to as a 'sawtooth-down'. The approximation is performed by truncating the Fourier series for a perfect sawtooth waveform after n harmonics according to equation (1) (with a positive sign for sawtooth-down),

$$V_{ac}(t) = \pm V_0 \sum_{k=1}^{n} \frac{1}{k} \sin(k\omega t), \tag{1}$$

where V_{ac} is the voltage applied to the powered electrode, ω is the fundamental angular frequency, and V_0 is adjusted to give the desired peak-to-peak voltage V_{PP}. Alternatively, 'sawtooth-up' waveforms (equation (1), minus sign) can be used. In a geometrically symmetric reactor this signal inversion produces a reversal of the roles of each electrode. The experiments were carried out at pressures of 400 mTorr, 900 mTorr, and 600 mTorr, for Ar, H_2, and CF_4, respectively. The exact value of the pressure has a negligible effect on the qualitative features shown hereafter. For example, it was shown in [21] that only a 5% increase of the ion flux ratio was obtained by increasing the pressure from 400 mTorr to 800 mTorr in Ar. The peak-to-peak voltage V_{PP} was set to 250 V, 350 V, and 400 V for Ar, H_2, and CF_4 respectively.

Figure 2 compares the experimental results and simulations for the three gases using a sawtooth-down waveform with n = 5 (corresponding to the waveform displayed in figure 1). The results for Ar, H_2, and CF_4 are displayed from top to bottom. The x-axis represents position (with the powered electrode at x = 0 cm and the grounded electrode at x = 2.5 cm), while the y-axis represents time, and spans one fundamental RF cycle. The first column shows the excitation rate, K_{exc}, of the respective observed emitting states, as derived from PROES measurements. The next three columns display simulation results. The second column shows the excitation rate, K_{exc}, while the third shows the electric field distribution, $E(x, t)$ divided by V_{PP}, outside of the sheath regions (which are shown in black). The same scale is used for the three gases, so that the normalized electric field values can be directly compared. The fourth column shows the electron power absorption, S_e. The sheath edge position was determined using Brinkmann's criterion [32], and the sheath edge position is shown as a white line in the second and fourth columns. For all three gases the simulations show fast sheath expansion and slow sheath contraction in front of the grounded electrode, and slow sheath expansion with fast sheath contraction in front of the powered electrode, as expected from the sawtooth-down waveform used.

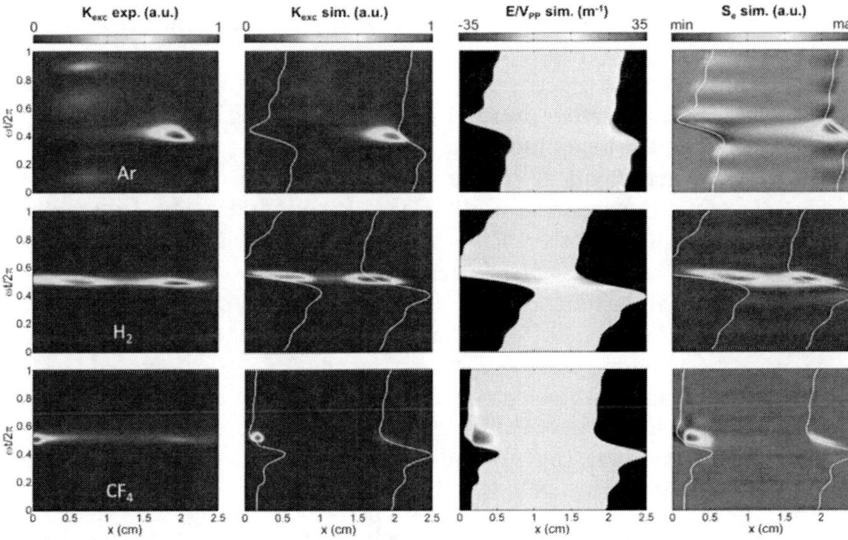

FIGURE 2. Excitation rate obtained from the experiments (first column), and from the simulations (second column), spatiotemporal plots of the electric field (third column), and electron power absorption rate (fourth column), from the simulations, for an argon discharge (top row), a hydrogen discharge (middle row), and a carbon tetrafluoride discharge (bottom row).

Looking at the excitation rate, one can see that an excellent agreement is found between experiments and simulations for all three gas chemistries, despite significantly different excitation features between them. In Ar, excitation is strongly localized close to the grounded electrode, and occurs during the fast sheath expansion (as indicated by the simulations in the second column). Only very weak excitation is observed close to the powered electrode during the (much slower) sheath expansion. In contrast, for H_2 two excitation peaks of similar amplitude occur almost simultaneously, the first one close to the grounded electrode where the sheath is expanding rapidly (sheath expansion heating) and a second peak close to the powered electrode where the sheath is contracting rapidly (due to a reversed electric field). Finally, in CF_4, weak excitation occurs close to the grounded electrode during the fast expansion, followed by a strong peak very close to the powered electrode during the fast sheath contraction. In order to understand the difference in the excitation features between the gas chemistries, both the electric field and the electron power absorption were analyzed.

In Ar, no significant electric field is observed within the bulk plasma outside of the sheath regions (the green color indicates zero field). In contrast, in H_2, a strong positive electric field occurs close to the powered electrode during the fast sheath contraction, corresponding to an electric field reversal. It should be noted that, although smaller, a field reversal also occurs at the grounded electrode when the sheath collapses. These field reversals occur because the motion of electrons is hindered by collisions (which are much more significant in H_2, a molecular gas, compared to Ar), and therefore a significant (reversed) electric field is necessary to accelerate the electrons to the electrodes and balance the escaping ion current [33]. The intensity of these field reversals therefore only depends on how fast the sheath contracts. When the sheath at the powered electrode is contracting rapidly the electron thermal flux is too low to ensure current balance, and a high, reversed, electric field is necessary. In contrast, the sheath contraction at the grounded electrode is slow, and therefore a weaker reversed electric field is sufficient to ensure current balance.

In CF_4, a very strong electric field (much stronger than in H_2) is observed during the fast sheath contraction. This electric field builds up because of the low electron density (because the majority of electrons have been converted to heavy F^- ions by dissociative attachment to CF_4), leading to low conductivity. In this discharge, the negative ion density (not shown), is particularly high close to the electrode where fast sheath contraction occurs (in this case the powered electrode), hence the electron density and the conductivity are particularly low. In contrast, close to the opposite (grounded) electrode the negative ion density is significantly lower and the electron density is higher, and therefore the conductivity is higher. As a consequence, a strong electric field is only observed close to the powered electrode, where fast sheath contraction occurs. Furthermore, on approaching the powered electrode, this positive electric field (corresponding to field reversal) is followed by a negative electric field, therefore creating an electric potential well, which acts as a trap for the electrons at one moment in time and limits their flux to the electrode.

The electron power absorption dynamics (shown in the fourth column) is clearly strongly affected by the gas properties. In all cases (for these relatively high gas pressure plasmas) the maximum in the electron power absorption occurs close to where the highest excitation rate is observed. This is because the electron mean free path is short compared to the discharge length. In Ar strong electron power absorption occurs close to the grounded electrode during the fast sheath expansion, while much lower electron power absorption occurs near the powered electrode during the slow sheath expansion. This is consistent with the hard wall model of Godyak [34, 35], according to which, the faster the sheath expansion rate, the greater the electron energy gain. In addition, because there is no significant electric field in the plasma bulk, and because secondary electron emission from the electrodes plays no significant role under these conditions, there is no other source of energy gain for the electrons. The discharge therefore operates in α-mode. In H_2, the situation is more complex: strong electron power absorption again occurs close to the grounded electrode during the fast sheath expansion, but is immediately followed by a second electron power absorption peak of similar amplitude, close to the powered electrode, due to the electric field reversal in this region. These two power absorption peaks are therefore the cause for the two excitation peaks.

In CF_4, the situation is again different. While some weak electron power absorption can be observed close to the grounded electrode during the fast sheath expansion, the largest power absorption peak is observed close to the powered electrode during the fast sheath contraction. Indeed, the large reversed electric field during this period strongly accelerates electrons from the plasma bulk towards the powered electrode. This leads to significant power absorption, accelerating electrons up to a mean energy of about 10 eV at this point. Because the electric fields in this phase can be split into drift and ambipolar components, this heating mode has been named the 'drift-ambipolar' (DA) mode [24]. This power absorption peak again coincides with a strong excitation peak. The latter is further strengthened by the formation of a potential well, which traps the energetic electrons and reduces their loss to the electrode, hence increasing the probability for these electrons to excite the neutral gas. In addition, this electron confinement increases the probability of electron attachment, and therefore the generation of negative ions. This further decreases the conductivity, and therefore increases the reversed electric field, forming a positive feedback loop.

In short, sawtooth waveforms induce a strong asymmetry in the excitation rate in both the Ar and CF_4 cases. However, the location of maximal excitation in CF_4 is opposite that of Ar, due to the change in the dominant electron power absorption mechanism. In the case of H_2, any asymmetry is cancelled out as both mechanisms occur almost simultaneously.

From the location of the electron excitation (figure 2, first and second columns) one would expect differences in the ion fluxes to the grounded and powered electrodes. Figure 3(a) shows the ratio β, of the ion flux to the powered electrode divided by that to the grounded electrode, for Ar (black squares), H_2 (red circles), and CF_4 (blue triangles) discharges, obtained from simulations, as a function of the number of applied harmonics n. This ratio exceeds unity for CF_4, because the (phase-integrated) excitation and ionization maxima are localized close to the powered electrode. In contrast, the ratio is below one for Ar, because the majority of the excitation and ionization occurs near the grounded electrode. It is worth noting that the asymmetry in the ion flux obtained for CF_4 is not only opposite compared to Ar, it is also significantly larger. As one can see, an ion flux ratio of about five can be obtained in a geometrically symmetric reactor with CF_4, while this drops to 0.6 (corresponding to an ion flux 1.6 times larger at the grounded electrode) in the case of Ar under these conditions. Finally, for H_2 the ion flux ratio remains close to unity (it is always between 0.9 and 1), since no clear asymmetry is observed in the spatial profile of the excitation dynamics.

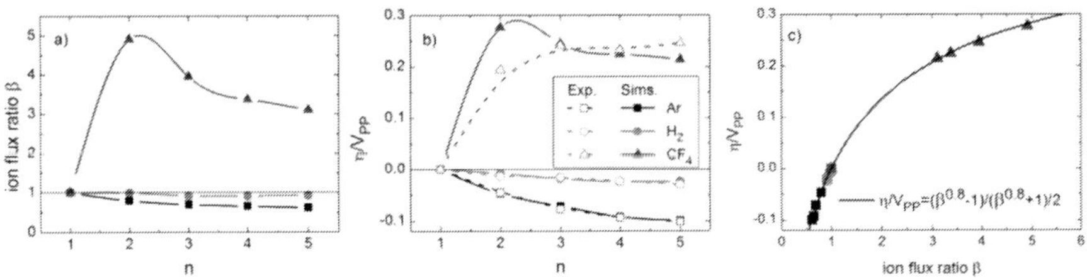

FIGURE 3. (a) Ion flux ratio β obtained by simulations for Ar (black squares), H_2 (red circles), and CF_4 (blue triangles) discharges, (b) normalized dc self-bias voltage, from simulations (full symbols) and from experiments (empty symbols), obtained with sawtooth-like waveforms composed of n frequencies. (c) Normalized dc self-bias voltage as a function of the ion flux ratio for the three gases.

In capacitively coupled plasmas, a dc self-bias voltage establishes such as to equalize the period-averaged electron and ion fluxes to the electrodes. Figure 3(b) shows the dc self-bias voltage, η, normalized to the peak-to-peak voltage V_{PP}, obtained from simulations (full symbols), and experiments (empty symbols), showing excellent agreement (with the exception of n = 2 in CF_4, for reasons not yet known). In CF_4, η is always positive, whereas it is always negative in Ar and H_2. It is worth noting that a strong correlation exists between the ion flux ratio, β, and η. The sign of η will be positive if $\beta > 1$, and negative if $\beta < 1$. Its amplitude is proportional to the difference of β from unity. This link can be understood by considering the expression for η given by Heil *et al* [1] applied to the case of sawtooth waveforms (which have no amplitude asymmetry):

$$\eta = -\frac{V_{PP}}{2} \times \frac{1-\varepsilon}{1+\varepsilon} \tag{2}$$

where ε is the symmetry parameter, which can be approximated by the ratio of the mean ion densities in the sheath at the powered and grounded electrode. Because the mean ion densities and the ion fluxes are strongly linked, we can expect a relationship between ε and β. Assuming a power law, $\varepsilon = \beta a$ (because we should have $\varepsilon = 1$ when $\beta = 1$, and $\varepsilon \rightarrow 1/\varepsilon$

when $\beta \rightarrow 1/\beta$), where a is a fitting parameter, and replacing ε by its expression as a function of β in equation (2), it is possible to obtain a relationship between the normalized dc self-bias voltage and the ion flux ratio. Figure 3(c)shows η/V_{PP} as a function of β, for the three gases and for different values of n. Also shown in black is the best fit obtained by replacing ε by its expression as a function of β in equation (2). We find a = 0.8, indicating that the ion flux ratio is always larger than the symmetry parameter. The agreement is excellent for all three gases and for all conditions, suggesting that this relationship may have general applicability.

In conclusion, we have observed that the charged particle dynamics occurring in plasmas of gases with different properties leads to fundamental differences in the spatial asymmetry of discharges excited with the same sawtooth waveform. The spatiotemporal distribution of the dominant electron energy gain mechanism is identified as the main cause for these asymmetries. When the discharge is in α-mode (Ar), most of the ionization occurs close to the electrode where the fast sheath expansion occurs, whereas in a strongly electro-negative gas where the DA mode dominates (CF_4) ionization is predominantly close to the electrode where fast sheath contraction occurs. Hydrogen discharges were found to operate in a mixed mode, where significant ionization is simultaneously observed near both sheath edges when they contract or expand quickly. The net effect for such H_2 discharges is an absence of asymmetry in the spatial profile of the time-averaged ionization. Finally, despite these differences, a power-law relationship is observed between the ion flux ratio and the normalized dc self-bias, and a single value of exponent fits the data for all three gas chemistries studied herein.

ACKNOWLEDGMENTS

The authors would like to thank F Farci and P Bulkin for their help with the reactor used in this study. Work described in this paper was made possible by financial support from the ANR projects APOCALYPSO (ANR-13-PRGE-0003-01), CleanGRAPH (ANR-13-BS09-0019), the UK EPSRC (EP/K018388/1 & EP/H003797/1), the York-Paris Low Temperature Plasma Collaborative Research Centre, and the Hungarian Fund for Scientific Research (OTKA K 105476). B Bruneau is also supported by the Doctoral School of Ecole Polytechnique (EDX grant, Ecole Doctorale de l'X). P Diomede and D J Economou are grateful to the Department of Energy, Office of Fusion Energy Science, contract DE-SC0001939 for financial support.

REFERENCES

1. Goto H H, Löwe H-D and Ohmi T 1992 Dual excitation reactive ion etcher for low energy plasma processing J. Vac. Sci. Technol. A **10** 3048–54
2. Kitajima T, Takeo Y, Petrović Z L and Makabe T 2000 Functional separation of biasing and sustaining voltages in two-frequency capacitively coupled plasma Appl. Phys. Lett. **77** 489–91
3. Boyle P C, Ellingboe A R and Turner M M 2004 Independent control of ion current and ion impact energy onto electrodes in dual frequency plasma devices J. Phys. Appl. Phys. **37** 697
4. Georgieva V, Bogaerts A and Gijbels R 2004 Numerical investigation of ion-energy-distribution functions in single and dual frequency capacitively coupled plasma reactors Phys. Rev. E **69** 026406
5. Turner M M and Chabert P 2006 Collisionless heating in capacitive discharges enhanced by dual-frequency excitation Phys. Rev. Lett. **96** 205001
6. Chabert P and Braithwaite N 2011 Physics of Radio-Frequency Plasmas (Cambridge: Cambridge University Press) [Online] available: www.cambridge.org/us/academic /subjects/physics/plasma-physics-and-fusion-physics/physics-radio-frequency-plasmas [Accessed: 3 April 2015]
7. Heil B G, Czarnetzki U, Brinkmann R P and Mussenbrock T 2008 On the possibility of making a geometrically symmetric RF-CCP discharge electrically asymmetric J. Phys. Appl. Phys. **41** 165202
8. Johnson E V, Verbeke T, Vanel J-C and Booth J-P 2010 Nanocrystalline silicon film growth morphology control through RF waveform tailoring J. Phys. Appl. Phys. **43** 412001
9. Lafleur T, Boswell R W and Booth J P 2012 Enhanced sheath heating in capacitively coupled discharges due to non-sinusoidal voltage waveforms Appl. Phys. Lett. **100** 194101
10. Lafleur T, Delattre P A, Johnson E V and Booth J P 2012 Separate control of the ion flux and ion energy in capacitively coupled radio-frequency discharges using voltage waveform tailoring Appl. Phys. Lett. **101** 124104
11. Coumou D J, Clark D H, Kummerer T, Hopkins M, Sullivan D and Shannon S 2014 Ion energy distribution skew control using phase-locked harmonic RF bias drive IEEE Trans. Plasma Sci. **42**1880–93
12. Zhang Q-Z, Zhao S-X, Jiang W and Wang Y-N 2012 Separate control between geometrical and electrical asymmetry effects in capacitively coupled plasmas J. Phys. Appl. Phys. **45** 305203
13. Zhang Y, Zafar A, Coumou D J, Shannon S C and Kushner M J 2015 Control of ion energy distributions using phase shifting in multi-frequency capacitively coupled plasmas J. Appl. Phys. **117** 233302
14. Gibson A R, Greb A, Graham W G and Gans T 2015 Tailoring the nonlinear frequency coupling between odd harmonics for the optimisation of charged particle dynamics in capacitively coupled oxygen plasmas Appl. Phys. Lett. **106** 054102
15. Schüngel E, Mohr S, Schulze J, Czarnetzki U and Kushner M J 2014 Ion distribution functions at the electrodes of capacitively coupled high-pressure hydrogen discharges Plasma Sources Sci. Technol.**23** 015001
16. Donkó Z, Schulze J, Czarnetzki U, Derzsi A, Hartmann P, Korolov I and Schüngel E 2012 Fundamental investigations of capacitive radio frequency plasmas: simulations and experiments Plasma Phys. Control. Fusion **54** 124003
17. Schüngel E, Brandt S, Donkó Z, Korolov I, Derzsi A and Schulze J 2015 Electron heating via self-excited plasma series resonance in geometrically symmetric multi-frequency capacitive plasmas Plasma Sources Sci. Technol. **24** 044009

18. Schulze J, Derzsi A and Donkó Z 2011 Electron heating and the electrical asymmetry effect in dual-frequency capacitive CF4 discharges Plasma Sources Sci. Technol. **20** 045008
19. Derzsi A, Korolov I, Schüngel E, Donkó Z and Schulze J 2013 Electron heating and control of ion properties in capacitive discharges driven by customized voltage waveforms Plasma Sources Sci. Technol. **22** 065009
20. Bruneau B, Novikova T, Lafleur T, Booth J P and Johnson E V 2014 Ion flux asymmetry in radiofrequency capacitively-coupled plasmas excited by sawtooth-like waveforms Plasma Sources Sci. Technol. **23** 065010
21. Bruneau B, Novikova T, Lafleur T, Booth J P and Johnson E V 2015 Control and optimization of the slope asymmetry effect in tailored voltage waveforms for capacitively coupled plasmas Plasma Sources Sci. Technol. **24** 015021
22. Bruneau B, Gans T, O'Connell D, Greb A, Johnson E V and Booth J-P 2015 Strong ionization asymmetry in a geometrically symmetric radio frequency capacitively coupled plasma induced by sawtooth voltage waveforms Phys. Rev. Lett. **114** 125002
23. Greb A, Niemi K, O'Connell D and Gans T 2013 The influence of surface properties on the plasma dynamics in radio-frequency drivenoxygen plasmas: measurements and simulations Appl. Phys. Lett.**103** 244101
24. Schulze J, Derzsi A, Dittmann K, Hemke T, Meichsner J and Donkó Z 2011 Ionization by drift and ambipolar electric fields in electronegative capacitive radio frequency plasmas Phys. Rev. Lett. **107**275001
25. Dittmann K, Küllig C and Meichsner J 2012 Electron and negative ion dynamics in electronegative cc-rf plasmas Plasma Phys. Control. Fusion **54** 124038
26. Lafleur T and Booth J P 2012 Control of the ion flux and ion energy in CCP discharges using non-sinusoidal voltage waveforms J. Phys. Appl. Phys. **45** 395203
27. Diomede P, Longo S, Economou D J and Capitelli M 2012 Hybrid simulation of a dc-enhanced radio-frequency capacitive discharge in hydrogen J. Phys. Appl. Phys. **45** 175204
28. Diomede P, Economou D J, Lafleur T, Booth J-P and Longo S 2014 Radio-frequency capacitively coupled plasmas in hydrogen excited by tailored voltage waveforms: comparison of simulations with experiments Plasma Sources Sci. Technol. **23** 065049
29. Franek J, Brandt S, Berger B, Liese M, Barthel M, Schüngel E and Schulze J 2015 Power supply and impedance matching to drive technological radio-frequency plasmas with customized voltage waveforms Rev. Sci. Instrum. **86** 053504
30. Patterson M M, Chu H-Y and Wendt A E 2007 Arbitrary substrate voltage wave forms for manipulating energy distribution of bombarding ions during plasma processing Plasma Sources Sci. Technol. **16**257–64
31. Gans T, O'Connell D, der Gathen V S and Waskoenig J 2010 The challenge of revealing and tailoring the dynamics of radio-frequency plasmas Plasma Sources Sci. Technol. **19** 034010
32. Brinkmann R P 2007 Beyond the step model: approximate expressions for the field in the plasma boundary sheath J. Appl. Phys. **102** 093303
33. Czarnetzki U, Luggenhölscher D and Döbele H F 1999 Space and time resolved electric field measurements in helium and hydrogen RF-discharges Plasma Sources Sci. Technol. **8** 230
34. Godyak V A 1972 Statistical heating of electrons at an oscillating plasma boundary Sov. Phys. - Tech. Phys. **16** 1073
35. Lieberman M A and Lichtenberg A J 2005 Principles of Plasma Discharges and Materials Processing(Hoboken, NJ: Wiley)

CHAPTER 19

Influence of Argon Gas Flow on Mechanical and Electrical Properties of Sputtered Titanium Nitride Thin Films

Kaykhosrow Khojier[1]*, Hadi Savaloni[2] , Ebrahim Shokrai[3] , Zohreh Dehghani[3] and Naser Zare Dehnavi[3]

[1] *Department of Physics, Chalous Branch, Islamic Azad University, Chalous 46615-397, Iran.*

[2] *Department of Physics, University of Tehran, North Kargar Street, Tehran 14395-515, Iran.*

[3] *Department of Physics, Faculty of Science, Central Tehran Branch, Islamic Azad University, Tehran 13185-786, Iran.*

ABSTRACT

Titanium nitrides have good mechanical, tribological, electrical, biomedical, and optical properties; therefore, they are used to harden and protect cutting and sliding surfaces, as semiconductor devices, and as a nontoxic exterior for biomedical applications. The dependence of the mechanical and electrical properties of titanium nitride thin films deposited on silicon substrates by direct-current reactive magnetron sputtering technique on argon gas flow (in the range of 8 to 20 sccm) was investigated. The crystallographic structure of the films was studied by X-ray diffraction (XRD), while surface morphology was studied using atomic force microscopy (AFM). Mechanical and electrical properties of these films were investigated by nanoindentation test and a four-point probe instrument, respectively. The XRD patterns showed titanium nitride (TiN) formation with a face-centered cubic structure for all samples. It was also observed that (111) crystallographic direction was the preferred orientation for TiN thin films which became more pronounced with increasing argon gas flow. The AFM images showed a granular structure for TiN layers. The hardness, crystallite/grain size (obtained from XRD and AFM), and surface roughness increased with the flow of argon gas, while elastic modulus and dislocation density in the films decreased. The study on electrical properties showed that the dependence of voltage with current for all samples was linear, and film resistivity was increased with argon gas flow.

Keywords: Titanium nitride Thin films Argon gas flow Hardness Elastic modulus Resistivity

INTRODUCTION

Because of their desirable mechanical, chemical, and electrical properties, such as high hardness, good corrosion, wear resistance, chemical and metallurgical stability, and low electrical resistivity, titanium nitride thin films have been widely used in different areas of semiconductor device technology, such as diffusion barriers, gate electrodes in field-effect transistors, contact layers in solar cells, and a replacement of polycrystalline Si in very large-scale integrated circuits [1, 2, 3, 4]. For the development of new materials in microelectronic applications, in addition to their electrical properties, their mechanical characteristics may play an important role on long-term reliability problems. In fact, new advanced metallization materials should combine good electrical properties, comparable to those currently used, with high hardness and elasticity, fine adhesion on Si, and low interface stresses [1].

Among the different methods of titanium nitride thin film preparation including physical vapor deposition, plasma focus, chemical vapor deposition, electrochemical and chemical nitridation, sputtering, and ion implantation [1, 5, 6, 7, 8, 9, 10, 11, 12, 13, 14, 15, 16, 17], direct-current (DC) magnetron sputtering is a simple and suitable method. However, the nanostructure and different properties of titanium nitride thin films in this method are strongly dependent on sputtering conditions such as bias voltage, substrate temperature, argon gas flow, and nitrogen content. The aim of this work is to investigate the influence of argon gas flow on the crystallographic structure, surface morphology, and both the mechanical and electrical properties of sputtered titanium nitride thin films. The results of this study can be of importance and used in different applications as mentioned throughout this section.

EXPERIMENTAL DETAILS

Film preparation

Titanium nitride thin films of 100 nm thick were deposited with a DC reactive magnetron sputtering system using a circular 76-mm-diameter and 1-mm-thick Ti target of 99.998% purity. The target-to-substrate distance was 10 cm. A continually variable DC power supply of 700 V and 0.3 A was used as power source for sputtering. The thickness and deposition rate of these titanium nitride films were checked *in situ* using a quartz crystal monitor (6-MHz gold, Inficon Company, East Syracuse, NY, USA) located near the substrate during the sputtering process. The substrates were 20 × 20 mm^2 Si (400) (n-type) and were ultrasonically cleaned in heated acetone then etalon. The substrate temperature was 373 K. The base pressure was lower than 2×10^{-6} Torr. Pure argon gas (99.998% purity) and nitrogen gas (99.998% purity) were used as sputter and reactive gases, respectively. Titanium nitride thin films were deposited under various argon gas flows. The flow rates of both argon and nitrogen gases were controlled individually using mass flow controllers. The argon gas flow rate was varied at 8, 12, 16, and 20 sccm during sputtering, while the nitrogen gas flow rate was kept constant at 10 sccm. Figure 1depicts the variations of deposition pressure and rate as a function of argon gas flow. This figure shows that deposition pressure and deposition rate increase with argon gas flow.

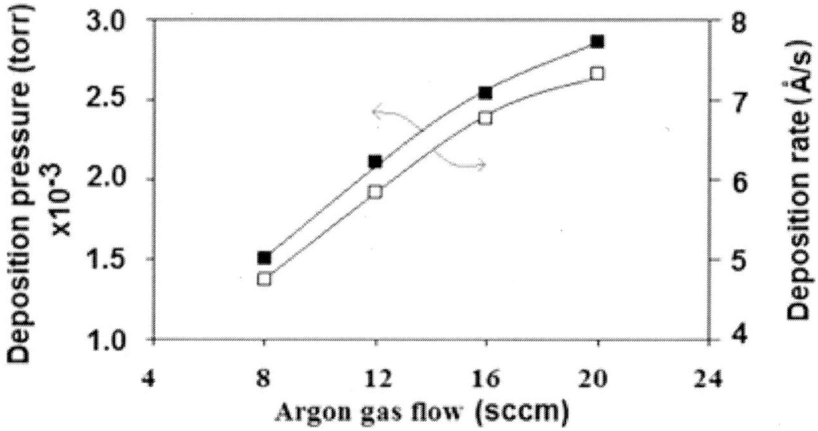

Figure 1. Variations of deposition pressure and rate as a function of argon gas flow.

Film characterization

The nanostructure and crystallographic orientation of the samples were obtained using a Philips X'pert MPD diffractometer (CuKα radiation; Amsterdam, The Netherlands) with a step size of 0.02° and count time of 1 s per step. The surface morphology and roughness of the samples were obtained using an atomic force microscope (AFM; Autoprobe PC, Park Scientific Instrument, Sunnyvale, CA, USA) with a scan size of 1 × 1 μm^2 and a scan rate of 1 Hz. A Hysitron Inc. (Eden Prairie, MN, USA) TriboScope® Nanomechanical Test Instrument with a 2D transducer, complete software, and Berkovich diamond indenter was used for mechanical test. Over three indentation tests were performed on each of the samples, and the average of the data obtained from these tests is presented for each sample. In this test, force, loading time, holding time, and unloading time were 600 μN, 30 s, 10 s, and 30 s, respectively. The electrical resistivity of the samples was also measured using a four-point probe instrument at room temperature.

RESULT AND DISCUSSIONS

Crystallographic structure

X-ray diffraction (XRD) patterns of all TiN/Si thin film samples prepared in this work are shown in Figure 2. The XRD pattern of the deposited film with 8-sccm argon gas flow in addition to the substrate peak (at 69.15° that is not presented in the figure) shows two weak peaks at 35.33° and 41.01° that can be related to the (111) and (200) crystallographic orientations of TiN with a face-centered cubic (FCC) structure, respectively (with reference to JCPDS card no. 02–1159; $2\theta = 35.317°$ and 41.00°, respectively). Titanium nitride with closed-packed FCC structure may form when nitrogen atoms occupy all the octahedral sites of titanium with hexagonal close-packed (HCP) or body-centered cubic (BCC) structures. In order to accommodate nitrogen atoms with small size (0.74 Å) in the interstitial sites of Ti with larger size (1.47 Å), the titanium structure should be transformed from HCP (at lower temperatures) and BCC (at higher temperatures) to FCC structure. By increasing the argon gas flow to 12 sccm, TiN(111) peak intensity is increased, while TiN(200) intensity is decreased. In the deposited films with a higher value of argon gas flow (i.e., 16 and 20 sccm), this process is repeated so that the TiN(111) diffraction line can be introduced as a preferred orientation. In fact, the result shows that an increase in argon gas flow causes the growth of TiN thin films in (111) crystallographic orientation.

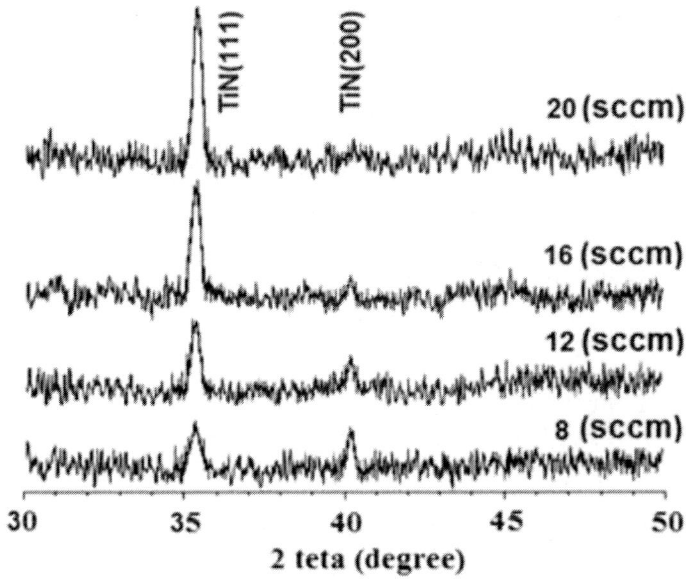

Figure 2. X-ray diffraction patterns of prepared tin thin films at different argon gas flows.

It is known that TiN films deposited by various PVD methods generally grow in three crystallographical orientations (i.e., (200), (111), and (220)) [18]. The preferred orientation of a film is due to its lowest overall energy, which is a result of competition between the surface energy, the strain energy, and the stopping energy of different lattice planes [19, 20]. In the case of TiN films, the (200) plane has the lowest surface energy, the (111) plane has the lowest strain energy, and the (220) plane has the lowest stopping energy. Only when the deposited ion energy is sufficiently high, the stopping energy becomes dominant, and the (220) preferred orientation can be observed [18]. In this work, due to the low deposition power and low substrate temperature, the (220) peak is not observed. Hence, one may propose that only surface energy and strain energy have affected the film growth process.

The absence of a titanium peak in the XRD pattern (different from previous similar literature [18]) indicates the absence of Ti in the structure of these films and complete nitride formation process, which can be related to the high value of nitrogen gas flow rate in the deposition process (10 sccm).

Crystallite size (coherently diffracting domains), D, was obtained by applying the Scherrer formula ($D = k\lambda/B\cos\theta$) to measure the full width at half maximum (FWHM) of the dominant peak TiN(111) in the XRD patterns of Figure 2[21],where $k = 0.9$, $\lambda = 1.54248$ Å, B = FWHM that is calculated from B2=B2M−B2S$B^2 = B_M^2 - B_S^2$ (B_M, FWHM of the sample and B_S, FWHM of the standard sample), and θ is the peak position in radian. The crystallite size (coherently diffracting domains) results are shown in Figure 3. The results show that the crystallite size gradually increases with increasing argon gas flow. This increase can be attributed to the fact that the increase in the argon gas flow enhances the energetic ion bombardment which promotes the mobility of adatoms and results in the increase of grain size. When the argon gas flow increases further, the ion density becomes large enough for crystal growth; thus, the grain size increases slightly. A similar kind of this behavior of change in grain size with the argon gas flow is also observed in previous literatures [18, 22].

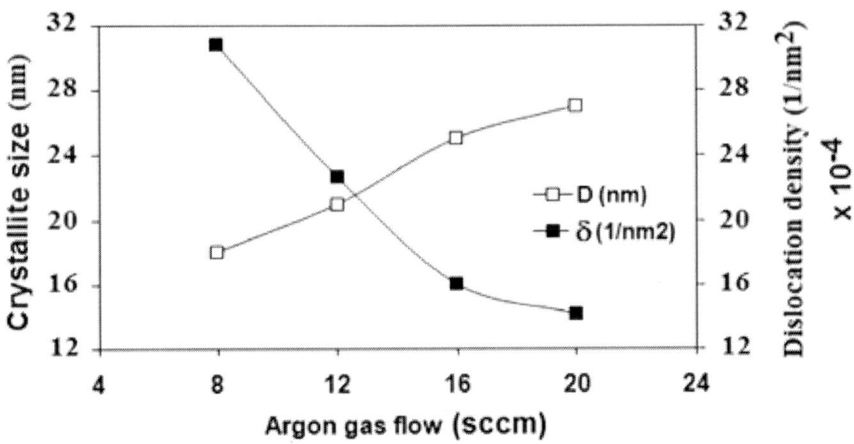

Figure 3. Variations of crystallite size and dislocation density as a function of argon gas flow. The crystallite size is obtained from XRD results.

Surface morphology

Two-dimensional (2D) and three-dimensional (3D) AFM images of TiN thin films prepared at different argon gas flows are shown in Figure 4, while variations of grain sizes (obtained from 2D pictures by JMicroVision code) and surface roughness as a function of argon gas flow are plotted in Figure 5. These images show a granular structure with small and sharp grains for deposited samples at low argon gas flows (i.e., 8 and 12 sccm), while at higher values of argon gas flow (i.e., 16 and 20 sccm), both grain size and surface roughness are increased. As mentioned, an increase in the argon gas flow increases the grain size which in turn increases the surface roughness.

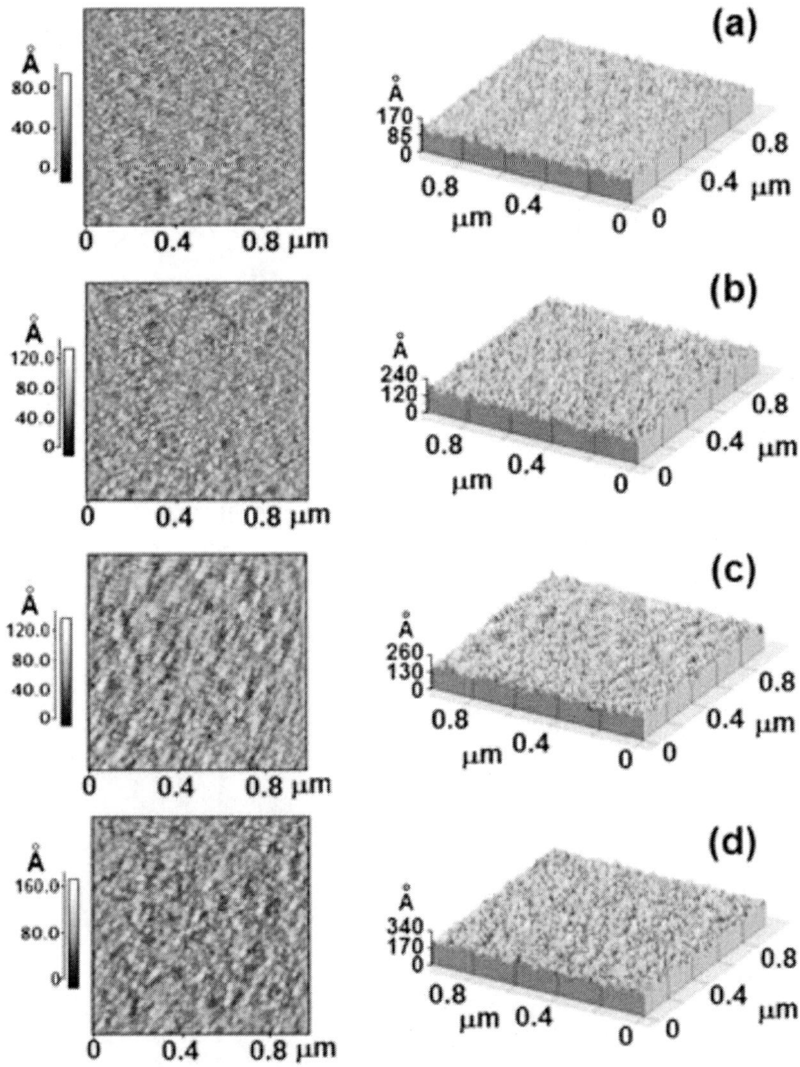

Figure 4. 2d and 3d afm images of deposited tin thin films at different argon gas flows. (A) 8, **(B)** 12, **(C)** 16, and **(D)** 20 sccm.

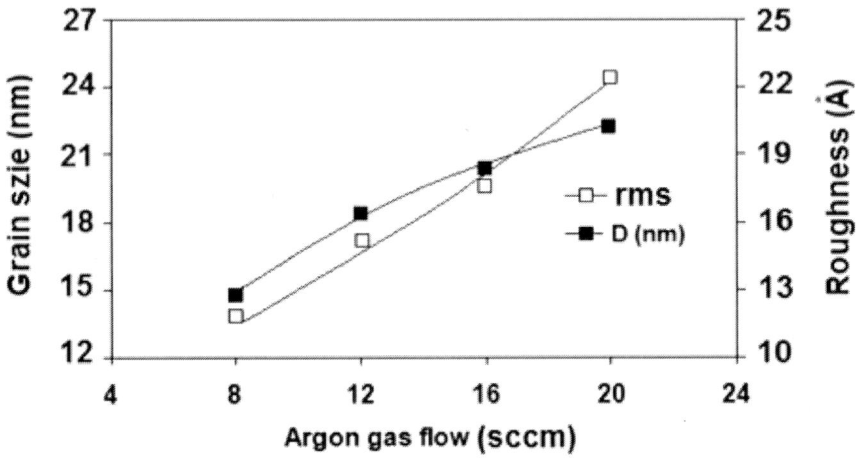

Figure 5. Variations of grain size and surface roughness as a function of argon gas flow. The grain size is obtained from 2D AFM images.

Mechanical properties

Figure 6 illustrates the variation of hardness of all TiN/Si thin films as a function of the argon gas flow. As can be seen, an increase in the argon gas flow causes the increase of this quantity. This change can be explained in the following paragraph.

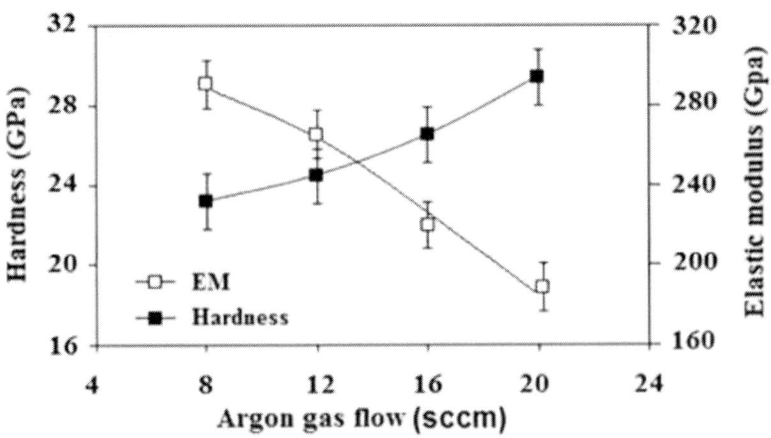

Figure 6. Variations of hardness and em of tin thin films as a function of argon gas flow.

The hardness and mechanical properties of films can be affected by different parameters such as grain size, crystallographic orientations, film density, lattice parameters, and stoichiometry [1, 10, 12, 23]. Despite the fact that variation of grain size with the argon gas flow is not large (remarkable), the increase of hardness can be attributed to (1) film growth in (111) crystallographic orientation, (2) the increase of density or elimination of voids, (3) and improvement of crystal quality with increasing argon gas flow. XRD results showed that the intensity of the TiN(111) diffraction line increased with the argon gas flow. TiN_x is an anisotropic material with $H\,111 > H\,200$ or $H\,220$ (where $H\,111$, $H\,200$, and $H\,220$ are the hardness along the (111), (200), and (220) crystallographic directions, respectively). On the other hand, TiN films with (111) preferred orientation possess the highest hardness [10, 12]. Figure 3 also depicts the variation of dislocation density, which represents the amount of defects and pores in the film. The dislocation density (δ) is determined from the formula $\delta = 1/D^2$[24], where D is the crystallite size and is calculated using Scherrer's formula. The results show that films with better crystallization quality can be obtained at higher argon gas flows, which result in decreased number of pores (or increased film density) in the structure of the film which in turn increases the hardness of the film.

Figure 6 also plots the variation of elastic modulus (EM) as a function of argon gas flow. The figure shows that the increase of argon gas flow results in the decrease of elastic modulus that can be attributed to the low elasticity of (111) crystallographic orientation.

Electrical properties

Figure 7 plots the dependence of voltage versus current obtained by the four-point probe analysis for all samples. In order to investigate the influence of possible (low frequency) charging effects at the electrical contacts and leads, current–voltage (*I-V*) curves were recorded, scanning the voltage both in increasing and decreasing increments. The results of measurements in the decreasing direction of the voltage are not shown as they matched the results in the increasing voltage direction and only clutter the figure unnecessarily. However, it should be noted that the current was kept low enough in order not to heat up the samples. Figure 8 depicts the variation of titanium nitride thin film resistivity as a function of argon gas flow. The results show that the resistivity of TiN thin films increases with the argon gas flow. XRD results showed that the intensity of the TiN(111) diffraction line also increased with the argon gas flow It has also been reported earlier that the film with a high (111) diffraction peak intensity has a high electrical resistivity [25]. Therefore, it may be deduced that there is a correlation between the intensity of the TiN(111) diffraction line and the resistivity of these TiN thin films, which is in agreement with the results reported in [18, 22].

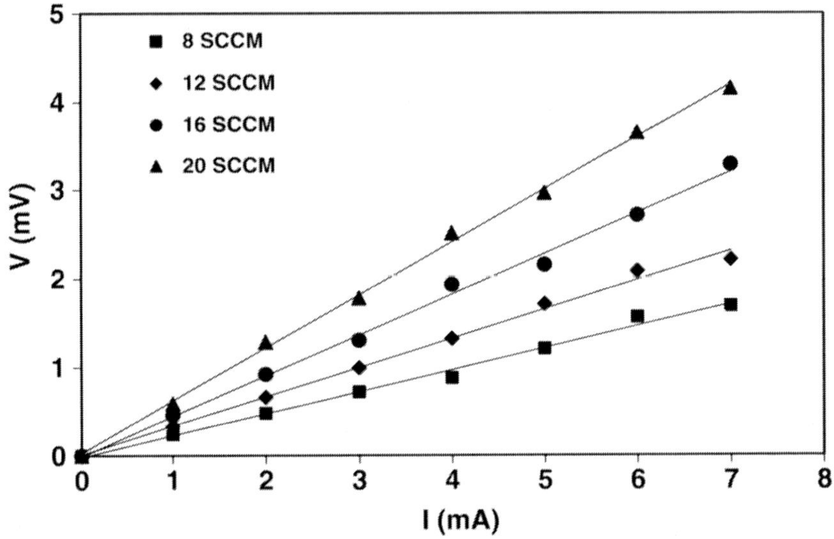

Figure 7.. Dependence of voltage with current of prepared tin thin films at different argon gas flows.

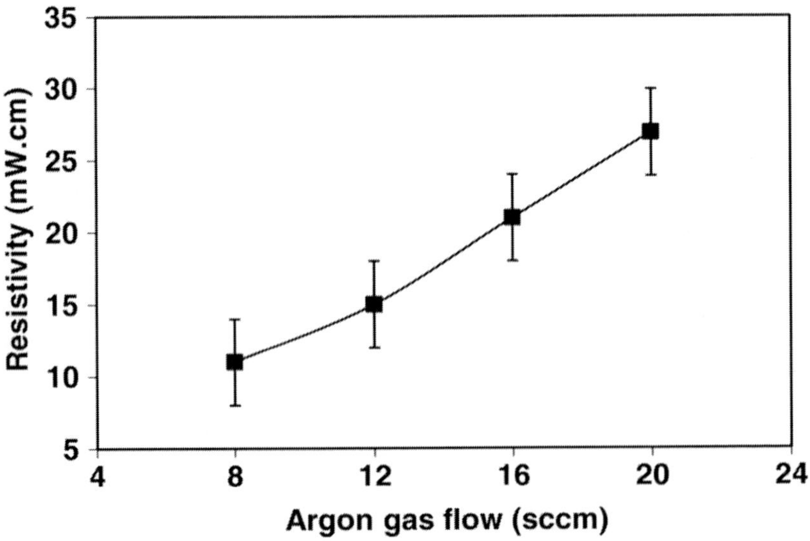

Figure 8. Variation of resistivity of tin thin films as a function of argon gas flow.

Furthermore, an increase in the intensity of titanium nitride peak may directly be related to the increase in the amount of nitrogen ions bonded to the Ti atoms in the structure of the thin film. In low bonded nitrogen content, the nitrogen atoms bonding with the titanium atoms are not enough, and the remainder amorphous titanium atoms act as donors and can provide free electrons. With these free electrons, the films have a high carrier density with high mobility and low resistivity. However, with the increase of the amount of bonded nitrogen content, there will be sufficient nitrogen atoms to integrate with the titanium atoms that cause only a few free electrons and result in higher resistivity.

CONCLUSION

Titanium nitride thin films were deposited on Si substrate by DC reactive magnetron sputtering system. The effect of argon flow rate (8 to 20 sccm in steps of 4 sccm) on the structural, morphological, mechanical, and electrical properties was studied using XRD, AFM, nanoindentation test, and four-point probe instrument, respectively. Structural investigation showed TiN with FCC structure for all samples. It was also observed that (111) crystallographic orientation was the preferred orientation for TiN thin film growth with increasing argon gas flow. The surface morphology of the samples depicted a granular structure with small grains, and the grain size and surface roughness were increased with the argon gas flow. The increase of grain size with argon gas flow was due to the energetic ion bombardment and was consistent with previous literatures. The mechanical and electrical studies showed that film surface hardness and resistivity increased with the argon gas flow, while film elastic modulus decreased with increasing argon gas flow. Structural analyses illustrated that these variations were related to the (111) crystallographic orientations and were consistent with previous similar works.

ACKNOWLEDGEMENTS

This work was carried out with the support of the Islamic Azad University, Chalous Branch and Central Tehran Branch. HS is grateful to the University of Tehran and to the Center of Excellence for Structural and Microscopic Properties of Matter, Department of Physics, University of Tehran for the partial support of this work.

COMPETING INTERESTS

The authors declare that they have no competing interests.

AUTHORS' CONTRIBUTIONS

KK, HS, ES, ZD, and NZD provided the same contributions in this article. All authors read and approved the final manuscript.

REFERENCES

1. Patsalas P, Charitidis C, Logothetidis S: **THE EFFECT OF SUBSTRATE TEMPERATURE AND BIASING ON THE MECHANICAL PROPERTIES AND STRUCTURE OF SPUTTERED TITANIUM NITRIDE THIN FILMS.** *Surf. Coat. Technol.* 2000, **125:** 335–340.
2. Travis EQ, Fiordalice RF: **MANUFACTURING ASPECTS OF LOW PRESSER CHEMICAL-VAPOR-DEPOSITED TIN BARRIER LAYERS.** *Thin Solid Films* 1993, **236:** 325–329.
3. Gagnon G, Currie JF, Beique C, Brebner JL, Gujrathi SG, Onllet L: **CHARACTERIZATION OF REACTIVELY EVAPORATED TIN LAYERS FOR DIFFUSION BARRIER APPLICATIONS.** *J. Appl. Phys.* 1994, **75:** 1565–1570.
4. Hara T, Yamanoue A, Iio H, Inoue K, Washidzu G, Nakamura S: **PROPERTIES OF TITANIUM NITRIDE FILMS FOR BARRIER METAL IN ALUMINUM OHMIC CONTACT SYSTEMS.** *Jpn. J. Appl. Phys.* 1991, **30:** 1447–1451.
5. Vaz F, Ferreira J, Ribeiro E, Rebouta L, Lanceros-Mendez S, Mendes JA, Alves E, Goudeau P, Riviere JP, Ribeiro F, Moutinho I, Pischow K, Rijk J: **INFLUENCE OF NITROGEN CONTENT ON THE STRUCTURAL, MECHANICAL AND ELECTRICAL PROPERTIES OF TIN THIN FILMS.** *Surf. Coat. Technol.* 2005, **191:** 317–323.
6. Eriksson L, Harju E, Korhonen AS, Pischow K: **FORMABILITY AND CORROSION RESISTANCE OF TIN-COATED STAINLESS STEEL SHEET.** *Surf. Coat. Technol.* 1992, **53:** 153–160.
7. Chou W-J, Yu G-P, Huang JH: **MECHANICAL PROPERTIES OF TIN THIN FILM COATINGS ON 304 STAINLESS STEEL SUBSTRATES.** *Surf. Coat. Technol.* 2002, **149:** 7–13.
8. Valvoda V: **STRUCTURE OF THIN FILMS OF TITANIUM NITRIDE.** *J. Alloy. Compd.* 1995, **219:**83–87.
9. Baker MA, Monclus MA, Rebholz C, Gibson PN, Leyland A, Matthews A: **A STUDY OF THE NANOSTRUCTURE AND HARDNESS OF ELECTRON BEAM EVAPORATED TIALBN COATINGS.** *Thin Solid Films* 2010, **518:** 4273–4280.
10. Hainsworth SV, Soh WC: **THE EFFECT OF THE SUBSTRATE ON THE MECHANICAL PROPERTIES OF TIN COATINGS.** *Surf. Coat. Technol.* 2003, **163–164:** 515–520.
11. Kumar A, Singh D, Kumar R, Kaura D: **EFFECT OF CRYSTALLOGRAPHIC ORIENTATION OF NANOCRYSTALLINE TIN ON STRUCTURAL, ELECTRICAL AND MECHANICAL PROPERTIES OF TIN/NITI THIN FILMS.** *J. Alloy. Compd.* 2009, **479:** 166–172.
12. Holleck H: **MATERIAL SELECTION FOR HARD COATINGS.** *J. Vac. Sci. Technol. A* 1986, **4:**2661–2669.

13. Savaloni H, Khojier K, Torabi S: **INFLUENCE OF N^+ ION IMPLANTATION ON THE CORROSION AND NANO-STRUCTURE OF TI SAMPLES.** *Corros. Sci.* 2010, **52:** 1263–1267.
14. Chatterjee S, Chanrashekhar S, Sudarshan TS: **DEPOSITION PROCESSES AND METAL CUTTING APPLICATIONS OF TIN COATINGS.** *J. Mater. Sci.* 1992, **27:** 3409–3423.
15. Russell SW, Rack MJ, Adams D, Alford TL: **TITANIUM NITRIDATION ON COPPER SURFACES.** *J. Electrochem. Soc.* 1996, **143:** 2349–2353.
16. Firouzi M, Savaloni H, Ghoranneviss M: **DEPENDENCE OF SURFACE NANO-STRUCTURAL MODIFICATION OF TI IMPLANTED BY N^+ IONS ON TEMPERATURE.** *Appl. Surf. Sci.* 2010, **256:** 4502–4511.
17. Fani N, Savaloni H: **INVESTIGATION ON THE FORMATION OF TITANIUM NITRIDE THIN FILMS ON 304 TYPE STAINLESS STEEL USING PLASMA FOCUS DEVICE.** *J. Theor. Appl. Phys.* 2012, **6:** 1–7.
18. Arshi N, Lu J, Joo YK, Lee CG, Yoon JH, Ahmed F: **STUDY ON STRUCTURAL, MORPHOLOGICAL AND ELECTRICAL PROPERTIES OF SPUTTERED TITANIUM NITRIDE FILMS UNDER DIFFERENT ARGON GAS FLOW.** *Mater. Chem. Phys.* 2012, **134:** 839–844.
19. Zhao JP, Wang X, Chen ZY, Yang SQ, Shi TS, Liu XH: **OVERALL ENERGY MODEL FOR PREFERRED GROWTH OF TIN FILMS DURING FILTERED ARC DEPOSITION.** *J. Phys. D: Appl. Phys.* 1997, **30:** 5–12.
20. Pelleg J, Zevin LZ, Lungo S: **REACTIVE-SPUTTER-DEPOSITED TIN FILMS ON GLASS SUBSTRATES.** *Thin Solid Films* 1991, **197:** 117–128.
21. Warren BE: *X-ray Diffraction*. London: Addison-Wesley; 1969.
22. Yu GQ, Tay BK, Lau SP, Prasad K, Pan LK, Chai JW, Lai D: **EFFECTS OF N ION ENERGY ON TITANIUM NITRIDE FILMS DEPOSITED BY ION ASSISTED FILTERED CATHODIC VACUUM ARC.** *Chem. Phys. Lett.* 2003, **374:** 264–270.
23. Wang TG, Jeong D, Liu Y, Wang Q, Iyengar S, Melin S, Kim KH: **STUDY ON NANOCRYSTALLINE CR_2O_3 FILMS DEPOSITED BY ARC ION PLATING: II. MECHANICAL AND TRIBOLOGICAL PROPERTIES.** *Surf. Coat. Technol.* 2012, **206:** 2638–2644.
24. Williamson GK, Smallman RE: **DISLOCATION DENSITIES IN SOME ANNEALED AND COLD-WORKED METALS FROM MEASUREMENTS ON THE X-RAY DEBYE-SCHERRER SPECTRUM.** *Philos. Mag.* 1956, **1:** 34–45.
25. Meng LJ, Dos Santos MP: **CHARACTERIZATION OF TITANIUM NITRIDE FILMS PREPARED BY D.C. REACTIVE MAGNETRON SPUTTERING AT DIFFERENT NITROGEN PRESSURES.** *Surf. Coat. Technol.* 1997, **90:** 64–70.

Index